Error Correcting Coding and Security for Data Networks

Error Correcting Coding and Security for Data Networks

Analysis of the Superchannel Concept

G. Kabatiansky
*Institute of Information Transmission Problems RAN, Russia
and INRIA-Projet CODES, France*

E. Krouk
St. Petersburg State University of Aerospace Instrumentation, Russia

S. Semenov
Nokia Technology Platforms, Finland

John Wiley & Sons, Ltd

Contents

Preface

This book provides a systematic approach to the problems involved in the application of error-correcting codes in data networks.

Over the last two decades the importance of coding theory has become apparent. Thirty years ago developers of communication systems considered error-correcting coding to be somewhat exotic. It was considered as an area of interest only for mathematical engineers or mathematicians involved in the problems of engineering. Today however, theory is an important part of any textbook on communications, and the results of coding theory have become standards in data communication. The increasing demand for communications quality and the progress in information technologies has led to the implementation of error-correcting procedures in practice and an expansion in the area of error-correcting codes applications.

Following the growth of coding theory in practice the number of publications on coding theory and its applications has also increased. However, most of these books are concerned with the problem of coding of physical or data-link network layers, which are the traditional application of error-control codes.

There are several classes of errors that are inherent in the process of information transfer over networks. One of these classes is formed by normal errors originating in communication links, and other classes are formed by special distortions, connected to the organisation of the network. However, in modern data networks the error-correcting (or controlling) codes are used only as a means of increasing the reliability of information during data transmission over different channels – no correlation between coding and other network procedures is considered. There is also a lack of research on the capability of codes to improve the operation of the network as a whole. Recently a number of non-traditional directions have appeared for applying coding at network layers higher than the data-link layer. In particular, the authors of this book have succeeded in showing the effectiveness of using error-correcting codes for reducing the delivery time of the message and increasing the reliability of the message itself. The desire to demonstrate these possibilities was the first reason for writing this book. However, during the preparation of the book it became clear that we had collected enough examples of using error-correcting codes at application and presentation layers of networks (coding for image compression and code-based cryptosystems for example) to enable us to apply the results of coding theory to all network layers.

The basic concept considers the data network as some superchannel (consisting of several layers of the network). This concept allows us to solve the problems of error-correcting coding in a data network as a whole. As most errors depend on protocols used at the corresponding network layer, the suggested 'global' approach to coding in a network is connected with other network procedures, and this fact determines the complexity and originality of coding problems in a network. Surprisingly, results indicate that coding in a network helps not only to increase the reliability of the

transmitted information, but can also be used to improve such important characteristics of a network as the mean message delay. We may also consider the encryption of messages just as a coding process at the presentation layer in a network. We can therefore distinguish different coding processes in different network layers and so it is necessary to consider the impact of coding in one layer on another layer. Thus, the problem of reconciliation of coding in different network layers arises. In this book we set out some solutions to this problem.

The importance of solving this problem and our perspectives on the possible solutions were especially emphasised by the reviewers of this book. We are very grateful for their helpful comments. It is the attempt to consider the problems of coding at higher network layers which, in our opinion, is the distinguishing feature of this book.

The theoretical material is accompanied by concrete recommendations for the use of codes in data networks and by calculations of the benefits that can be obtained with the help of error-correcting coding. The organisation of the book (from the problems to the theory and from the theory to the problems, and the relative independence of chapters from each other) is chosen in such a way as to facilitate reading for engineers who would like to familiarise themselves with new approaches to the use of error-correcting codes in data networks. Since a significant part of the material is new or is not reflected in the literature, we hope that this book will be of interest to readers from different disciplines who are interested in issues of data communication and applications of coding theory.

The book is organised as follows. Chapter 1 introduces the problems of coding is different network layers. Chapter 2 presents the main algebraic structures used in coding theory and one of the most studied class of codes: linear block codes. Chapter 3 covers the different methods of linear codes decoding and introduces some new results obtained by authors in this field. Chapter 4 describes the very widely used codes: Hamming codes, BCH codes, and Reed-Solomon codes. The decoding of these helps to demonstrate very important and comprehensive results of coding theory. Chapter 5 introduces the problems of LDPC codes decoding. Chapter 6 presents another very widely-used class of codes: convolutional codes and turbo codes, and covers some problems of iterative decoding. Chapter 7 is devoted to the new area of application of error-correcting codes: transport coding. In this chapter the possibility of using error-correcting codes to control such important data characteristics as mean message delay is demonstrated. Chapter 8 covers coding methods used in cryptography. Chapter 9 analyses the problems of reconciliation of coding in different network layers. In this chapter some solutions based on the superchannel approach are considered.

Additional research results including some new constructions of LDPC codes, joint error-control coding and synchronization, Reed-Muller codes and their list decoding can be obtained from the book's companion website at ftp://ftp.wiley.co.uk/pub/books/kabatiansky.

The problem of error control coding in data networks is very wide and not yet fully defined today so the authors do not claim to provide a full solution of the problem but are hoping that this book can become a first step to further research in the field.

ACKNOWLEDGMENTS

We would like to acknowledge the assistance of our colleagues in writing this book. Peter Trifonov contributed to Section 4.5, Andrey Ovchinnikov contributed to Chapter 5, and Section 9.4 is based on the Ph.D. thesis of Andrey Belogolovy. Evgeny Linskii provided significant help in writing Chapter 8. We are also grateful to Alexander Barg, Ilya Dumer, Grigorii Evseev, Ulrich Sorger, and many people at Nokia for their help in choosing the material for this book.

We would like to thank Sarah Hinton at John Wiley & Sons, Ltd for her help in co-ordinating the writing process.

Special thanks to our families who supported us in this work.

1
Problems Facing Error Control Coding in Data Networks

1.1 INTERNATIONAL RECOMMENDATIONS ON USING ERROR CONTROL CODING AT DIFFERENT NETWORK LAYERS

The aim of any data network is to provide reliable and effective (fast) transmission of information between the network users. The international standards aim to fulfill these two conflicting requirements independently, increasing the reliability by means of inserting redundancy in the transmitted data to detect or correct the errors, and increasing the speed by developing the 'economy' procedures of retransmission, initialisation, connection and disconnection, and so on. The information theory approach to the problem of coding in a network is based on the fact that all the actions in the particular network layer can be regarded as some method of transmission of 'messages' over the 'channel', considering this network layer as a channel.

The Open Systems Interconnection Reference Model (OSIRM) developed by the International Organisation for Standardisation (ISO) contains 7 network layers:

1. **Physical Layer.** The physical layer (L1) provides transparent transmission of a bit stream across the physical interconnections of the network elements. The different modulation techniques are used at this layer.

2. **Data Link Layer.** The primary function of the data link layer (L2) is to establish a reliable protocol interface across the physical layer (L1) on behalf of the network layer (L3). This means that the link layer performs error detection or error correction. It is the most common area of coding applications.

3. **Network Layer.** The main function of the network layer (L3) is to provide the delivery of protocol data between transport layer entities. There is one network layer process associated with each node and with each network site of the network. All these processes are peer processes and all work together in implementing routing and flow control for the network.

4. **Transport Layer.** The main functions of the transport layer (L4) are segmentation, (re)assembly of messages into packets, and multiplexing over a single L3 interface. If the

Error Correcting Coding and Security for Data Networks   G. Kabatiansky, E. Krouk and S. Semenov
© 2005 John Wiley & Sons, Ltd   ISBN: 0-470-86754-X

network layer is unreliable, the transport layer might achieve reliable end-to-end communication. End-to-end flow control is often done at the transport layer.

5. **Session Layer.** The session layer's (L5) main function is to provide the user's interface to the network. Sessions usually provide connections between a user and a host. Other session layer functions include flow control, control over the direction of data transfer, and transaction support.

6. **Presentation Layer.** The presentation layer (L6) determines how data is presented to the user. The main functions of the presentation layer are data encryption, data conversion, and code conversion.

7. **Application Layer.** The application layer (L7) manages the program or device generating the data to the network.

Data flows down from L7 at the originating end system to L1 and onto the physical medium, where it is transmitted, and back up to L7 of the destination end system, as shown in Figure 1.1. In accordance with the information theory concept, the channel corresponding to the ith network layer (i-channel) is characterised by its own alphabet (the symbols of this alphabet are the data blocks of the layer $i - 1$), and by the error types that depend on the organisation of lower layers. Notice that upper layers (higher than L2) have non-binary alphabets and quite specific types of errors: duplicates, drops and 'overtaking'. With this kind of approach, error control coding is a universal means of providing not only the required reliability of data transmission but also reduction in message delay.

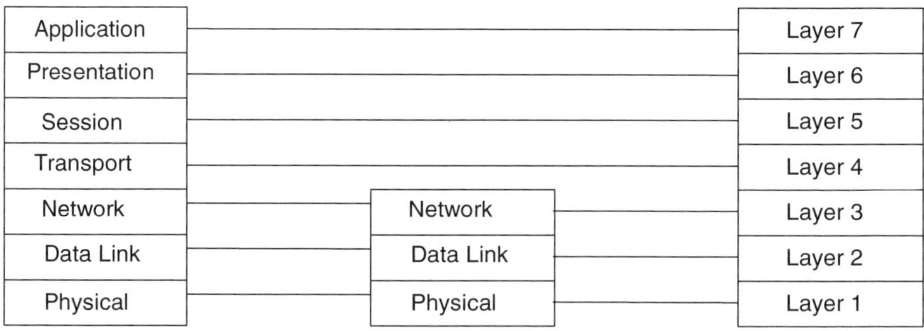

Figure 1.1   The open systems interconnection reference model

The main area of application of coding in a network is the data link layer (DLL). The wide range of coding techniques, from the simplest parity checks to more comprehensive codes, is represented in different DLL protocols.

The binary cyclic block codes are used for the calculation of the 16, 24 or 32-bit Cyclic Redundancy Check (CRC) in almost all protocols. The 16-bit CRC was already being used in the Digital Data Communications Message Protocol (DDCMP). The same 16-bit CRC is used in one of most popular protocols–the High-level Data Link Control (HDLC) or ISO4335. Also 16-bit CRC is used in the Link Access Procedure (LAP) protocols, which can be regarded as subsets of HDLC. In particular, the LAPB protocol is used in the famous X.25 standard. The IEEE 802.3 CSMA/CD (Carrier Sense Multiple Access with Collision Detection) or

Ethernet protocol, invented in the early 1970s and adopted in the 1980s, already contains 32-bit CRC for the protection of the data in the frame. The IEEE 802.4 Token Bus, the IEEE 802.5 Token Ring and the Fibre Distributed Data Interface (FDDI) protocols also contain the same 32-bit CRC [1]. The error protection of data in ATM is provided with the help of 10-bit CRC and the ATM header data is protected with 8-bit CRC. The interesting feature of this 8-bit CRC is that it is sometimes used to correct errors rather than just to detect them [2]. More comprehensive coding techniques are used in wireless networks. In GSM standards, except the 8-bit CRC, the convolutional code of constraint length 5 and coding rate 1/3, and block cyclic Fire code, are used. In the ECSD (Enhanced Circuit Switch Data), part of GSM, the shortenings of systematic Reed-Solomon (255,243) code over $GF(2^8)$ are used [3]. In UMTS, except the 24-bit CRC, the convolutional codes with constraint length 9 and coding rates 1/3 and 1/2, and turbo code with coding rate 1/3 (two 8-state constituent encoders) are used [4]. IEEE 802.16 Broadband Wireless Access uses the shortened (255,239,17) Reed-Solomon code whilst the shortened (15,10) Hamming code is used in BlueTooth specification.

Some of these coding techniques are also used at higher than L2 layers of a network. For example, error detection is quite often used in the transport level protocols. The Transmission Control Protocol (TCP), which is the L4 protocol, uses 16-bit CRC to detect the errors in TCP frame. Usually the information protection at L4 is restricted to the calculation of CRC and the initialisation of retransmission procedure if needed.

Unfortunately, very often the protection of information in one layer does not interact with the information protection in another layer or with other procedures in the same layer. A more or less acceptable solution of this problem is the interaction of L1 and L2. However, the main technique in layer 1 is modulation rather than coding. As an example of the lack of interaction of different information procedures we can consider HDLC. One of the most important problems in HDLC (and also to LAP, which was based on the HDLC Set Asynchronous Response Mode (SARM)) is the interaction of the procedure of frame (block) synchronisation with the error detection with the help of CRC (see Figure 1.2). In HDLC and LAP the border of the transmitted message is marked by a flag of form '01111110' in addition to the procedure of bit stuffing, which during the transmission inserts in the frame bit '0' after each five successive bits '1'. Thus, the frame, after bit stuffing, never contains more than five consecutive 1s, and the flag at the end of the frame is uniquely recognisable. At the receiver end, the first 0 after each string of five consecutive 1s is deleted. If a string of five consecutive 1s is followed by bit '1', the frame is declared to be finished. After this the CRC is checked.

8 bits	8 bits	8 bits	variable	16 bits	8 bits
Flag	Address field	Control	User data (information)	FCS	Flag

Figure 1.2 HDLC frame format

It is assumed therefore, that the cyclic code should detect the errors in some extended discrete channel (EDC), which is the combination of the discrete channel (DC) and the procedures of bit stuffing and flag addition, as shown in Figure 1.3. Due to the errors in DC there are two possible failures of block synchronisation, i.e. the incorrect detection of the end of the frame. The first case is when the false flag appears inside the frame, and the frame

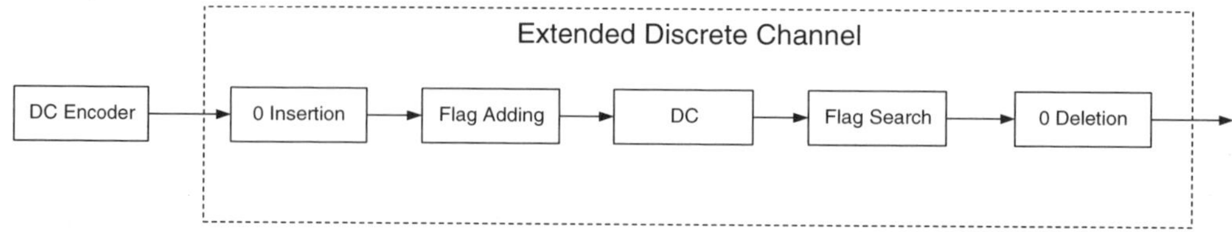

Figure 1.3 Extended discrete channel in HDLC frame synchronisation procedure

is split into two frames with deletion of bits on the position of the false flag. In the second case, due to the corruption of the flag, two (or more) frames are glued together with the insertion of bits at the position of the corrupted flag. Thus, such errors lead to a change of frame borders with insertion or deletion of 7 or 8 bits. This can be explained by the fact that a single error in DC leads to one of the following transformations in EDC: single error and insertion of bit '0', single error and deletion of bit '0', single error. Unfortunately, the detection capability of the binary cyclic codes relative to these specific errors in EDC is quite poor, while in an additive channel they can detect any $d - 1$ errors, where d is the minimum distance of the code.

One of the aims of this book is to show the possibility of better interaction of coding at different network layers.

1.2 CLASSIFICATION OF PROBLEMS ON CODING IN NETWORKS

Increasing noise immunity is one of the central problems in designing data communications networks. To solve this problem it is necessary to take into account the fact that information loss in data networks is caused not only by distortions in communication channels, but also by errors resulting from particular features of the organisation of data networks as a message delivery service.

Normally the coding is only used to provide transmission reliability in the communication channel. Meanwhile, every layer of a data network can be considered as a data channel with its own data quantum (bit, frame, packet, message), transmission method and specific distortions. From this point of view the problem of coding becomes a global problem relative to the data network as a whole, since it requires analysis of not only physical error sources, but also the protocol supplying the data network; solving this task is connected with developing non-typical coding methods, providing co-ordination of codes on different network layers.

The use of redundant coding on higher network layers was always limited by the idea that increasing network loading by coding leads to an unconditional increase in transmitted messages delay.

In a number of works [5,6,7,8,9] the principal use of coding not only to increase the network reliability, but also to improve the reliability of its functioning parameters, in particular, to decrease the average message delay, has been demonstrated. Later in this book both traditional methods of coding usage in physical and data link layers, and the problems of coding in the transport layer of data network are considered. The possibilities of using error-correcting codes in the application layer are investigated to a lesser extent. Public-key cryptosystems based on error-correcting codes appeared at the same time as number-theoretic systems. Although the number-theoretic cryptosystems became the basis of cryptographic standards, cryptosystems based on error-correcting codes remain the object of many investigations as being the most serious alternative to number-theoretic standards. In this book the possibilities of code-based cryptosystem development are considered. However, it seems that prospects of code-based cryptosystems are determined not so much by their own advantages, as by the development of error-protection integrated systems intended to combat not only natural distortions, but also artificial impacts. These implications are very uncertain at the present time, but it seems that the material of this book can be the basis of elucidating this.

Let us formulate the problems solved by means of error-correcting coding (all these tasks are considered below):

- Coding providing reliable transmission in data communication channels (physical layer)

- Coding in feedback systems (data link layer)

- Coding in the transport layer

- Coding providing unauthorized access protection (application layer)

- Coding for data compression with losses (application layer)

All these tasks have different levels of readiness. Of course the traditional coding in a physical layer is investigated to a much greater extent than are the other mentioned tasks. However, examples of solving tasks in layers higher than physical allow consideration of error-correction coding as a universal tool suitable for application in a wide area of network tasks.

Moreover, the material collected in this book allows the formulation of the task of joint coding usage on different data network layers to provide optimal distribution of algorithmical redundancy in a data network.

REFERENCES

1. Spohn, D. L. *et al.* (2002). *Data Network Design*. McGraw Hill, New York.
2. Bertsekas, D. and Gallager, R. (1992). *Data Networks*. Prentice Hall, New Jersey.
3. 3GPP TS 45.003. *Channel Coding*, V6.0.0 (2003–08).
4. 3GPP TS 25.212. *Multiplexing and channel coding (FDD)*, V5.6.0 (2003–09).
5. Kabatiansky, G. A. and Krouk, E. A. (1993). Coding Decreases Delay of Messages in Networks. *IEEE International Symposium on Information Theory. Proceedings.*
6. Maxemchuk, N. F. (1975). Dispersity routing, *IEEE Conf. Commun.* **3** San Francisco.
7. Krouk, E. and Semenov, S. (2002). Application of Coding at the Network Transport Level to Decrease the Message Delay. *Proceedings of Third International Symposium on Communication Systems Networks and Digital Signal Processing.* 15–17 July 2002 Staffordshire University, UK.
8. Krouk, E. and Semenov, S. (2004). Transmission of Priority Messages with the Help of Transport Coding. *Proceedings of 10th International Conference on Telecommunications.* Papeete, Tahiti French Polynesia.
9. Byers, J. Luby, M. Mitzenmacher, M. and Rege, A. (1998). Digital Fountain Approach to Reliable Distribution of Bulk Data, in *International Computer Science Institute Technical Reports*.

2
Block Codes

This chapter introduces the theory of block codes. Here we describe the features of block codes for correcting independent errors. The importance of this class of code is shown by its significance for practical applications and by the fact that analyses of these codes demonstrates the main methods and results of coding theory.

2.1 MAIN DEFINITIONS

Let us consider in accordance with Shannon [1] the following model of a data transmission system (Figure 2.1). A data source generates messages $\mathbf{u}_1, \ldots, \mathbf{u}_M$ and a receiver would like to receive them correctly (with high reliability). The data source and the receiver are connected by a *channel* allowing the transmission of symbols from an input alphabet (set) A in a sequential way. However, due to some noise in the channel the output sequence may differ from the input one. Moreover, in general terms, the input alphabet A and the output alphabet B do not coincide. The probabilistic model of the channel is given by transition probabilities P(b|a) that an output sequence (of symbols) is b under condition that an input sequence was a. We restrict our consideration to the case of the well-explored coding theory q-ary *memoryless channels*, for which:

- input and output alphabets coincide;

- cardinal number of the input (and the output) alphabet equals to q;

- statistic characteristics of the output symbol are fully defined by the input symbol (i.e. there is no memory in the channel);

- statistic characteristics of the output symbol do not depend on the time.

An important instance of such a channel is a binary symmetric channel (BSC), where the probability of the output (binary) symbol coinciding with the input symbol is equal to Q and the probability of the output symbol differing from the input symbol is equal to $P = 1 - Q$.

To provide reliable data transmission, messages should be encoded, i.e. each message $\mathbf{u}_i$ corresponds to a finite sequence $\mathbf{v}_i = (v_{i1}, \ldots, v_{in_i})$ of symbols of alphabet A. This sequence is called the *codeword*. The codeword is transmitted over the channel. The set of all codewords

Error Correcting Coding and Security for Data Networks   G. Kabatiansky, E. Krouk and S. Semenov
© 2005 John Wiley & Sons, Ltd   ISBN: 0-470-86754-X

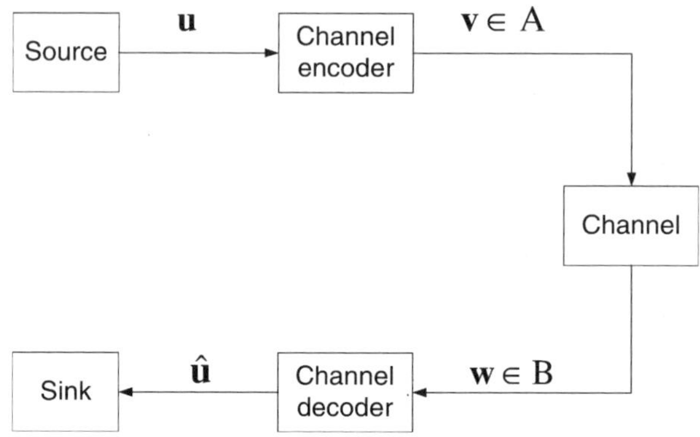

Figure 2.1   Model of a data communication system

is called a *code* and the mapping $\varphi : \mathbf{u} \to \mathbf{v}$ is called an *encoding procedure*. If codewords consist of the same number of symbols ($n_i = n$), then a code is called a *block code* of length n. We assume that encoding maps different messages to different codewords. Hence a block code is an arbitrary subset V (of cardinality M) of the vector space E_q^n of all q-ary words of length n.

The receiver tries to restore the source message $\mathbf{u}$ relying on the received output sequence $\mathbf{w}$. The corresponding mapping $\Psi : \mathbf{w} \to \hat{\mathbf{u}}$ is called the *decoding procedure*. This mapping is to some extent the reverse mapping to encoding. Devices that realise encoding and decoding procedures, are called the *encoder* and the *decoder* respectively. Due to the presence of noise in the channel, $\mathbf{w}$ may differ from $\mathbf{v}$. In this case message $\Psi(\mathbf{w}) = \hat{\mathbf{u}}$ may differ from source message $\mathbf{u}$. This event is called the *decoding error*. Since there is reciprocation between the messages and the codewords, it is possible to consider mapping $\psi = \varphi \circ \Psi : \mathbf{w} \to \hat{\mathbf{v}} = \varphi(\hat{\mathbf{u}})$ instead of mapping Ψ. Mapping ψ also is called the decoding procedure. It is more convenient to consider mapping ψ, because the definition of ψ is equivalent to a partition of the set A^n of all words of length n on *decision regions* Ω_i, so that $A^n = \cup_{i=1}^{M} \Omega_i$; $\Omega_i \cap \Omega_j = \emptyset, i \neq j$, where $\Omega_i = \{\mathbf{w} \in A^n : \psi(\mathbf{w}) = \mathbf{v}_i\}$. It is intuitively clear that to minimise the probability of decoding error it is necessary to include in Ω_i words of A^n, which are 'close enough' to $\mathbf{v}_i$, where the measure of 'closeness' should be agreed with the channel in the sense that the closer two words are, the more probable it is that one word will be received from the output of the channel if another word was fed to the channel input. Such measures of 'closeness' for BSC is the *Hamming distance* [2].

The Hamming distance $d(\mathbf{a}, \mathbf{b})$ between words $\mathbf{a} = (a_1, \ldots, a_n) \in A^n$ and $\mathbf{b} = (b_1, \ldots, b_n) \in A^n$ is defined as the number of positions where these words are different.

It is easy to check that Hamming distance is the metric, i.e.

$$\left. \begin{array}{l} d(\mathbf{a}, \mathbf{b}) > 0, \quad \mathbf{a} \neq \mathbf{b}, \quad d(\mathbf{a}, \mathbf{a}) = 0; \\ d(\mathbf{a}, \mathbf{b}) = d(\mathbf{b}, \mathbf{a}); \\ d(\mathbf{a}, \mathbf{c}) \leq d(\mathbf{a}, \mathbf{b}) + d(\mathbf{b}, \mathbf{c}) \end{array} \right\} \qquad (2.1)$$

By definition, the Hamming distance between transmitted and received words is equal to the number of errors that occurred during the transmission over the channel. Therefore we can say that a decoding procedure ψ of a code V corrects t errors, if the decoding result is always correct on condition that there were no more than t errors during the data transmission over the channel, i.e., $\psi(\mathbf{w}) = \mathbf{v}$, if $d(\mathbf{w}, \mathbf{v}) \leq t$. One of the most important decoding procedures is the *minimum distance (MD) decoding* which, for a given received word $\mathbf{w}$, outputs the closest (in Hamming distance) to its codeword $\hat{\mathbf{v}}$ (or if there are several such codewords, then to any of them).

The following notion in many ways characterises the capability of a code to correct errors. The *minimum distance* (shortly, distance) $d(V)$ of a code V is the minimum of pairwise Hamming distances between different codewords, i.e.

$$d(V) = \min_{\mathbf{v}, \mathbf{v}' \in V; \mathbf{v} \neq \mathbf{v}'} d(\mathbf{v}, \mathbf{v}'). \tag{2.2}$$

Let us denote a code of length n, cardinal number M and distance d as (n, M, d) code.

It is easy to check that a decoding procedure ψ corrects t errors if, and only if, for any $i \in 1, \ldots, M$ the decision region Ω_i contains the n-dimensional (hyper)sphere

$$A_t^n(\mathbf{v}_i) = \{\mathbf{x} \in A^n : d(\mathbf{x}, \mathbf{v}_i) \leq t\}. \tag{2.3}$$

of radius t and with the centre at the point $\mathbf{v}_i$. Since decision regions Ω_i and Ω_j do not intersect for $i \neq j$, then the corresponding spheres do not intersect either; and this fact, by virtue of properties of metric (2.1), is equivalent to the property: $d(\mathbf{v}_i, \mathbf{v}_j) > 2t$ for all $i \neq j$. So we derive one of the fundamental results of coding theory: code V corrects t errors if, and only if, $d(V) \geq 2t + 1$.

For many applications it is convenient to extend the definition of encoding procedure by allowing 'denial decoding', i.e. to define one more decision region Ω_* consisting of received words for which no decision about the transmitted word was done. This kind of decoding is called *partial decoding* (as opposed to the previously described *full decoding*) and it is the mapping $\psi : A^n \to A^n \cup \{*\}$. The most important example of partial decoding is the *error detection* ψ_0, which refuses to output a codeword if a given received word is not from the code (i.e., produces an error detection mark *) and in case a received word is a codeword it assumed that this word was uncorrupted. Thus, $\psi_0(\mathbf{v}_i) = \mathbf{v}_i$ and $\psi_0(\mathbf{w}) = *$ for $\mathbf{w} \notin V$. The generalisation of error detection is the decoding procedure ψ_t, which corrects no more than t errors. For this method of decoding, the decision regions Ω_i coincide with spheres $A_t^n(\mathbf{v}_i)$ of radius t around $\mathbf{v}_i$ and $\Omega_* = A^n \backslash \{\cup_{i=1}^M A_t^n(\mathbf{v}_i)\}$, i.e., an error will be detected if the distance between the received word and an arbitrary codeword is more than t.

We can say that the decoding procedure ψ of code V is capable of correcting t errors and detecting s errors ($s > t$) if the decoding result is the transmitted codeword if there are no more than t errors occurring during transmission, or the decoding result is the error detection if more than t but no more than s errors occur. This is equivalent to the following condition: $\Omega_i = A_t^n(\mathbf{v}_i)$ and $A_t^n(\mathbf{v}_i) \cap A_s^n(\mathbf{v}_j) = 0$ for all $i \neq j$. The obtained condition, in turn, is equivalent to $d(V) > t + s$. Thus the code V is capable of correcting t errors and detecting s errors $s > t$ if, and only if, its distance $d(V) \geq t + s + 1$.

Example 2.1 Consider a binary code $V = \{\mathbf{v}_0 = (00000), \mathbf{v}_1 = (11111)\}$ of length 5 and cardinal number 2. The distance of this code is 5. The decision region

$\Omega_0 = A_2^5(\mathbf{v}_0)$ for MD (minimum distance) decoding consists of those binary sequences of length 5 in which the number of zeros is more than the number of ones; and, vice versa, $\Omega_1 = A_2^5(\mathbf{v}_1)$ consists of sequences in which the number of ones is more than the number of zeros. In this case the MD decoding is the majority decoding, i.e. $\psi_2(\mathbf{w}) = \mathbf{v}_i$, where $i = \text{maj}(w_1, w_2, w_3, w_4, w_5)$. The decoding procedure ψ_1 with the decision regions $\Omega_0 = A_1^5(\mathbf{v}_0) = \{(00000), (10000), (01000), (00100), (00010), (00001)\}$ and $\Omega_1 = A_1^5(\mathbf{v}_1) = \{(11111), (01111), (10111), (11011), (11101), (11110)\}$ is capable of correcting 1 error and detecting 2 and 3 errors.

From the above it follows that the greater the code distance, the more errors the code can correct. Thus, one of the main tasks of coding theory is to find optimal codes. An (n, M, d) code is called optimal if for two fixed parameters (of n, M and d) it is impossible to 'improve' third ones, i.e., to increase the cardinal number, to increase the distance or to decrease the length. The more important task is to find the code with the maximum cardinal number $M = m_q(n, d)$ for a given code length n and distance d.

Note that little is known about optimal codes or about the behaviour of the function $m_q(n, d)$ (see sec. 2.5).

The capability of a code to correct and/or to detect errors is connected to *redundancy*. Since for transmission of M messages over the noiseless channel it is enough to use q-ary k-tuples $k = \lceil \log_q M \rceil$, where $\lceil x \rceil$ denotes the least integer greater or equal to x) then the value $r = n - k$ symbols is called the redundancy of a code.

The code rate, defined as $R = R(V) = \log_q M / n$, is an important parameter, which characterises the 'slowing down' of the information transmission due to redundancy. The fundamental result of information theory – *Shannon's noisy channel coding theorem* – states that for any rate R, less than the *channel capacity C*, the probability of decoding error for best codes (in fact for almost all codes) tends (exponentially) to zero with increasing code length; and vice versa, in case $R > C$ the probability of decoding error greater than some constant $a = a(R, C)$ for any code. Thus, for a given acceptable probability of decoding error $P_{acc.}^*$ and code rate R^* ($R^* < C$), it is possible to search for the code with a minimum code length of a set of codes $\mathbf{V}$ such that for any code $V \in \mathbf{V}$, $P(V) \leq P_{acc.}^*$ and $R(V) \leq R^*$.

Note that this problem is close to the above mentioned extreme problem $\min\limits_{V:|V|=M, d(V)=d} n(V)$; and the minimum code length means decreasing the decoding time, connected to accumulation of all n symbols in the decoder. However, from the practical point of view the complexity of the encoding and decoding procedures is the more important factor.

Until now we have considered codes as arbitrary sets of codewords without any restrictions on code structure. It is obvious that the utilisation of codes, defined in such a way, is very restricted. For example, even implementation of an encoding procedure, which usually has much less complexity than decoding, requires that the table of mapping messages to codewords is stored in memory. In cases when $k \geq 50$, the size of such a table becomes unacceptably large. Because of this, great attention is given to codes that have some algebraic structure, providing the opportunity to simplify the realisation of these codes as well as their construction. The most important class of such codes is the class of linear codes. To describe these codes we need some information about algebraic structures, which are stated in the following section.

2.2 ALGEBRAIC STRUCTURES

In this section we briefly describe such algebraic structures as groups, fields and vector spaces. This knowledge will useful for an understanding of the subsequent sections. For more detailed information see [3] or any major textbook on modern algebra. One of the simplest algebraic structures is the *semigroup*, defined as a set M with *binary operation* $\circ$ that assigns to each pair of elements $a, b \in M$ a uniquely defined element, denoted as $a \circ b$. The binary operation $\circ$ should be *associative*, i.e. for any $a, b, c \in M$

$$(a \circ b) \circ c = a \circ (b \circ c) \tag{2.4}$$

A *group* is defined as a semigroup G which, at first, contains an *identity* element e such that, for any $a \in G$,

$$a \circ e = e \circ a = a; \tag{2.5}$$

and, secondly, for any element $a \in G$, there exists a uniquely *inverse* element, denoted as a^{-1} such that

$$a \circ a^{-1} = a^{-1} \circ a = e. \tag{2.6}$$

If in addition the following condition is satisfied

$$a \circ b = b \circ a \tag{2.7}$$

for all $a, b \in G$, then a group G is said to be *commutative* or *Abelian*.

The binary operation on a group is usually called (by convention) the multiplication or the addition and is denoted by $\cdot$ or by $+$ respectively. The cardinal number of group G (the number of elements in group) is called the *order* of the group and is denoted by $|G|$. A mapping φ of group G to group G' is called a *homomorphism* if, for all $g_1, g_2 \in G$,

$$\varphi(g_1 \circ g_2) = \varphi(g_1) \circ \varphi(g_2). \tag{2.8}$$

If, moreover, φ is one-to-one mapping, then it is called *isomorphism*; and groups G and G' are called *isomorphic groups* (i.e. algebraic identical).

The important example of a group is the group $S(X)$ of one-to-one mappings of the set X to itself with the superposition of mappings as the binary operation $\circ$, i.e. $(f \circ g)(x) = f(g(x))$, $x \in X$. Let $X = \{1, 2, \ldots, n\}$ be the finite set of n elements, then the group $S(X)$ is called a *symmetric group* of order n; and its elements, i.e. mappings $\sigma : X \to X$ are called *permutations* and denoted as tables:

$$\begin{pmatrix} 1 & 2 & \ldots & n \\ i_1 & i_2 & \ldots & i_n \end{pmatrix}, \qquad \text{where } i_k = \sigma(k).$$

For group G with 'multiplication' it is possible to raise the elements to integer power, that is $g^0 = e$, $g^i = g \circ g^{i-1}$ for $i > 0$, $g^i = (g^{-1})^{-i}$ for $i < 0$. A group is called *cyclic* if each of its elements is a power of some element a, which is called a *generator element*, i.e. $G = \{a^i : i \in Z\}$. For example, set Z of integers is the cyclic group under addition to the generator element 1 (or -1). Another example of a cyclic group is group Z_q^+ of elements, which are integers (residues) $0, 1, \ldots, q - 1$; and the binary operation addition on modulo q is defined as:

$$(i + j) \bmod q = \begin{cases} i + j, & \text{if } i + j < q \\ i + j - q, & \text{otherwise.} \end{cases} \qquad (2.9)$$

Any cyclic group G is either isomorphic to Z if $|G| = \infty$, or to Z_q^+ where $|G| = q$.

A subset $H \in G$ is called a *subgroup* of G if $h_1 \circ h_2 \in H$ for any $h_1, h_2 \in H$, i.e., the set H is a group relative to the binary operation $\circ$.

Example 2.2 The subset $\langle g \rangle = \{g^i : i \subset Z\}$ is always the subgroup, which is called the subgroup generated by the element g; and the order of $\langle g \rangle$ is said to be the order of element g. If this order is finite, then it is equal to the minimal positive integer n such that $g^n = e$.

A subset $g \circ H = \{g \circ h : h \in H\}$ is called a left *coset* of group G on subgroup H. Any two cosets either coincide or do not intersect each other, i.e., they define the partition of G; and all of them have the same order $|H|$. Hence, $|H|$ is the divisor of $|G|$, i.e. the order of any subgroup is a divisor of the order of the group. This is the statement of the famous Lagrange theorem. Therefore, the order of any element is the divisor of the order of the group; and, for all $g \in G$,

$$g^{|G|} = e. \qquad (2.10)$$

A subgroup H is said to be *normal* if $g \circ h \circ g^{-1} \in H$, for all $g \in G, h \in H$; or in the equivalent statement, the left coset $g \circ H$ coincides with the right coset $H \circ g$ for all g. It follows from the second definition that any subgroup of a commutative group is normal. For the normal subgroup H the binary operation on the group G induces the binary operation on the set G/H of the cosets, i.e. $g_1 H \circ g_2 H = (g_1 \circ g_2)H$; and relevant to this operation the set G/H is the group, called *factorgroup* (or *quotient group*). For instance, let G be the group Z of all integers with addition as the binary operation; and let G be the group Z of all integers divisible by q. Then the corresponding factor-group is isomorphic to the above-mentioned group Z_q^+.

In some early works on coding theory the so-called binary group codes were considered. These codes are defined as arbitrary subgroups of the group Z_2^n of binary sequences of length n with the binary operation of symbol-by-symbol addition by modulo 2. The attempts to transpose these results to the case of an arbitrary finite alphabet of q elements, for example by changing the set E_q^n of q-ary sequences of length n to the group with the binary operation of symbol by symbol addition on modulo q, showed that the group structure is not enough; and that set E_q^n should be regarded (if possible) as an n-dimension vector space on the field F_q of q elements and the linear subspaces should be chosen as the codes.

It is possible to consider a *field* as a set in which it is possible to add, subtract, multiply and divide (by non-zero elements) while preserving the usual properties of these operations. Well-known example fields are field Q of rational numbers; R of real numbers; and C of complex numbers.

On the other hand, if we define in the set $\{0, 1\}$ the addition on modulo 2 and conjunction as the binary operations of addition and multiplication respectively, then we obtain the field F_2 consisting of two elements. This field plays a very important role in discrete mathematics, close to the role of the fields R and C in classic mathematics. To make certain that F_2 is a field despite being unlike fields Q, R or C, let us give a formal definition.

A set K of more than one element is called a field if for any two elements $a, b \in K$ there are defined their sum $a + b \in K$ and their product $a \cdot b \in K$ with the following properties:

$$\left.\begin{array}{ll}
1.1 & (a + b) + c = a + (b + c) \\
1.2 & a + b = b + a \\
1.3 & \text{there exists element } 0 \text{ such that } a + 0 = a \\
1.4 & \text{there exists element } (-a) \text{ such that } a + (-a) = 0 \\
2.1 & (a \cdot b) \cdot c = a \cdot (b \cdot c) \\
2.2 & a \cdot b = b \cdot a \\
2.3 & \text{there exists element } 1 \text{ such that } a \cdot 1 = a, \quad a \neq 0 \\
2.4 & \text{there exists element } a^{-1} \text{ such that } a \cdot a^{-1} = 1, \quad a \neq 0 \\
3 & (a + b) \cdot c = a \cdot c + b \cdot c
\end{array}\right\} . \tag{2.11}$$

The axioms of the first section mean that a field is a commutative group relative to addition (see 2.4–2.7); and the axioms of the second section mean that a field without element '0' is a commutative group relative to multiplication.

If we relax the axioms of the second section by excluding conditions 2.2 and 2.4, then we obtain the definition of an associative *ring* with unit. We can define the homomorphism (isomorphism) for the rings, as was done for groups, with the natural demand that condition (2.8) should be valid for operation of both addition and multiplication.

A finite field of q elements is denoted by F_q or $GF(q)$. In particular, the set of residues by modulo p, where p is the prime number forms the field F_p (or in other notation Z_p). The field F_p consists of integers $0, 1, \ldots, p - 1$; and to add or to multiply two elements of F_p means to add (or to multiply) these two elements just as integers, and then to find the remainder after division by p (this remainder is called the residue by module p).

If the equation $n \cdot 1_K = 0_K$ $(n \in Z)$ has only zero solution $n = 0$ in a field K, then the field K is said to be a field of zero characteristic. Otherwise, the field K is said to be a field of characteristic p, where p is a minimal positive integer such that $n \cdot 1_K = 0_K$. Thus, p is the order of the element 1_K as an element of the additive group of field K. It is easy to show that p should be a prime number. A field K of characteristic p contains the subfield $K(1) = \{n \cdot 1_K, n \in Z\}$, which is isomorphic to F_p; and a field K of zero characteristics contains the subfield, which is isomorphic to Q. In particular, a finite field of q elements exists if, and only if, q is a power of prime number p, where p is the characteristic of that field, and the field is unique (up to isomorphism).

For any field of characteristic p the following unusual identity ('truncated' Newton identity)

$$(a + b)^p = a^p + b^p \tag{2.12}$$

is true. Moreover, by virtue of the Lagrange theorem another useful identity is valid. Namely, for any $a \in F_q, a \neq 0$:

$$a^{q-1} = 1. \tag{2.13}$$

This is known as Fermat's 'small theorem'. This identity is equivalent to

$$a^q = a, \tag{2.14}$$

for any $a \in F_q$.

Many applications of finite fields in coding theory are based on the fact that it is possible to regard the finite field F_{p^m} as m-dimension *vector space* on the field F_p; and vice versa.

A set V is called *vector (linear) space* over a field K if

1. V is an Abelian group under addition.

2. For any $\mathbf{v} \in V$ and $\lambda \in K$, the *multiplication of vector by scalar* (or *scalar multiplication*) $\lambda \cdot \mathbf{v} \in V$ is defined. Moreover,

   2.1 $\lambda \ (\mathbf{v}_1 \mid \mathbf{v}_2) - \lambda \cdot \mathbf{v}_1 + \lambda \cdot \mathbf{v}_2,$

   2.2 $(\lambda_1 + \lambda_2) \cdot \mathbf{v} = \lambda_1 \cdot \mathbf{v} + \lambda_2 \cdot \mathbf{v};$

   2.3 $(\lambda_1 \cdot \lambda_2) \cdot \mathbf{v} = \lambda_1 \cdot (\lambda_2 \cdot \mathbf{v});$

   2.4 $1_K \cdot \mathbf{v} = \mathbf{v}.$

As an example of a vector space we can consider the so-called 'n-dimension *coordinate (arithmetic) space* K^n', the elements of which are the sequences $\mathbf{a} = (a_1, \ldots, a_n)$, $a_i \in K$; and the operations of the addition and the scalar multiplication are defined as follows:

$$\begin{aligned}
\mathbf{a} + \mathbf{b} &= (a_1 + b_1, \ldots, a_n + b_n) \\
\lambda \cdot \mathbf{a} &= (\lambda \cdot a_1, \ldots, \lambda \cdot a_n).
\end{aligned} \tag{2.15}$$

Vector $\sum_{i=1}^{n} \lambda_i \cdot \mathbf{v}_i$ is called a linear combination of vectors $\mathbf{v}_1, \ldots, \mathbf{v}_n$; and λ_i are coefficients of linear combination. The *basis* of vector space V over field K is the set of vectors $\mathbf{v}_1, \ldots, \mathbf{v}_m \in V$ such that any vector $\mathbf{x} \in V$ can be represented uniquely as the linear combination

$$\mathbf{x} = \sum_{i=1}^{m} \lambda_i \cdot \mathbf{v}_i, \qquad \lambda_i \in K.$$

The coefficients λ_i are called coordinates of vector $\mathbf{x}$ in the basis $\{\mathbf{v}_i, i = 1, \ldots, m\}$. All bases of given vector space V consist of the same number of vectors, referred to as the dimension of vector space V and denoted by $\dim V$. The vector space V is called m-dimension vector space, where $m = \dim V$. We consider only vector spaces of finite dimension $\dim V < \infty$.

A set of vectors $\{\mathbf{v}_i\}$ is said to be *linear independent* if a linear combination of these vectors is equal to 0 only if all coefficients of the linear combination are zeros. Otherwise the set $\{\mathbf{v}_i\}$ is said to be *linear dependent*. Another definition of the basis is that it is the maximal (in the sense of cardinal number) set of linear independent vectors.

The set of vectors $\{\mathbf{f}_i, \quad i = 1, \ldots, m\}$ with the coefficients f_{ij} in the basis $\mathbf{v}_1, \ldots, \mathbf{v}_n$ (i.e. $\mathbf{f}_i = \sum f_{ij} \cdot \mathbf{v}_j$) is the basis of the space if and only if $\det(f_{ij}) \neq 0$; calculation of the determinant should be done in the field K.

A mapping $A : V \to U$ of vector space V to vector space U is said to be linear if for any $\mathbf{v}, \mathbf{v}', \mathbf{v}'' \in V, \lambda \in K$,

$$A(\mathbf{v}' + \mathbf{v}'') = A(\mathbf{v}') + A(\mathbf{v}'');$$
$$A(\lambda \cdot \mathbf{v}) = \lambda \cdot A(\mathbf{v}). \tag{2.16}$$

A linear mapping is a homomorphism over the field K and, as in a case with groups and rings, the vector spaces V and U are called the isomorphic vector spaces if there exists one-to-one mapping $A : V \to U$. Since there exists one-to-one mapping of any vector $\mathbf{v} \in V$ to coordinates of this vector $\mathbf{v}$ in some fixed basis, an n-dimension vector space over the field K is isomorphic to the n-dimension coordinate space. Therefore, all vector spaces of the same dimension over the same field are isomorphic. In particular, any n-dimension vector space over the field F_q consists of $|K^n| = q^n$ vectors.

Let $\mathbf{v}_1, \ldots, \mathbf{v}_n$ be the basis of V and $\mathbf{u}_1, \ldots, \mathbf{u}_m$ be the basis of U, then every linear mapping $A : V \to U$ corresponds to $(m \times n)$ matrix $[a_{ij}]$, $i = 1, \ldots, m; j = 1, \ldots, n$, which coefficients are defined by the following equation:

$$A(\mathbf{v}_j) = \sum_{i=1}^{m} a_{ij} \cdot \mathbf{u}_i, \qquad j = 1, \ldots, n. \tag{2.17}$$

An important case of a linear mapping is a *linear functional* $f : V \to K^1$, which in accordance with (2.17) can be represented as $f(\mathbf{x}) = \sum f_i \cdot x_i$, where $f_i = f(\mathbf{v}_i)$, $\mathbf{x} = \sum x_i \cdot \mathbf{v}_i$. The set of all linear functionals forms a vector space V^* under the operations of addition of functionals and their multiplication by elements of the field K. This vector space is said to be dual to V and it has the same dimension: $\dim V^* = \dim V$.

Another important case is the linear mapping $A : V \to V$, called a *linear operator* over V. In this case $\mathbf{u}_i = \mathbf{v}_i$ (only single basis is used); and the linear operator A corresponds to a square $(n \times n)$ matrix $[a_{ij}]$ such that

$$A(\mathbf{v}_j) = \sum_{i=1}^{n} a_{ij} \cdot \mathbf{v}_i, \qquad j = 1, \ldots, n. \tag{2.18}$$

It is possible not only to add linear operators and multiply them by an element of the field but also to multiply operator by operator (as mappings). In this case the operator $E : E(\mathbf{v}) = \mathbf{v}$, referred to as a unit operator, is the unit (the neutral element under multiplication), since for any operator A, $EA = AE = A$. The matrix of operator E in any basis can be written as $\mathbf{E}_n = [\delta_{ij}]$,

$$\mathbf{E}_n = [\delta_{ij}],$$
$$\text{where } \delta_{ij} = \begin{cases} 1 & \text{if } i = j; \\ 0 & \text{if } i \neq j. \end{cases} \tag{2.19}$$

The existence of the inverse operator $A^{-1}: A \cdot A^{-1} = A^{-1} \cdot A = E$ is equivalent to any one of the following conditions:

- $\text{Ker } A = \{\mathbf{v} \in V : \quad A(\mathbf{v}) = 0\} = \{0\} (\text{Nonsingular});$

- $\text{Im } A = \{A(\mathbf{v}) : \quad \mathbf{v} \in V\} = V;$

- $\det(a_{ij}) \neq 0.$

The set of nonsingular operators forms the group under addition, called the general linear group and denoted by $GL(n, K)$. If a metric is defined on V, then the subset of nonsingular linear operators such that any of them preserves the distance between points of V, forms the subgroup. Such an operator is called an isometry operator, and the corresponding group is called the isometry group. For a Hamming metric this subgroup consists of linear operators $\sigma \cdot \Delta_A$, where σ is the linear operator of permutation $(\sigma(\mathbf{e}_i) = \mathbf{e}_{\sigma(i)})$; $\Delta_A(\mathbf{e}_j) = \lambda_j \cdot \mathbf{e}_j$ is the 'diagonal' operator; and $\mathbf{e}_i$ is the i th row of matrix $\mathbf{E}_n$, $\mathbf{e}_i$ has only one nonzero component at i th position.

The subset $L \in V$ is said to be a *linear subspace* of the space V if L is the subgroup of V under addition, and for any $\mathbf{l}, \mathbf{l}_1, \mathbf{l}_2 \in L$, $\lambda \in K$, the following statements are correct: $\mathbf{l}_1 - \mathbf{l}_2 \in L$ and $\lambda \cdot \mathbf{l} \in L$. In other words, L is the linear space under the operations of vector addition and multiplication by scalar, defined on the whole set V. Therefore, the basis $\mathbf{l}_1, \ldots, \mathbf{l}_k$ exists in L, where $k = \dim L$. Thus we obtain the definition of subspace L as follows:

$$L = \left\{ \mathbf{x}; \, \mathbf{x} = \sum_{i=1}^{k} \lambda_i \cdot \mathbf{l}_i \right\}. \tag{2.20}$$

An important fact is that any basis of a subspace could be extended up to the basis of the full space.

A linear subspace can be also described as solutions of some system of linear equations. Define the subspace L^* of the dual space V^*, which consists of all linear functionals such that any of these functionals equal to zero for any vector of L:

$$L^* = \{ f \in V^* : f(\mathbf{l}) = 0, \quad \mathbf{l} \in L \}. \tag{2.21}$$

This subspace is said to be dual to L. Then for any basis $f_1, \ldots, f_r$ of space L^*, where $r = \dim L^* = n - \dim L$, L is the set of solutions of system of r linear equations:

$$L = \{ \mathbf{x} \in V : f_i(\mathbf{x}) = 0, \quad i = 1, 2, \ldots, r \}. \tag{2.22}$$

Define the scalar product of two vectors $\mathbf{x} = \sum x_i \cdot \mathbf{v}_i$ and $\mathbf{y} = \sum y_i \cdot \mathbf{v}_i$, $x_i, y_i \in K$ as

$$(\mathbf{x}, \mathbf{y}) = \sum_{i=1}^{n} x_i \cdot y_i. \tag{2.23}$$

Then an arbitrary linear functional f can be represented as

$$f(\mathbf{x}) = (\mathbf{F}, \mathbf{x}), \tag{2.24}$$

where $\mathbf{F} = (f(\mathbf{v}_1), \ldots, f(\mathbf{v}_n))$. The equation (2.24) establishes the isomorphism between V and V^*. In this case dual subspace L^* corresponds to a so-called orthogonal complement to L, denoted by $\bar{L}$ and defined as

$$\bar{L} = \{ \mathbf{x} \in V : (\mathbf{x}, \mathbf{l}) = 0 \} \textit{ for all } \mathbf{l} \in L. \tag{2.25}$$

Note that for the fields of finite characteristic the 'usual' (for fields of zero characteristic, like the field of complex numbers) property $L \cap \bar{L} = 0$ is not correct; and it is possible that L intersects $\bar{L}$. Furthermore, L could belong to $\bar{L}$ and such a subspace is called *self-orthogonal*.

Example 2.3 Consider 5-dimensional vector space V of all 5-tuples over the finite field F_2. The operations of vector addition and scalar multiplication are defined according to (2.15) and the scalar product of two vectors is defined by (2.23). The following four vectors form a 2-dimensional subspace L of V:

$$(0\,0\,0\,0\,0), (0\,0\,0\,1\,1), (0\,1\,1\,0\,0), (0\,1\,1\,1\,1).$$

The orthogonal complement $\bar{L}$ consists of the following 8 vectors:

$$\begin{matrix} (0\,0\,0\,0\,0), & (0\,0\,0\,1\,1), & (0\,1\,1\,0\,0), & (0\,1\,1\,1\,1), \\ (1\,0\,0\,0\,0), & (1\,0\,0\,1\,1), & (1\,1\,1\,0\,0), & (1\,1\,1\,1\,1). \end{matrix}$$

The dimension of $\bar{L}$ is 3. Clearly $L \cap \bar{L} \neq 0$. Moreover $L \subset \bar{L}$, and hence, the subspace L is the self-orthogonal subspace.

For any subspace L of any vector space V, $\bar{\bar{L}} = L$ and $\dim L + \dim \bar{L} = \dim V$.

For any linear mapping $A : V \to U$, the dimension of the kernel $\mathrm{Ker}\, A \subset V$ is connected with the dimension of the image $\mathrm{Im}\, A \subset U$ by the following equation:

$$\dim \mathrm{Ker}\, A + \dim \mathrm{Im}\, A = \dim V.$$

Now we will pay a special attention to the ring of polynomials and its quotient rings. The most important example of associative rings is the ring of integers Z. Another very important example, which is very close to the previous one, is the ring $K[x]$ of the polynomials with coefficients from the field K. The elements of the $K[x]$ are the polynomials, i.e. the sequences $f = (f_0, f_1, \ldots)$, where $f_i \in K$. The maximal m such that $f_m \neq 0$ is called the *degree of the polynomial* and denoted by $\deg f(x)$, where $f(x)$ is more usually representative of the polynomial with one *variable* x

$$f(x) = f_0 + f_1 x + \cdots + f_m x^m. \tag{2.26}$$

If $f_m = 1$ then a polynomial is called *normalised*. Two polynomials can be added and multiplied in accordance with standard formulas:

$$f(x) + g(x) = (f_0 + g_0) + (f_1 + g_1)x + \cdots + (f_i + g_i)x^i + \cdots, \tag{2.27}$$

$$f(x) \cdot g(x) = h_0 + h_1 x + \cdots + h_i x^i + \cdots, \text{ where } h_k = \sum_{i+j=k} f_i \cdot g_j, \tag{2.28}$$

and

$$\deg(f(x) + g(x)) \leq \max(\deg f(x), \deg g(x)), \quad \deg(f(x) \cdot g(x)) = \deg f(x) \cdot \deg g(x). \tag{2.29}$$

Example 2.4

$$f(x) = x^4 + 3x^2 + 4, \quad g(x) = 2x^2 + x + 3, \quad K = F_5$$
$$f(x) + g(x) = x^4 + (3+2)x^2 + x + (4+3) = x^4 + x + 2$$
$$f(x) \cdot g(x) = (1 \cdot 2)x^6 + (1 \cdot 1)x^5 + (1 \cdot 3 + 3 \cdot 2)x^4 + (3 \cdot 1)x^3$$
$$+ (3 \cdot 3 + 4 \cdot 2)x^2 + (4 \cdot 1)x + 4 \cdot 3$$
$$= 2x^6 + x^5 + 4x^4 + 3x^3 + 2x^2 + 4x + 2$$

It is easy to verify that zero element and unit element of the ring $K[x]$ are polynomials $\mathbf{0} = (0, \ldots, 0)$ and $\mathbf{1} = (1, 0, \ldots, 0)$ respectively, and $f(x) \cdot g(x) = \mathbf{0}$ if and only if $f(x) = \mathbf{0}$ or $g(x) = \mathbf{0}$. The latter property means that there are no divisors of zero in $K[x]$. Moreover, the ring $K[x]$ as well as the ring Z is an example of the *Euclidean* ring. The commutative ring L is said to be Euclidean if there is a nonnegative integer function $\mu()$ defined on $L \backslash \{0\}$ such that

1. $\mu(a \cdot b) \geq \mu(a)$ for all $a, b \neq 0$ from L;

2. for any $a, b \in L$, $\quad b \neq 0$ there exist $q, r \in L$ (the quotient and the reminder) such that

$$a = q \cdot b + r, \qquad \text{where } \mu(r) < \mu(b) \text{ or } r = 0.$$

The reminder r is also called the *residue* of a by modulo b and denoted by $a \bmod b$. For Z this function is $|a|$ and for $K[x]$ $\mu(f(x)) = \deg f(x)$. The property 2 can be realized by the usual algorithm of division of polynomials.

Example 2.5

$$a(x) = x^7 - 2x^5 + 4x^3 + 2x^2 - 2x + 2$$
$$b(x) = 2x^5 + 3x + 4, \qquad K = F_5$$

$$
\begin{array}{r}
3x^2 - 1 \\
2x^5 + 3x + 4 \overline{)\; x^7 - 2x^5 + 4x^3 + 2x^2 - 2x + 2} \\
x^7 + 4x^3 + 2x^2 \\
\hline
- 2x^5 - 2x + 2 \\
-2x^5 - 3x - 4 \\
\hline
x + 1
\end{array}
$$

As the result of the division $a(x)$ by $b(x)$ we obtain the quotient $q(x) = 3x^2 - 1$ and the remainder $r(x) = x + 1$, i.e.

$$a(x) = q(x) \cdot b(x) + r(x) = x^7 - 2x^5 + 4x^3 + 2x^2 - 2x + 2$$
$$= (3x^2 - 1) \cdot (2x^5 + 3x + 4) + x + 1$$

A subgroup $I \subset L$ of the additive group of the ring L is called an *ideal* of the commutative ring L if for all $a \in L$, $v \in I$,

$$a \cdot v \in I.$$

The set L/I of cosets $\{I + a\}$ is the ring under the operations of addition and multiplication, defined on the ring L: $\quad \{a+I\} + \{b+I\} = \{a+b+I\}; \{a+I\} \cdot \{b+I\} = \{a \cdot b+I\}$. The ring L/I is called the *quotient ring* or the *residue class ring* by modulo of ideal I. The simplest example of an ideal is the *principal ideal* $V = \{a \cdot v : a \in L\}$, generated by an element $v \in L$. Any Euclidean ring is the ring of principal ideals, i.e., there are no ideals except the principal ones; and the elements of the quotient ring L/V can be represented as elements $r \in L$ such that $\mu(r) < \mu(v)$ or $r = 0$ if we define operations as follows:

$$(r_1 + r_2)_{L/V} = (r_1 + r_2) \bmod v,$$
$$(r_1 \cdot r_2)_{L/V} = (r_1 \cdot r_2) \bmod v. \tag{2.30}$$

For example, for the ring $K[x]$ and the principal ideal generated by $g(x)$, elements of the quotient ring $K[x]/g(x)$ are the polynomials of the degree less than $n = \deg g(x)$. These elements can be added as usual polynomials; and the multiplication of polynomials by modulo $g(x)$ is chosen as the operation of multiplication.

It is a well-known fact that any positive integer can be uniquely represented as the product of prime numbers. This statement can be generalised to any ring of principal ideals, in particular to the Euclidean rings. Let us restrict our attention to the case of the ring $K[x]$. The polynomial $f(x)$, $\deg f(x) \geq 1$ is said to be *irreducible* over the field K if it cannot be represented as a product of two polynomials (with coefficients of K) of nonzero degree.

Example 2.6 $f_1(x) = x^2 - 2$ is the irreducible polynomial over Q, $f_2(x) = x^2 + 1$ is the irreducible polynomial over R, $f_3(x) = x^2 + x + 1$ is the irreducible polynomial over F_2. Notice that the field of the coefficients is significant. For instance, $f_1(x)$ is reducible polynomial over R, $f_2(x)$ can be reduced over C and $f_3(x)$ can be reduced over F_4.

Theorem 2.1 Any polynomial $f(x) \in K[x]$ can be uniquely represented as the product of the element of the field K and the irreducible normalised polynomials.

Notice that for the ring $K[x]$ there are known simple algorithms (with the polynomial complexity) of factorisation of polynomials to irreducible polynomials, as distinguished from the case of Z [3].

Consider calculations over finite fields since it is a very important issue for the main part of codes constructed with the help of algebraic coding theory. Let us outline several useful (for calculations) properties of finite fields. First of all, the field F_{p^m} can be represented as the m-dimension vector space over the field F_p, where p is prime number. That means the addition of the elements in F_{p^m} can be regarded as the addition by modulo p of the m-tuples. Secondly, the multiplicative group of the field F_q consist of $q-1$ elements and it is the cyclic group, i.e., there is at least one *primitive element* $\gamma \in F_q$ such that $a = \gamma^i$, $\quad 0 \leq i < q-1$, for any $a \in F_q$, $\quad a \neq 0$; the number i is called the logarithm of a to the base γ and denoted by $\log_\gamma a$. In fact, there are $\varphi(q-1)$ primitive elements of the field F_q, where $\varphi()$ is the Euler's function. This property allows us to use the 'logarithmic' representation of the elements in the process of the multiplication:

$$\log(a \cdot b) = (\log a + \log b) \bmod (q-1). \tag{2.31}$$

One more useful (but unusual in comparison with fields Q, R and C) property of the finite field F_{p^m} was mentioned above (2.12), (2.14):

$$(a + b)^p = a^p + b^p, \tag{2.32}$$

for any $a, b \in F_{p^m}$;

$$(\lambda \cdot a)^p = \lambda \cdot a^p, \tag{2.33}$$

for any $a \in F_{p^m}, \lambda \in F_p$. Therefore, the mapping $\sigma_p : F_{p^m} \to F_{p^m}$ $(a \to a^p)$ is a linear operator on F_{p^m} regarding as the m-dimension vector space over the field F_p. Moreover, the mappings σ_{p^i}, $i = 0, 1, \ldots, m - 1$ are the automorphisms of the field F_{p^m} and form the group.

2.3 LINEAR BLOCK CODES

Let us return to the model of a reliable data communications system, described in section 2.1. A discrete memoryless channel (we restrict consideration to this class of channels) is defined by the crossover probability $p(x/y)$ that is the (conditional) probability of receiving a q-ary symbol y as channel's output if a q-ary symbol x was transmitted. Let an additive group structure be defined on the q-ary channel alphabet A (for instance, consider A as the group Z_q^+ of residues by modulo q). If the received word $\mathbf{y} = (y_1, \ldots, y_n)$ does not coincide with the transmitted word $\mathbf{x} = (x_1, \ldots, x_n)$ then it is said that the error (or error vector) occurs during the data transmission over the channel. If

$$P(\mathbf{y}|\mathbf{x}) = P(\mathbf{y} - \mathbf{x}|\mathbf{0}) = P(\mathbf{y} - \mathbf{x}), \tag{2.34}$$

then such kind of channel is called a *channel with additive noise* (or *additive channel*).

It is natural to use for a channel with additive noise some codes capable of correcting some set of errors E. A code $V \subset A^n$ can correct a set of errors $E = \{\mathbf{0}, \mathbf{e}_1, \ldots, \mathbf{e}_m\}$ if any equation $\mathbf{v} + \mathbf{e} = \mathbf{v}' + \mathbf{e}'$, where $\mathbf{v}, \mathbf{v}' \in V$, $\mathbf{e}, \mathbf{e}' \in E$ has the unique solution $\mathbf{v} = \mathbf{v}'$ and $\mathbf{e} = \mathbf{e}'$. The choice of a set of correctable errors E should depend on the probability distribution of errors $\mathbf{P}$. Since the code V with a set of correctable errors E guarantees the decoding error probability P_e no more than $1 - \sum_{\mathbf{e} \in E} P(\mathbf{e})$ then a suitable set E is usually formed in such a way as to include the most probable error patterns. Therefore the problem of the construction of the corresponding optimal code, i.e. the code with maximum cardinal number (or with maximum code rate), should be investigated. Let us note however, that this choice of E is not necessarily the best in the sense of maximum code rate for a given decoding error probability P_e.

The *Hamming weight* of vector $\mathbf{x} = (x_1, \ldots, x_n)$, denoted by $wt(\mathbf{x})$, is defined as the number of nonzero components of $\mathbf{x}$. If a group under addition is defined on alphabet A, and a set A^n of all words of length n is regarded as a group under component-wise addition, then the relation between Hamming distance and Hamming weight can be written as follows

$$d(\mathbf{x}, \mathbf{y}) = wt(\mathbf{x} - \mathbf{y}). \tag{2.35}$$

Consider codes which are the subgroups of A^n and called group *codes*. To calculate the distance of a group code by virtue of property (2.35) it is enough to find the minimum weight of its nonzero codewords, i.e.

$$d(V) = \min_{\mathbf{v}\in V, \mathbf{v}\neq\mathbf{0}} wt(\mathbf{v}). \tag{2.36}$$

The group structure on A is not enough to construct good codes in A^n; the main results of coding theory are obtained in the case when q is a prime power when the alphabet A can be regarded as the finite field F_q, and A^n is regarded as n-dimension vector space F_q^n over F_q. By definition, a q-ary *linear block* (n,k) *code* is an arbitrary k-dimension subspace of vector space F_q^n.

Since a linear block code is a group code, equation (2.36) is correct for any linear block code. Notice that in the case $q = p$, where p is a prime number, the definition of a linear code coincides with the definition of a group code.

Since the number of vectors in the arbitrary k-dimension subspace of vector space F_q^n over the field F_q is equal to q^k, the number of messages M that it is possible to transmit by a q-ary (n,k) code is the same. It is convenient to represent these $M = q^k$ messages as k-dimension vectors $\mathbf{u}_i = (u_i^1, \ldots, u_i^k)$, $i = 1, \ldots, M$, from F_q^k, i.e. $\{\mathbf{u}_1, \ldots, \mathbf{u}_M\} = F_q^k$.

In the previous section two methods of description of linear subspaces were presented. Let us start from the first of them. Consider an (n,k) code V and $k \times n$ matrix $\mathbf{G}$, which rows are vectors $\mathbf{v}_1, \ldots, \mathbf{v}_k$ forming a basis of the subspace V, i.e.

$$\mathbf{G} = [g_{ij}], \quad \text{where } (g_{i1}, \ldots, g_{in}) = \mathbf{v}_i$$

Matrix $\mathbf{G}$ is called a *generator matrix* of the code. Every (n,k) code has exactly $\prod_{i=0}^{k-1} (q^k - q^i)$ bases and, therefore, the same number of generator matrices. Each generator matrix defines the encoding procedure $\varphi_{\mathbf{G}} : \quad F_q^k \to V$ by the following formula

$$\varphi_{\mathbf{G}}(\mathbf{u}) = \varphi_{\mathbf{G}}(u_1, \ldots, u_k) = \mathbf{u} \cdot \mathbf{G} = \sum_{i=1}^{k} u_i \cdot \mathbf{v}_i, \tag{2.37}$$

which is a linear mapping. Let $\mathbf{G}$ be some generator matrix of an (n,k) code V. Then an arbitrary generator matrix $\mathbf{G}'$ of this code can be represented as $\mathbf{G}' = \mathbf{C} \cdot \mathbf{G}$, where $\mathbf{C}$ is a nonsingular $k \times k$ matrix. Let us split the generator $k \times n$ matrix into matrices $\mathbf{G}_1$ and $\mathbf{G}_2$

$$\mathbf{G} = [\mathbf{G}_1 | \mathbf{G}_2], \tag{2.38}$$

where $\mathbf{G}_1$ is $k \times k$ matrix, and $\mathbf{G}_2$ is $k \times (n - k)$ matrix. If $\mathbf{G}_1$ is a nonsingular matrix, then matrix $\mathbf{G}'$

$$\mathbf{G}' = \mathbf{G}_1^{-1} \cdot \mathbf{G} = [\mathbf{I}_k | \mathbf{G}_2'], \quad \mathbf{G}_2' = \mathbf{G}_1^{-1} \cdot \mathbf{G}_2 \tag{2.39}$$

is also a generator matrix of the code V and defines in accordance with (2.37) the encoding procedure

$$\varphi_{\mathbf{G}}(u_1, \ldots, u_k) = (v_1, \ldots, v_k, v_{k+1}, \ldots, v_{k+r}) = (\mathbf{u}, \mathbf{u} \cdot \mathbf{G}_2'). \tag{2.40}$$

Such an encoding procedure is called *systematic encoding*, because the first k symbols of any codeword coincide with the corresponding symbols of an uncoded message ($v_i = u_i$, $i = 1, \ldots, k$). A code with generator matrix $\mathbf{G}'$ is called a systematic *code*. Not every linear code is systematic, because matrix $\mathbf{G}_1$ in (2.38) may appear to be a singular matrix. However, it is always possible to find k linear independent columns of the matrix $\mathbf{G}$ (since rank over the columns coincides with rank over the rows). Therefore, it is possible to transform the code V to a systematic form by some permutation of coordinates, i.e. any (n, k) code is *equivalent* to a systematic one. Hereafter, we often assume that considered (n, k) codes are systematic.

Let code V be a systematic code and matrix $\mathbf{G}'$ have the same form as in (2.39), then as previously mentioned, V can be defined as the set of solutions of the following system of linear equations

$$\mathbf{H} \cdot \mathbf{v}^T = \mathbf{0}, \tag{2.41}$$

where $\mathbf{H} = [-\mathbf{G}'_2 | \mathbf{I}_r]$. It means that the matrix $\mathbf{H}$ is a generator matrix of the subspace $\overline{V} = \{\mathbf{x} \in L : (\mathbf{x}, \mathbf{v}) = 0, \quad \mathbf{v} \in V\}$, which is called the *dual code*. This statement immediately follows from the substitution of equation (2.40) in (2.41), which shows that equation (2.41) is correct for any codeword and from the comparison of the dimensions ($\dim V + \dim \overline{V} = n$). The matrix $\mathbf{H}$ satisfying (2.41) is called a *parity-check matrix* of the code V. The equation (2.41) is the equation of linear dependence between those columns $\mathbf{h}_i$ of matrix $\mathbf{H}$, where $v_i \neq 0$. It leads immediately to the following useful result.

Lemma 2.1 (Bose criterion). The minimum distance of a code V is no less than d if any $d - 1$ columns of its parity-check matrix $\mathbf{H}$ are linear independent.

It follows from lemma 2.1 that to construct the code capable of correcting single errors, the matrix $\mathbf{H}$ with non-collinear columns should be constructed. For instance, a maximal (in number of columns) matrix $\mathbf{H}$ can be constructed by induction:

$$\mathbf{H}_r = \left[\begin{array}{ccc|cccc} 1 & \cdots & 1 & 0 & 0 & \cdots & 0 \\ & * & & & \mathbf{H}_{r-1} & & \end{array} \right]$$

or, what is the same,

$$\mathbf{H}_r = \left[\begin{array}{ccc|cccc} 1 & \cdots & 1 & 0 & 0 & \cdots & 0 \\ & & & 1 & 1 & & \vdots \\ & & & & & & 0 \\ & F_q^{r-1} & & & F_q^{r-2} & & 1 \end{array} \right], \tag{2.42}$$

The equation (2.42) allows detection of t errors in a very simple manner. Namely, it is enough to calculate vector $\mathbf{S}$ called a *syndrome*

$$\mathbf{S} = \mathbf{H} \cdot \mathbf{b}^T \text{ or } \mathbf{S} = \mathbf{b} \cdot \mathbf{H}^T, \tag{2.43}$$

where $\mathbf{b}$ is the received vector (depending on whether the type of calculation $\mathbf{S}$ is vector-row or vector-column), and check if $\mathbf{S}$ is equal to zero or not since $\mathbf{S} = 0$ if, and only if, $\mathbf{b}$

belongs to the code. Notice that the value of the syndrome depends not only on the vector **b** but also on the form of the parity-check matrix of the code. This fact we will use later when considering decoding algorithms of linear codes.

Of course, nontrivial code (i.e. code which consists of more than one word) cannot correct any errors. In particular, if errors that occurred in the channel form a codeword, then the received vector **b** is a codeword but not the transmitted one. Such kinds of error cannot be detected because the syndrome of the received vector is equal to zero. Let us introduce the concept of a *standard array* to describe errors, which can be corrected and detected by the code.

Let V be an (n, k) linear binary code $(n - k = r)$. Let $\mathbf{v}_0, \mathbf{v}_1, \ldots, \mathbf{v}_{2^k-1}$ be all codewords of the code

$$V = \{\mathbf{v}_0, \mathbf{v}_1, \ldots, \mathbf{v}_{2^k-1}\},$$

where $\mathbf{v}_0$ is the all-zero word. Let us form the table of 2^k columns and 2^r rows as follows. Any row consists of 2^k vectors. The first row we constrain all codewords with $\mathbf{v}_0$ as the first element of the row. Then we take any n-vector $\mathbf{e}_1$, which does not belong to the code; the second row consists of elements that are the sum $\mathbf{e}_1 + \mathbf{v}_i$, $i = 0, \ldots 2^k - 1$. Then we choose an element $\mathbf{e}_2$, which does not belong to the first and the second row and form the third from the sums $\mathbf{e}_2 + \mathbf{v}_i$. We continue this process until all vector space is exhausted. As a result of this procedure we obtain an array, which is called a standard array:

$$
\begin{array}{ccccc}
\mathbf{v}_0 & \mathbf{v}_1 & \mathbf{v}_2 & \cdots & \mathbf{v}_{2^k-1} \\
\mathbf{e}_1 + \mathbf{v}_0 & \mathbf{e}_1 + \mathbf{v}_1 & \mathbf{e}_1 + \mathbf{v}_2 & \cdots & \mathbf{e}_1 + \mathbf{v}_{2^k-1} \\
\cdots\cdots & \cdots\cdots & \cdots\cdots & \cdots & \cdots\cdots \\
\mathbf{e}_{2^r-1} + \mathbf{v}_0 & \mathbf{e}_{2^r-1} + \mathbf{v}_1 & \mathbf{e}_{2^r-1} + \mathbf{v}_2 & \cdots & \mathbf{e}_{2^r-1} + \mathbf{v}_{2^k-1}
\end{array}
\qquad (2.44)
$$

It is obvious that different rows of this array do not contain the same elements. Therefore, the number of rows is equal to 2^r. The syndromes of all vectors in the same row are identical:

$$\mathbf{H} \cdot (\mathbf{e}_i + \mathbf{v}_{j_1})^{\mathrm{T}} = \mathbf{H} \cdot (\mathbf{e}_i + \mathbf{v}_{j_2})^{\mathrm{T}} = \mathbf{H} \cdot \mathbf{e}_i^{\mathrm{T}};$$

and the syndromes of the elements from the different rows are different.

The standard array is the method of writing the whole n-dimension vector space. There can occur any error vector in the channel, but the code can correct only one received vector from the row of the standard array, because the vectors, placed in the same row, have the identical syndromes. The rows of the standard array are usually called the *cosets* of the code and the elements in the first column are called *coset leaders*. Any element in the row (in the coset) can be used as the coset leader.

A binary linear code can correct only 2^r vectors, which is significantly less than the overall number of possible error vectors 2^n. However, in most channels the different error vectors have different probabilities. In any channel it is necessary to choose the most probable error vectors as the coset leaders to realise the decoding on *maximum likelihood*. In particular, in the channel with independent errors the vectors with minimum Hamming weight should be chosen as the coset leaders.

If $\mathbf{e}_0, \mathbf{e}_1, \ldots, \mathbf{e}_{2^r-1}$ ($\mathbf{e}_0$ is the all-zero vector) are the coset leaders of code V, then the decoding error probability, provided by this code P_e is

$$P_e = 1 - \sum_{i=0}^{2^r-1} P(\mathbf{e}_i), \qquad (2.45)$$

where $P(\mathbf{e}_i)$ is the probability of vector $\mathbf{e}_i$ being the error vector in the channel. A code can be used in the channel if

$$1 - P_{e.acc.} < \sum_{i=0}^{2^r-1} P(\mathbf{e}_i),$$

where $P_{e.acc.}$ is the acceptable error probability.

To calculate the error probability with the help of formula (2.45) it is necessary to calculate 2^r probabilities, which is, as usual, a problem of very high complexity. Notice that the coding theorems of information theory show that there should be a subset of coset leaders among those sufficiently long codes with a code rate less than the channel capacity, which includes the set of the most probable channel error vectors. That is, there exists a code that provides an arbitrary small value of error probability P_e. The formula (2.45) defines the exact value of error probability provided by the code in the channel with independent errors with minimum distance decoding. The estimations of error probability, based on the use of minimum distance, can be obtained for the case of decoding in the hypersphere of radius ν. Let us find out the size of radius ν to show the decoding in the hypersphere is very close (in the sense of error probability) to the minimum distance decoding.

Let A^n be the set of n-tuples with symbols from the alphabet A; and let E_r be the set of q^r most probable error vectors $\mathbf{e} \in A^n$. Let V be the (n, k)-code ($n - k = r$) over A. Let E_V be the set of leader cosets of code V, and let $P(\mathrm{B})$ be the probability of error vector in the channel be a vector from some set B.

Lemma 2.2 [6]:

$$P(A^n \backslash (E_V \cap E_r)) \leq 2P(A^n \backslash E_V).$$

Proof. Since the number of elements in E_r is equal to the number of elements in E_V, then $|E_r \backslash (E_V \cap E_r)| = |E_V \backslash (E_V \cap E_r)|$. Therefore, in accordance with the definition of set E_r,

$$P(E_r \backslash (E_V \cap E_r)) \geq P(E_V \backslash (E_V \cap E_r)). \qquad (2.46)$$

Then from the obvious inclusion:

$$A^n \backslash E_V \supseteq E_r \backslash (E_V \cap E_r),$$

and in accordance with (2.46) it follows that

$$P(E_V \backslash (E_V \cap E_r)) \leq P(A^n \backslash E_V). \qquad (2.47)$$

And from the equation

$$A^n \backslash (E_V \cap E_r) = (A^n \backslash E_V) \cup (E_V \backslash (E_V \cap E_r)),$$

and from the inequality (2.47) we obtain:

$$P(A^n \backslash (E_V \cap E_r)) = P(A^n \backslash E_V) + P(E_V \backslash (E_V \cap E_r)) \leq 2P(A^n \backslash E_V).$$

Lemma 2.2 shows that decoding only those coset leaders, which belong to the set E_r (instead of decoding all error vectors, that can be corrected by the code) leads to the situation that the error probability $P(A^n \backslash (E_V \cap E_r))$ will not exceed the double error probability for decoding on the maximum likelihood $2P(A^n \backslash E_V)$.

Decoding in the hypersphere of radius ν means that the received vector is decoded to the nearest codeword at a distance no more than ν from the received vector. Moreover, the received vector is compared only with coset leaders of weight no more than ν. Therefore, to make it possible that the error probability for decoding in the hypersphere does not exceed more than two times the error probability for maximum likelihood decoding, it is necessary to choose the minimum value of ν, satisfying

$$A_\nu^n(\mathbf{0}) \supseteq E_r \cap E_V,$$

where $A_\nu^n(\mathbf{0})$ is the hypersphere of radius ν and with the center in all-zero vector.

In particular, it is enough $A_\nu^n(\mathbf{0}) \supseteq E_r$; and for BSC it means that $|A_\nu^n(\mathbf{0})| = \sum_{i=0}^{\nu} \binom{n}{i} \geq 2^r$.

Notice that given proof does not depend on the error model, i.e. this proof is applicable to any additive channel. Moreover, the proof does not depend on the method of full decoding, i.e. the proof is correct for any full decoding algorithm, not only for the maximum likelihood decoding.

2.4   CYCLIC CODES

Cyclic codes form the most explored subclass of linear codes. The majority of known good codes are also cyclic codes. There is a simple encoding procedure for these codes and there are also simple decoding procedures for many of the cyclic codes.

Definition   A linear code is called a *cyclic code* if every cyclic shift of a codeword is also a codeword.

Thus, if $\mathbf{a} = (a_0, a_1, \ldots, a_{n-1})$ is the codeword of the cyclic code of length n, then the cyclic shift of this codeword $T(\mathbf{a}) = (a_{n-1}, a_0, a_1, \ldots, a_{n-2})$ is the codeword of the same code. Let each n-dimension vector $\mathbf{f} = (f_0, f_1, \ldots, f_{n-1})$, $f_i \in K$ correspond to the polynomial $f(x) = f_0 + f_1 x + \ldots + f_{n-1} x^{n-1} \in K[x]$. Then each n-tuple corresponds to the polynomial of degree of no more than $n - 1$. Hereafter we will not distinguish between vector and the corresponding polynomial.

Let $a(x)$ be the codeword of the cyclic code of length n. Consider the polynomial $xa(x) \bmod (x^n - 1)$:

$$xa(x) = a_{n-1}x^n + a_{n-2}x^{n-1} + \ldots + a_1x^2 + a_0x,$$

and the residue of $xa(x)$ on modulo $(x^n - 1)$ is equal to

$$xa(x) \bmod (x^n - 1) = a_{n-2}x^{n-1} + \ldots + a_1x^2 + a_0x + a_{n-1}. \tag{2.48}$$

The right side of the equation (2.48) is the cyclic shift of codeword $a(x)$. Therefore, $xa(x) \bmod (x^n - 1)$ is the codeword of the cyclic code. Considering the cyclic shifts of vector $a(x)$: $xa(x) \bmod (x^n - 1)$, $x^2a(x) \bmod (x^n - 1)$, etc. obtain that any polynomial $x^ia(x) \bmod (x^n - 1)$ is the codeword. Since the cyclic code is the linear code, each linear combination of it codewords is also the codeword, i.e. all polynomials

$$\sum_{i,j} \lambda_i \cdot x^j a(x) \bmod (x^n - 1), \qquad \lambda_i \in K \tag{2.49}$$

are the codewords. Thus, the set of codewords is an ideal in the ring $K[x]/(x^n - 1)$. As mentioned above $K[x]/f(x)$ is the ring of principal ideals. Therefore, there exists the element $g(x) \in K[x]/(x^n - 1)$ such that $I = \langle g(x) \rangle$, i.e. this element generates the cyclic code I. It is convenient to choose a nonzero normalised polynomial of minimum degree as the element $g(x)$. Then it is easy to verify that any codeword $v(x)$ of the code I can be represented uniquely as

$$v(x) = m(x) \cdot g(x), \qquad \deg m(x) < n - \deg g(x). \tag{2.50}$$

Let us consider the division of $v(x)$ by $g(x)$:

$$v(x) = m(x) \cdot g(x) + r(x),$$

where $\deg r(x) < \deg g(x)$ or $r(x) = 0$. The first statement cannot be correct since in that case $r(x) = v(x) - m(x) \cdot g(x) \in I$, i.e. $r(x)$ is the codeword (polynomial) of degree less than degree of $g(x)$; and this contradicts the choice of $g(x)$. The fact that the polynomial $m(x) \cdot g(x)$ belongs to the code follows from the properties of an ideal. The uniqueness of the representation (2.50) follows from the fact that there are no divisors of zero in the ring of polynomials. The polynomial $g(x)$ is called the *generator polynomial* of the code. Notice that the generator polynomial $g(x)$ is the divisor of the polynomial $x^n - 1$. Since the degree of the polynomial $x^{n-\deg g(x)}g(x)$ is equal to n, then it can be represented as

$$x^{n-\deg g(x)}g(x) = x^n - 1 + r(x), \tag{2.51}$$

where $r(x) = (x^{n-\deg g(x)}g(x)) \bmod (x^n - 1)$. In accordance with (2.49) $r(x)$ is the codeword, i.e. $g(x)$ is the divisor of $r(x)$. Then from (2.51) it follows that $x^n - 1$ is also divisible by $g(x)$. We showed that all codewords could be represented as (2.50); the number of such words is equal to the number of possible choices of the information polynomial $m(x)$, i.e. $q^{n-\deg g(x)} = q^k$. The number of information symbols of the code $k = n - \deg g(x)$. The generator matrix $\mathbf{G}$ of the cyclic code can be formed in accordance with (2.50) by the cyclic shifts of $g(x)$:

$$\mathbf{G} = \left.\begin{bmatrix} g_0 & \cdot & \cdot & \cdot & \cdot & g_r & & & \\ & g_0 & & & & & g_r & & \\ & & \cdot & & & & & \cdot & \\ & & & \cdot & & & & & \cdot \\ & & & g_0 & \cdot & \cdot & \cdot & \cdot & g_r \end{bmatrix}\right\}k, \qquad (2.52)$$

where $r = n - k$. Any cyclic code is defined by the corresponding generator polynomial $g(x)$, which is the divisor of $x^n - 1$. The opposite is also true, i.e. if we choose the polynomial $g(x)$ and form the code from the words of form (2.50) then we obtain the cyclic (n, k) code, where n is the positive integer such that $g(x)$ is the factor of $x^n - 1$ and $k = n - \deg g(x)$.

The results given above can be formulated as the following theorem [4].

Theorem 2.2 Any q-ary cyclic (n, k) code is generated by the normalised polynomial $g(x)$ over $GF(q)$ of degree $n - k$, which $g(x)$ is the factor of $x^n - 1$. And vice versa, any normalised polynomial $g(x)$ over $GF(q)$ of degree $n - k$, which $g(x)$ is the factor of $x^n - 1$, generates the cyclic (n, k) code.

Let polynomial $h(x)$ be

$$h(x) = \frac{x^n - 1}{g(x)}. \qquad (2.53)$$

Then the multiplication of any codeword $v(x) = m(x) \cdot g(x)$ by $h(x)$ is equal

$$v(x) \cdot h(x) = m(x) \cdot h(x) \cdot g(x) = m(x) \cdot (x^n - 1) = 0 \bmod (x^n - 1).$$

This equation defines the parity-check sums for codewords, and the polynomial $h(x)$ is called the *parity polynomial*. The parity-check matrix of the cyclic code can be represented with the help of $h(x)$ as follows:

$$\mathbf{H} = \left.\begin{bmatrix} h_0 & \cdot & \cdot & \cdot & \cdot & h_k & & & \\ & h_0 & & & & & h_k & & \\ & & \cdot & & & & & \cdot & \\ & & & \cdot & & & & & \cdot \\ & & & h_0 & \cdot & \cdot & \cdot & \cdot & h_k \end{bmatrix}\right\}n-k, \qquad (2.54)$$

A minimum distance of the cyclic code can be found using the parity-check matrix with the help of the lemma 2.1.

Example 2.9 Consider the polynomial $g(x) = x^{10} + x^8 + x^5 + x^4 + x^2 + x + 1$. It is easy to verify that the minimal n, for which $x^n - 1$ is divisible by $g(x)$, is equal to 15.

Then the polynomial $g(x)$ generates $(15,5)$ cyclic code over F_2, and $h(x) = \dfrac{x^{15} - 1}{g(x)} = x^5 + x^3 + x + 1$. Therefore

$$
\mathbf{G} = \left.\begin{bmatrix}
1 & 1 & 1 & 0 & 1 & 1 & 0 & 0 & 1 & 0 & 1 & 0 & 0 & 0 & 0 \\
0 & 1 & 1 & 1 & 0 & 1 & 1 & 0 & 0 & 1 & 0 & 1 & 0 & 0 & 0 \\
0 & 0 & 1 & 1 & 1 & 0 & 1 & 1 & 0 & 0 & 1 & 0 & 1 & 0 & 0 \\
0 & 0 & 0 & 1 & 1 & 1 & 0 & 1 & 1 & 0 & 0 & 1 & 0 & 1 & 0 \\
0 & 0 & 0 & 0 & 1 & 1 & 1 & 0 & 1 & 1 & 0 & 0 & 1 & 0 & 1
\end{bmatrix}\right\}5, \tag{2.55}
$$

$$
\mathbf{H} = \left.\begin{bmatrix}
1 & 0 & 1 & 0 & 1 & 1 & 0 & 0 & 0 & 0 & 0 & 0 & 0 & 0 & 0 \\
0 & 1 & 0 & 1 & 0 & 1 & 1 & 0 & 0 & 0 & 0 & 0 & 0 & 0 & 0 \\
0 & 0 & 1 & 0 & 1 & 0 & 1 & 1 & 0 & 0 & 0 & 0 & 0 & 0 & 0 \\
0 & 0 & 0 & 1 & 0 & 1 & 0 & 1 & 1 & 0 & 0 & 0 & 0 & 0 & 0 \\
0 & 0 & 0 & 0 & 1 & 0 & 1 & 0 & 1 & 1 & 0 & 0 & 0 & 0 & 0 \\
0 & 0 & 0 & 0 & 0 & 1 & 0 & 1 & 0 & 1 & 1 & 0 & 0 & 0 & 0 \\
0 & 0 & 0 & 0 & 0 & 0 & 1 & 0 & 1 & 0 & 1 & 1 & 0 & 0 & 0 \\
0 & 0 & 0 & 0 & 0 & 0 & 0 & 1 & 0 & 1 & 0 & 1 & 1 & 0 & 0 \\
0 & 0 & 0 & 0 & 0 & 0 & 0 & 0 & 1 & 0 & 1 & 0 & 1 & 1 & 0 \\
0 & 0 & 0 & 0 & 0 & 0 & 0 & 0 & 0 & 1 & 0 & 1 & 0 & 1 & 1
\end{bmatrix}\right\}10. \tag{2.56}
$$

The generator and the parity matrices can be reduced to the systematic form:

$$
\mathbf{G} = \left.\begin{bmatrix}
1 & 0 & 0 & 0 & 0 & 1 & 1 & 1 & 0 & 1 & 1 & 0 & 0 & 1 & 0 \\
0 & 1 & 0 & 0 & 0 & 0 & 1 & 1 & 1 & 0 & 1 & 1 & 0 & 0 & 1 \\
0 & 0 & 1 & 0 & 0 & 1 & 1 & 0 & 1 & 0 & 1 & 1 & 1 & 1 & 0 \\
0 & 0 & 0 & 1 & 0 & 0 & 1 & 1 & 0 & 1 & 0 & 1 & 1 & 1 & 1 \\
0 & 0 & 0 & 0 & 1 & 1 & 1 & 0 & 1 & 1 & 0 & 0 & 1 & 0 & 1
\end{bmatrix}\right\}5, \tag{2.57}
$$

$$
\mathbf{H} = \left.\begin{bmatrix}
1 & 0 & 1 & 0 & 1 & 1 & 0 & 0 & 0 & 0 & 0 & 0 & 0 & 0 & 0 \\
1 & 1 & 1 & 1 & 1 & 0 & 1 & 0 & 0 & 0 & 0 & 0 & 0 & 0 & 0 \\
1 & 1 & 0 & 1 & 0 & 0 & 0 & 1 & 0 & 0 & 0 & 0 & 0 & 0 & 0 \\
0 & 1 & 1 & 0 & 1 & 0 & 0 & 0 & 1 & 0 & 0 & 0 & 0 & 0 & 0 \\
1 & 0 & 0 & 1 & 1 & 0 & 0 & 0 & 0 & 1 & 0 & 0 & 0 & 0 & 0 \\
1 & 1 & 1 & 0 & 0 & 0 & 0 & 0 & 0 & 0 & 1 & 0 & 0 & 0 & 0 \\
0 & 1 & 1 & 1 & 0 & 0 & 0 & 0 & 0 & 0 & 0 & 1 & 0 & 0 & 0 \\
0 & 0 & 1 & 1 & 1 & 0 & 0 & 0 & 0 & 0 & 0 & 0 & 1 & 0 & 0 \\
1 & 0 & 1 & 1 & 0 & 0 & 0 & 0 & 0 & 0 & 0 & 0 & 0 & 1 & 0 \\
0 & 1 & 0 & 1 & 1 & 0 & 0 & 0 & 0 & 0 & 0 & 0 & 0 & 0 & 1
\end{bmatrix}\right\}10. \tag{2.58}
$$

To detect the errors in the received word $b(x)$ it is enough to check the condition

$$b(x) \cdot h(x) = 0 \bmod (x^n - 1). \tag{2.59}$$

We show that the generator polynomial of the cyclic code is the factor of $(x^n - 1)$. Therefore, it is necessary to consider all combinations of the factors of the polynomial $(x^n - 1)$ in order to enumerate all cyclic codes of length n. It is well known that if the characteristic p of the field is not the divisor of n, then the polynomial $(x^n - 1)$ can be factored by the irreducible divisors

$$x^n - 1 = f_1(x) \cdot \ldots \cdot f_l(x).$$

Therefore, it is possible to choose any polynomial of form

$$g(x) = f_{i_1}(x) \cdot \ldots \cdot f_{i_s}(x), \qquad i_1 < i_2 < \ldots < i_s, \qquad s < l$$

as the generator polynomial. Each of these polynomials corresponds to the code with some values of k and d; and there are $2^l - 2$ nontrivial cyclic codes of length n at all.

Example 2.10 Construct all binary cyclic codes of length 7. The polynomial $x^7 - 1$ can be factored as follows:

$$x^7 - 1 = (x + 1) \cdot (x^3 + x + 1) \cdot (x^3 + x^2 + 1).$$

The corresponding six cyclic codes of length 7 are defined by the following polynomials:

$$g_1(x) = (x + 1), g_2(x) = (x^3 + x + 1), g_3(x) = (x^3 + x^2 + 1)$$
$$g_4(x) = (x + 1) \cdot (x^3 + x + 1), g_5(x) = (x + 1) \cdot (x^3 + x^2 + 1)$$
$$g_6(x) = (x^3 + x + 1) \cdot (x^3 + x^2 + 1).$$

It is easy to verify (for instance, by enumerating the codewords) that the codes corresponding to these polynomials have the following parameters:

$$
\begin{array}{ll}
G_1 : & k = 6, d = 2; \\
G_2, G_3 : & k = 4, d = 3; \\
G_4, G_5 : & k = 3, d = 4; \\
G_6 : & k = 1, d = 7.
\end{array}
$$

The parameters of some binary cyclic codes are listed in Table 2.1.

One of the most important operations for the implementation of the cyclic codes is the calculation of the remainder resulting from dividing one polynomial by another.

Table 2.1 Parameters of some cyclic codes

n	k	d	Generator polynomial $g(x)$
7	4	3	13[a]
15	11	3	23
	9	3	171
	7	5	721
	5	7	2467
31	26	3	45
	21	5	3551
	16	7	107657
	11	11	5423325
	6	15	313365047
63	57	3	103
	51	5	12471
	45	7	1701317
	39	9	166623567
	30	13	157464165547
	16	23	6331141367235453
	7	31	5231045543503271737
127	120	3	211
	85	13	130704476322273
	71	19	6255010713253127753
	22	47	12337607040472252243544562663764 7043
255	247	3	435
	187	19	52755313540001322236351
	139	31	4614017320601755615707227302474535 67445
	47	85	25335420170626465630330413774062331751233341454460 45005066024552543173

[a]All generator polynomials are given in the octal format, e.g. $13_8 = 1011 = x^3 + x + 1$.

This operation can be executed with the help of a tapped filter, i.e. a device containing delay elements with taps, adders in field $GF(q)$ and multipliers in field $GF(q)$ (see Figure 2.2).

The state of the delay element is $s^{(t)} = s$ at the time t if at the moment t we obtain symbol $s \in GF(q)$ at the output of this delay element. Let r be the number of delay elements in the filter. Then vector $\mathbf{s}^{(t)} = (s_1^{(t)}, \ldots, s_r^{(t)})$, where $s_i^{(t)}$ is the state of the ith element at time t, is called the state of the filter at time t.

Hereafter we will consider the filters with one input and one output. The input and output signals of the filters will be the sequences of symbols from $GF(q)$.

Notice that this kind of filter is called the *linear filter*, i.e. the filter response (the output signal) to the sum of input signals is the sum of filter responses to each input signal (sum in $GF(q)$).

Let $\mathbf{s} = (s_1, \ldots, s_r)$ be the preceding and $\mathbf{s}' = (s'_1, \ldots, s'_r)$ be the succeeding states of the filter. Then to define the filter it is necessary to define the following equations

$$s'_1 = c_{11} \cdot s_1 + \ldots + c_{1r} \cdot s_r;$$
$$\cdots\cdots\cdots\cdots\cdots\cdots\cdots\cdots\cdots\cdots \tag{2.60}$$
$$s'_r = c_{r1} \cdot s_1 + \ldots + c_{rr} \cdot s_r,$$

where c_{ij} is the coefficient defined by the structure of the filter; the addition and the multiplication is carried out in $GF(q)$. That means the preceding state fully defines the succeeding state in case there is no signal at the filter input.

Assume that the input signal appears at the input of the first delay element. Then if the input signal is equal to α, filter from the state $\mathbf{s} = (s_1, \ldots, s_r)$ goes to the state $\mathbf{s}' = (s'_1, \ldots, s'_r)$ defined by the equations

$$s'_1 = c_{11} \cdot s_1 + \ldots + c_{1r} \cdot s_r + \alpha;$$
$$\cdots\cdots\cdots\cdots\cdots\cdots\cdots\cdots\cdots\cdots \tag{2.61}$$
$$s'_r = c_{r1} \cdot s_1 + \ldots + c_{rr} \cdot s_r$$

Let us consider the synthesis of the calculator of the remainder. The $(r \times r)$ matrix $\mathbf{C} = [c_{ij}]$ is called the transfer matrix. This matrix defines the filter. The general view of the filter defined by matrix $\mathbf{C}$ is shown in Figure 2.2.

Put in correspondence to the jth column of matrix $\mathbf{C}$ the polynomial $c_j(x) = \sum_{i=1}^{r} c_{ij} x^{i-1}$.

Theorem 2.3 Let $c_i(x) = x^i \bmod g(x)$ be the i th column of the matrix $\mathbf{C}$, where $g(x)$ is an arbitrary polynomial of degree r. Then if signal α appears at the filter input,

$$s'(x) = \alpha + xs(x) \bmod g(x), \tag{2.62}$$

where $s'(x) = \sum_{i=1}^{r} s'_i x^{i-1}$.

Proof It follows from (2.61) that

$$s'(x) = s_1 \cdot c_1(x) + s_2 \cdot c_2(x) + \ldots + s_r \cdot c_r(x) + \alpha$$

Then substituting expressions for $c_i(x)$ we obtain

$$s'(x) = \alpha + s_1 x \bmod g(x) + s_2 x^2 \bmod g(x) + \ldots + s_r x^r \bmod g(x) =$$
$$= (\alpha + s_1 x + s_2 x^2 + \ldots + s_r x^r) \bmod g(x) = (\alpha + x(s_1 + s_2 x + \ldots + s_r x^{r-1})) \bmod g(x)$$
$$= (\alpha + xs(x)) \bmod g(x). \quad Q.E.D$$

Thus, for the arbitrary polynomial $g(x)$ and the input signal α in Theorem 2.3 the filter goes to the state s' defined by the equation (2.62). Let the initial state of the filter be all-zero and the elements $a_{n-1}, a_{n-2}, \ldots, a_0$, which are the coefficients of the polynomial $a(x) = \sum_{i=0}^{n-1} a_i x^i$, consecutively appears at the input of the filter. Then the filter will consecutively go to the states:

$$s^{(0)}(x) = 0, \quad s^{(1)}(x) = a_{n-1}, \quad s^{(2)}(x) = (a_{n-2} + a_{n-1} x) \bmod g(x),$$
$$s^{(3)}(x) = (a_{n-3} + a_{n-2} x + a_{n-1} x^2) \bmod g(x), \ldots,$$
$$s^{(n)}(x) = (a_0 + a_1 x + \ldots + a_{n-1} x^{n-1}) \bmod g(x),$$

but $s^{(n)}(x)$ is the remainder resulting from dividing $a(x)$ by $g(x)$.

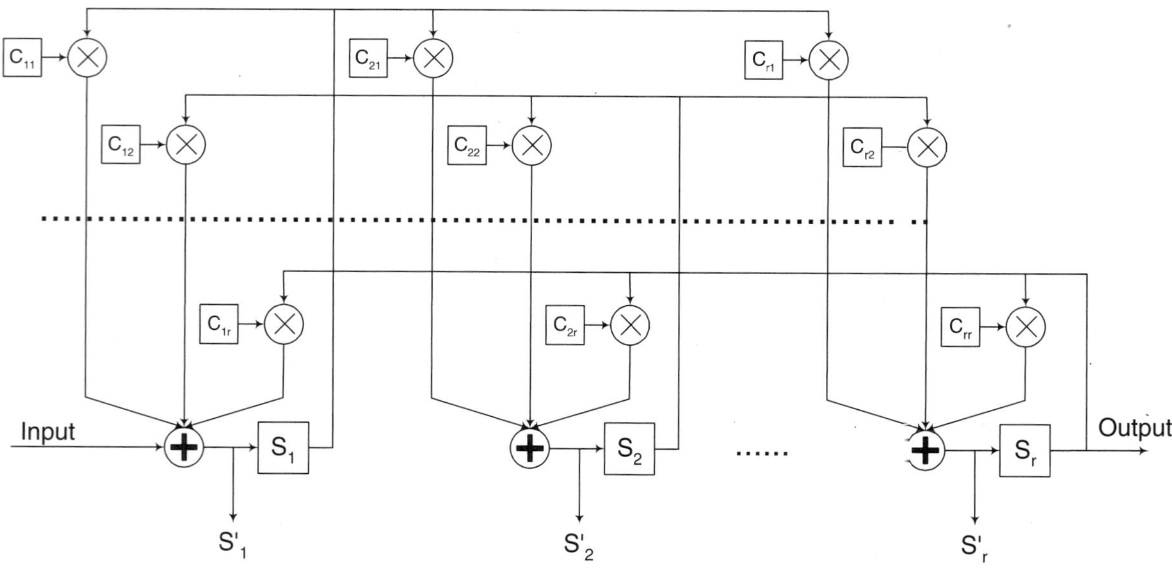

Figure 2.2 General structure of the filter

Example 2.11 $g(x) = x^3 + x + 1$, $a(x) = x^6 + x^3 + 1$, $GF(2)$. The transfer matrix is

$$C = \begin{bmatrix} 0 & 0 & 1 \\ 1 & 0 & 1 \\ 0 & 1 & 0 \end{bmatrix}$$

The filter defined by the matrix C is shown in Figure 2.3. The states of the filter are

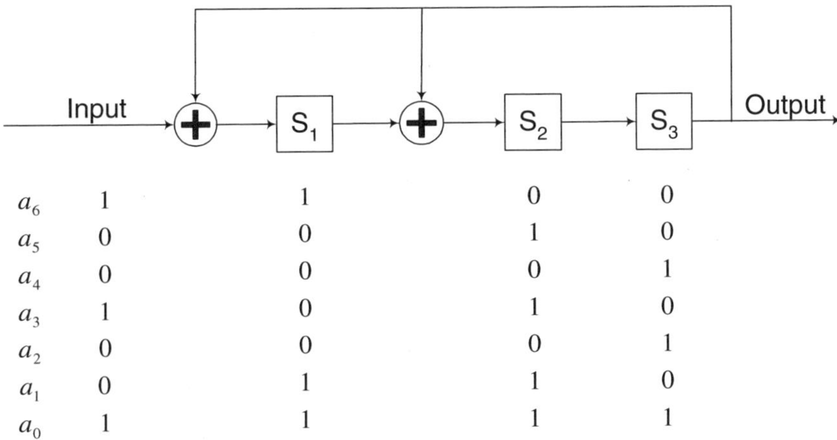

		S₁	S₂	S₃
a_6	1	1	0	0
a_5	0	0	1	0
a_4	0	0	0	1
a_3	1	0	1	0
a_2	0	0	0	1
a_1	0	1	1	0
a_0	1	1	1	1

Figure 2.3   Filter for the division by $g(x) = x^3 + x + 1$

$$s^{(0)}(x) = 0, \quad s^{(1)}(x) = a_{n-1} = a_6 = 1, \quad s^{(2)}(x) = a_5 + a_6 x = x, \quad s^{(3)}(x) = x^2,$$
$$s^{(4)}(x) = x, \quad s^{(5)}(x) = x^2, \quad s^{(6)}(x) = x + 1, \quad s^{(7)}(x) = x^2 + x + 1.$$

The remainder resulting from dividing $a(x)$ by $g(x)$ is equal to $x^2 + x + 1$.

All codewords of a cyclic code can be represented in the form of (2.50), where $m(x)$ is the information message and $m(x) \cdot g(x)$ is the corresponding codeword. Such encoding procedures correspond to the generator matrix of form (2.52). There is no necessity to keep the whole matrix (2.52) for the encoding. It is enough to keep only the first row of this matrix, i.e. the generator polynomial. Thus, the realisation of the encoding reduces to the realisation of the multiplication of two polynomials. However, the encoding procedure

$$m(x) \rightarrow m(x) \cdot g(x) \tag{2.63}$$

defines the nonsystematic code. It is impossible to select the information symbols in the codeword $m(x) \cdot g(x)$; and it is necessary to divide the codeword by $g(x)$ in order to obtain the information message $m(x)$. Here we will show two methods of systematic encoding of the cyclic codes, for which the information symbols are the coefficients of the most significant powers of the polynomial corresponding to the codeword.

Let $a(x) = \sum_{i=0}^{n-1} a_i x^i$ be the codeword and $h(x) = \sum_{i=0}^{k} h_i x^i$ be the parity polynomial of the (n,k) cyclic code. Consider the product of $a(x)$ by $h(x)$

$$
\begin{aligned}
f(x) = a(x) \cdot h(x) &= m(x) \cdot g(x) \cdot h(x) = x^n m(x) - m(x) \\
&= m_{k-1} x^{n+k-1} + \ldots + m_0 x^n - m_{k-1} x^{k-1} - \ldots - m_0.
\end{aligned}
\tag{2.64}
$$

It follows from the definition of the parity polynomial that the coefficients of the polynomial $f(x) = \sum_{i=0}^{n+k-1} f_i x^i$ for x^i, $k \leq i \leq n-1$ are equal to zero. Now substituting coefficients of $m(x)$ and $h(x)$ to f_i we obtain

$$
f_i - \sum_{j=0}^{k} h_j\, a_{i-j}, \quad k \leq i \leq n-1
\tag{2.65}
$$

Since $h_k = 1$, then we can derive the following equation

$$
a_{i-k} = -\sum_{j=0}^{k} h_j \cdot a_{i-j}, \quad k \leq i \leq n-1
\tag{2.66}
$$

The equation (2.66) defines the recurrent formula for the sequential calculation of $a_{n-1-k}, a_{n-1-k-1}, \ldots, a_0$ using the information symbols $a_{n-1}, a_{n-2}, \ldots, a_{n-k}$. Thus, the equation (2.66) defines the method of systematic encoding for cyclic code.

The circuit implementing the calculation by the formula (2.66) is shown in Figure 2.4. The operation of the circuit can be described as follows:

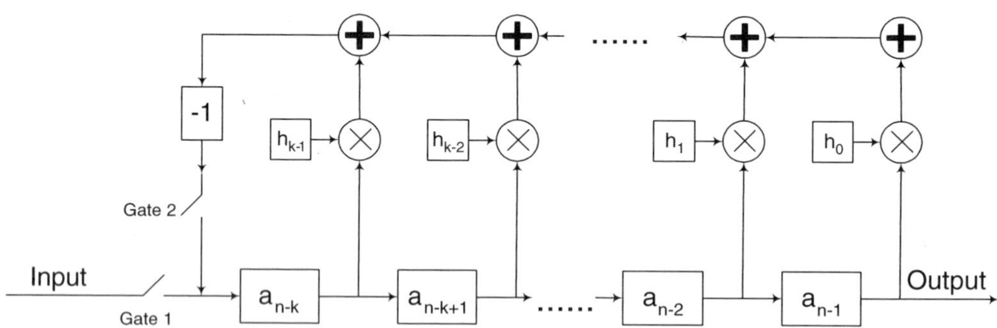

Figure 2.4   The encoder of cyclic code with k delay elements

Initially gate 1 is switched on and gate 2 is switched off. The information symbols $a_{n-1}, a_{n-2}, \ldots, a_{n-k}$ are sequentially shifted into the register. As soon as the k information symbols enter the shift register, gate 1 is switched off and gate 2 is switched on. Symbol a_{n-1} appears at the output of the encoder; and the new symbol is fed to the input. This new symbol is the inverted sum of the products of the symbols from the register elements by the corresponding coefficients h_i, i.e. as follows from (2.66) a_{n-1-k}. During the next shifts the symbols $a_{n-2}, a_{n-3}, \ldots$ will appear at the output of the encoder; and the symbols $a_{n-k-2}, a_{n-k-3}, \ldots$ will be fed to the input of the register. After n shifts we obtain the whole codeword at the output of the encoder. Notice that multiplication and addition should be executed in the field $GF(q)$, so we have to have special devices for these operations.

The encoder considered above uses k delay elements (or k-stage shift register). Let us consider the encoder with $(n - k)$ delay elements. It is obvious that such a kind of encoder will be more economic in case $k > n - k$.

Let us represent the information symbols as the polynomial $a_{n-1}x^{n-1} + a_{n-2}x^{n-2} + \ldots + a_{n-k}x^{n-k}$. In accordance with the algorithm of division

$$a_{n-1}x^{n-1} + a_{n-2}x^{n-2} + \ldots + a_{n-k}x^{n-k} = m(x) \cdot g(x) + r(x), \tag{2.67}$$

where $\deg r(x) < \deg g(x) = n - k$. It follows from (2.67) that the polynomial

$$a_{n-1}x^{n-1} + a_{n-2}x^{n-2} + \ldots + a_{n-k}x^{n-k} - r(x)$$

is the codeword. Therefore, it is enough to obtain the remainder which results from dividing $a_{n-1}x^{n-1} + a_{n-2}x^{n-2} + \ldots + a_{n-k}x^{n-k}$ by $g(x)$ in order to calculate the parity-check symbols. The device implementing this algorithm is shown in Figure 2.5

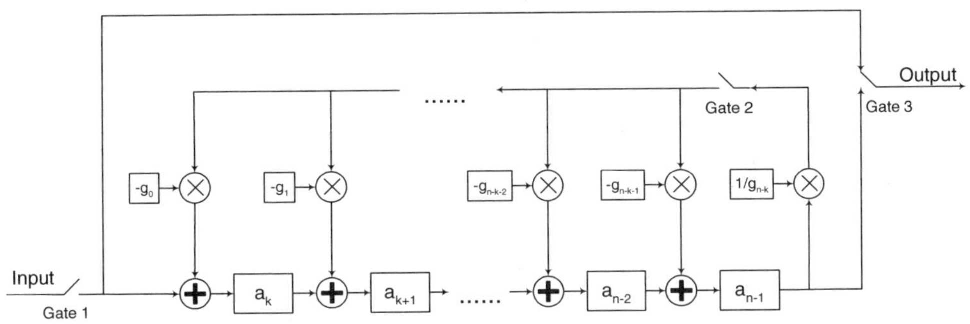

Figure 2.5   The encoder of cyclic code with $(n-k)$ delay elements

The device is operating as follows. Initially gate 1 and gate 2 are switched on and gate 3 is switched off. The information symbols $a_{n-1}, a_{n-2}, \ldots, a_{n-k}$ are fed into the input of the shift register and appear at the output of the encoder simultaneously. After $n - k$ shifts, symbol a_{n-1} appears at the output of the shift register and is multiplied by the coefficients $g_{n-k-1}, g_{n-k-2}, \ldots, g_0$ which is subtracted from the elements $a_{n-2}, a_{n-3}, \ldots, a_{k-1}$ correspondingly. This step corresponds to the first operation in the algorithm of the division of polynomials. After k shifts, gate 1 is switched off. After n shifts from the start of operation the remainder $r(x)$ we obtain the whole codeword at the output of the encoder.

Notice that in case of binary code the described circuits contain adders on modulo 2 and binary memory elements. The multiplication by '1' is realised by the presence, and the multiplication by '0' by the absence, of feedback.

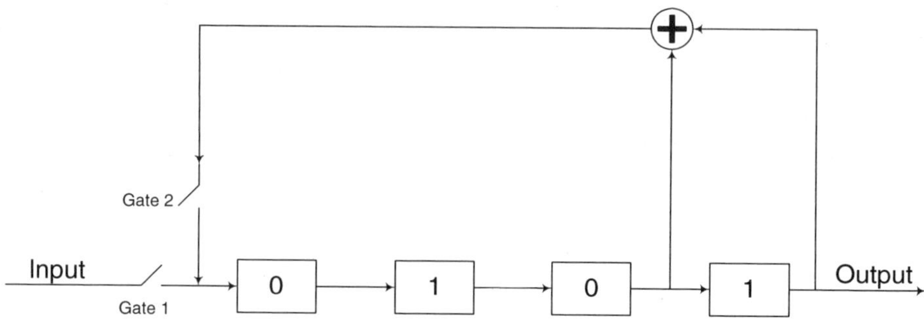

Figure 2.6   The k-stage encoder of the $(15,4)$ code

Example 2.12   Consider the $(15,4)$ binary cyclic code with parity polynomial $h(x) = x^4 + x + 1$. The k-stage encoder is shown in Figure 2.6.

Let $a_{n-1} = a_{n-3} = 1$, $a_{n-2} = a_{n-4} = 0$. The work of the encoder is shown in Table 2.2.

Example 2.13   Consider the $(15,11)$ binary cyclic code with generator polynomial $g(x) = x^4 + x^3 + 1$. The $(n-k)$-stage encoder of this code is shown in Figure 2.7.

Let $a_{14} = a_{13} = a_{10} = a_9 = a_5 = a_4 = 1$, $a_{12} = a_{11} = a_8 = a_7 = a_6 = 0$. The work of the encoder is shown in Table 2.3.

In the example being considered the division by the polynomial $g(x)$ only begins when the symbol a_{n-1} occupies the last (right) delay element in the shift register. On the other hand it

Table 2.2   The work of the encoder of (15,4) code

N of shift	State	Feedback symbol	Output symbol	
0	0 0 0 0	0	0	
1	1 0 0 0	0	0	
2	0 1 0 0	0	0	
3	1 0 1 0	0	0	
4	0 1 0 1	0	0	
5	1 0 1 0	1	1	
6	1 1 0 1	1	0	
7	1 1 1 0	1	1	
8	1 1 1 1	1	0	
9	0 1 1 1	0	1	
10	0 0 1 1	0	1	
11	0 0 0 1	0	1	
12	1 0 0 0	1	1	The result of the encoding
13	0 1 0 0	0	0	
14	0 0 1 0	0	0	
15	1 0 0 1	1	0	
16	1 1 0 0	1	1	
17	0 1 1 0	0	0	
18	1 0 1 1	1	0	
19	0 1 0 1	0	1	

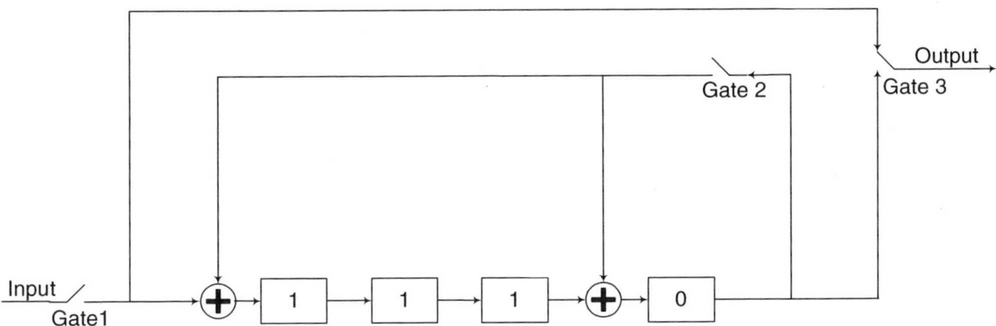

Figure 2.7 The $(n\text{-}k)$-stage encoder of the (15,11) code

takes an additional four cycles to obtain the calculated parity symbols at the output of the encoder. Hence, the overall encoding time is equal to 19 cycles (the same as for the scheme shown in Figure 2.6). It is possible to decrease the encoding time if we start the division by the generator polynomial simultaneously with feeding the information symbols. This kind of encoder is shown in Figure 2.8.

This encoder executes the multiplication of the information polynomial by x^{n-k} (which is equivalent to $(n-k)$ shifts of the information symbols in the shift register) due to feeding the information symbols at point A. The encoding time of the encoder in Figure 2.8 is equal to n cycles.

In order to detect error with the help of the cyclic code it is only necessary to calculate the remainder resulting from the division of the received word $b(x)$ by the generator polynomial $g(x)$ and to compare it (the remainder) with zero.

Table 2.3 The work of the encoder of (15,11) code

N of shift	State	Feedback symbol
0	0 0 0 0	0
1	1 0 0 0	0
2	1 1 0 0	0
3	0 1 1 0	0
4	0 0 1 1	0
5	0 0 0 0	1
6	1 0 0 0	0
7	0 1 0 0	0
8	0 0 1 0	0
9	0 0 0 1	0
10	0 0 0 1	1
11	0 0 0 1	1
12	1 0 0 1	1
13	1 1 0 1	1
14	1 1 1 1	1
15	1 1 1 0	1

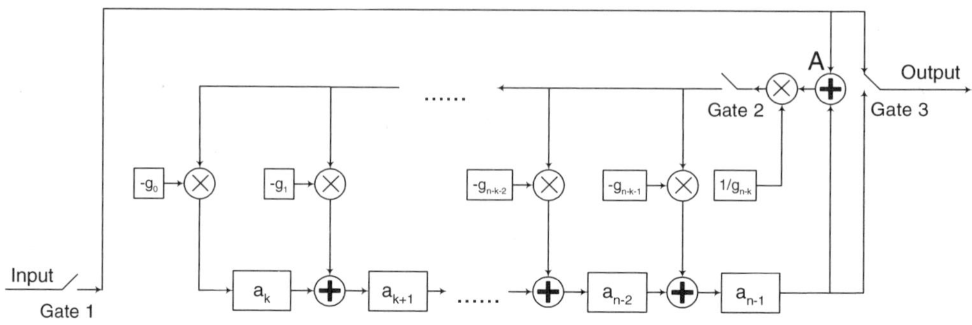

Figure 2.8   The $(n\text{-}k)$-stage encoder of cyclic code with the premultiplication by x^{n-k}

The remainder resulting from the division of the received word $b(x)$ by the generator polynomial $g(x)$ is called a *syndrome*. Let us denote the syndrome by $S(x)$. Then

$$S(x) = b(x) \bmod g(x). \qquad (2.68)$$

If the syndrome of the received word is equal to zero, then the received word is regarded as a codeword and it is assumed that no error can occur during the transmission of this word over the channel (or the undetectable error occurs). If the syndrome is not equal to zero the error is detected.

The syndrome can be calculated with the help of $(n-k)$-stage circuit with premultiplication by x^{n-k} used for the calculation of parity-check symbols.

Example 2.14   Consider the circuit for error detection for binary (15,9) cyclic code with the generator polynomial $g(x) = x^6 + x^5 + x^4 + x^3 + 1$. The circuit is shown in Figure 2.9. In 15 shifts the remainder resulting from division of the received word by the generator polynomial $g(x)$ will be stored in the shift register, i.e. the syndrome of the received word. If at least one coefficient of the syndrome is not equal to zero, then the signal of error detection appears at the output of the OR gate.

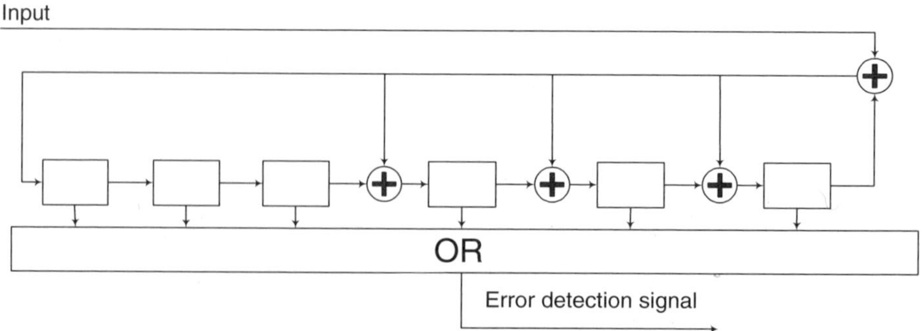

Figure 2.9   The error detector for the (15,9) code

The correction of the error pattern also can be done with the help of the syndrome. In most cases, to solve this problem it is necessary to keep in memory the table, in which every syndrome corresponds to the error pattern.

For cyclic codes the solution is not so complex, because it is possible to keep in memory only those syndromes that correspond to error patterns containing a nonzero symbol in the first position. Using the cyclic shifts of the received word $b(x)$ we obtain word $b'(x)$ with the error symbol in the first position and the syndrome of this word is stored in memory and can be used to correct the error pattern.

Let us describe the process of the error correction for cyclic code in detail. Let S be the set of syndromes of correctable error patterns with nonzero symbols in the first position. Let us denote by $S^{(\nu)}$ the subset of S, which contain the syndromes of error patterns with symbol ν ($\nu \neq 0, \nu \in GF(q)$) in the first position. It is obvious that $S = \cup_{\nu=1}^{q-1} S^{(\nu)}$. The algorithm of decoding the cyclic code consists of the following steps:

1. For received vector $\mathbf{b}$ there are computed syndromes $S_i (i = 0, \ldots, n-1)$ of the cyclic shift of vector $\mathbf{b}$ by i positions.

2. For each S_i the sets $S^{(\nu)}$ are formed $\nu = 1, \ldots, q-1$ such that $S_i \in S^{(\nu)}$. Then the ith element of error pattern e_i is assumed to be equal to v. If for none of v the condition $S_i \in S^{(\nu)}$ is correct, then e_i is assumed to be zero.

3. When all e_i, $(i = 0, \ldots, n-1)$ are obtained vector $\mathbf{b}$ is added with computed vector $\mathbf{e}$.

Step 2 is usually called the *selection* and the device executing this operation is called the *selector*. The most difficult part of realisation of the syndrome decoding is the design of the selector. As usual the complexity of this device is very high. Usually the syndrome decoding is used for short codes or in the case where only a small number of errors are to be corrected. For instance, the selector for correction of one error is a very simple device.

The list of cyclic codes is restricted by the fact that for all values of n and k there does not exist the polynomial $g(x)$ of degree $n - k$, which is the factor of $x^n - 1$. Moreover, for some values of n all cyclic codes are inefficient. For example, for all even n the (n, k) cyclic code has the distance $d = 2$ independently of the value of k. The list of codes may be extended if we consider the shortened cyclic codes.

The *shortened cyclic* $(n - i, k - i)$ *code* is the code that is constructed from the cyclic (n, k) code by the rejection of i high-order information symbols of each codeword. The length of the obtained code is $n - i$ and the number of information symbols is $k - i$. The distance of the shortened code is no less than the distance of the original cyclic code.

The shortened code is not the cyclic code because it contains the words that are not the cyclic shift of some codeword. However, all codewords of the shortened code are divisible by the generator polynomial of the original code, because the codewords of the shortened code are the codewords of the original code, for which the i high-order information symbols are equal to zero. Because of this fact the same circuits as those employed by the original cyclic code can accomplish the encoding of the shortened cyclic code.

It is also possible to use the decoder of the original (n, k) cyclic code for error detection and error correction of the shortened $(n - i, k - i)$ code. It is only necessary to add i zeros to the word of the shortened code to do it. However, the decoder of the original code needs to calculate the syndrome in n shifts and the decoding of the shortened code can be done with $n - i$ shifts.

Let symbols $\underbrace{0, 0, \ldots, 0}_{i}, b_{n-1-i}, b_{n-2-i}, \cdots, b_0$ consecutively appears at the input of the syndrome calculator. This device is executing the division of polynomial $b(x) = 0 \cdot x^{n-1} + 0 \cdot x^{n-2} + \cdots + 0 \cdot x^{n-i} + b_{n-1-i}x^{n-1-i} + b_{n-2-i}x^{n-2-i} + \cdots + b_0$ by the polynomial $g(x)$; and the operation takes n shifts. However, since the i high-order coefficients of $b(x)$ are equal to zero the feedback signal of the shift register is equal to zero during the first i shifts and in fact, there is no operation of the division of polynomials executed during these first i shifts. It is possible to avoid this inefficient operation by the preliminary shifting of $b(x)$ i times, which corresponds to the premultiplication of $b(x)$ by x^i on modulo $g(x)$.

Consider the premultiplication by x^{n-k}, which is necessary to compute the syndrome of the original code in n shifts. The calculator of the syndrome of the shortened $(n-i, k-i)$ code needs preliminary multiplication of the received word $b(x)$, by x^{n-k+i},

Let polynomial $f(x)$ be $f(x) = f_{n-k-1}x^{n-k-1} + f_{n-k-2}x^{n-k-2} + \cdots + f_0 = x^{n-k+i} \bmod g(x)$, then the premultiplication of $b(x)$ by $f(x)$ corresponds to the multiplication by x^{n-k+i}. This operation can be executed with the help of circuit shown in Figure 2.10.

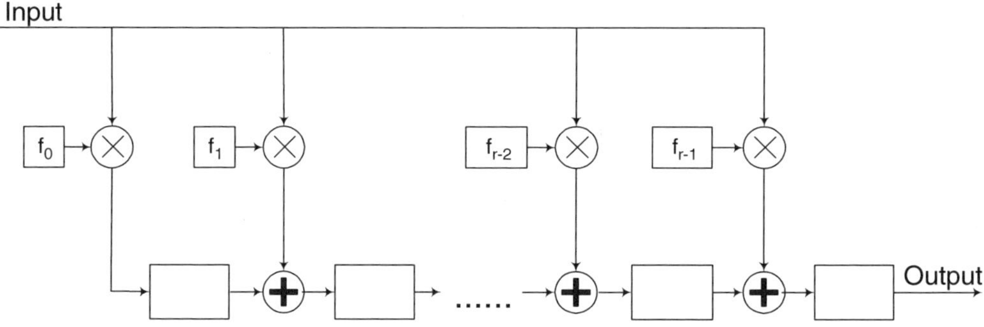

Figure 2.10   The circuit for multiplication by $f(x)$

The syndrome calculator for the shortened cyclic code executing the multiplication by $f(x)$ and the division by $g(x)$ is shown in Figure 2.11.

Example 2.15   Consider the syndrome calculator of (10,6) code obtained by shortening by 5 symbols the original cyclic (15,11) code with the generator polynomial $g(x) = x^4 + x + 1$.

$f(x) = x^{15-11+5} \bmod g(x) = x^9 \bmod (x^4 + x + 1) = x^3 + x$. Figure 2.12 shows the syndrome calculator for this code.

The cyclic codes can be used for burst-error correction. Here we will show the simple burst-error-correcting decoder.

Let $a(x)$ be the transmitted codeword of (n, k) cyclic code, $e(x)$ be the polynomial corresponding to error vector. Then the received word $b(x)$ can be represented as

$$b(x) = a(x) + e(x). \tag{2.69}$$

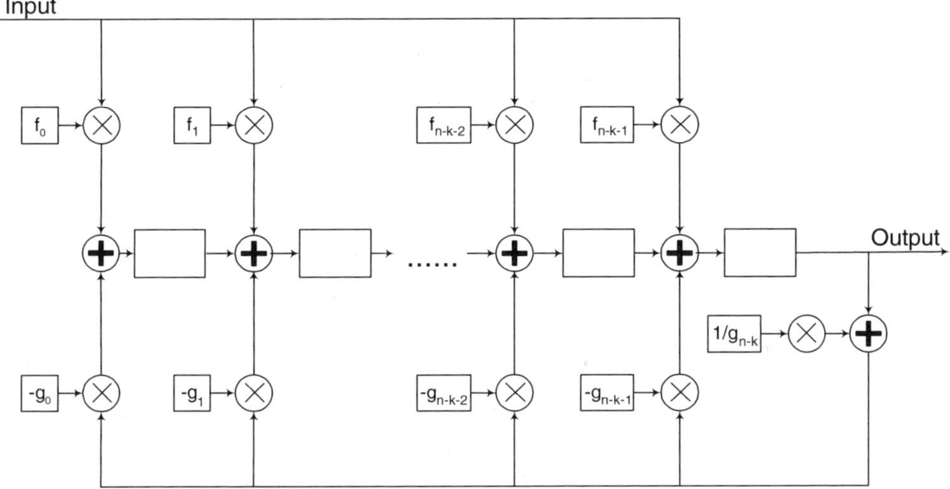

Figure 2.11   The circuit for calculation of shortened cyclic code syndrome

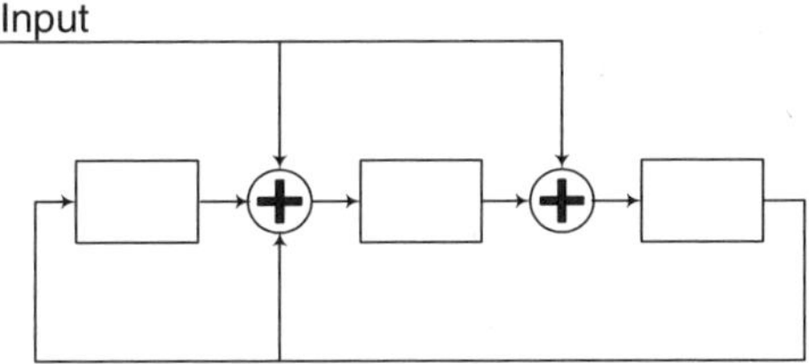

Figure 2.12   The syndrome calculator of shortened (10,6) code

Notice that if the burst-error occur on the parity-check symbols of the decoding word, then $\deg e(x) \leq n - k - 1$ (the parity-check symbols correspond to less significant digits of the word). Then $S(x) = b(x) \bmod g(x) = a(x) \bmod g(x) + e(x) \bmod g(x) = e(x) \bmod g(x) = e(x)$, since $a(x) \bmod g(x) = 0$ and $\deg e(x) < \deg g(x)$. In this case it is only necessary to calculate the syndrome $S(x)$ and to add it with $b(x)$:

$$a(x) = b(x) + S(x). \tag{2.70}$$

If the burst-error occur on the information symbols it is possible to sift it to the parity-check symbols by shifting the received codeword. Let

$$b(x) = a(x) + e(x) \cdot x^i,$$

where $\deg e(x) < \deg g(x) = n - k$ and i satisfies the relation

$$\deg(e(x) \cdot x^i) = i + \deg e(x) \geq n - k.$$

Then the polynomial

$$(x^{-i} \cdot b(x)) \bmod (x^n - 1) = (x^{-i} \cdot a(x) + e(x)) \bmod (x^n - 1)$$

is the sum of the codeword $a'(x) = (x^{-i} \cdot a(x)) \bmod (x^n - 1)$ and the burst-error $e(x)$ on the parity-check symbols positions.

Now shifting the received word we shift the burst-error on the parity-check symbols positions, and then by adding $(x^{-i} \cdot b(x)) \bmod (x^n - 1)$ to the corresponding syndrome we obtain the polynomial $x^{-i} \cdot a(x)$. Then we obtain the codeword $a(x)$ with the help of i cyclic shifts. However, it is not clear on which cyclic shift of the received word $b(x)$ the burst-error appears at the parity-check positions.

Theorem 2.4 Let the code correct the burst-errors of length b or less and let the decoding word $f(x)$ be the sum of the codeword $a(x)$ and burst-error $e(x)$ of length no more than b, $f(x) = a(x) + e(x)$. In order to ensure that the degree of $e(x)$ be no more than $b - 1$ it is necessary that the degree of the syndrome of word $f(x)$ be no more than $b - 1$.

Proof The necessity immediately follows from the fact that if $\deg e(x) \leq b - 1 \leq n - k - 1$, then $S(x) = e(x)$.

The sufficiency we will prove by contradiction. Let $\deg e(x) \leq b$ and $\deg S(x) \leq b$. In accordance with definition of $S(x)$

$$f(x) = m_1(x) \cdot g(x) + S(x),$$

on the other hand

$$f(x) = m_2(x) \cdot g(x) + e(x).$$

Therefore,

$$e(x) \bmod g(x) = S(x) \bmod g(x). \tag{2.71}$$

The equation (2.71) means that $e(x)$ and $S(x)$ belong to the same coset. But this fact, in turn means that the code is capable of correcting only one burst-error from two of the lengths of no more than b ($S(x)$ or $e(x)$). And this fact contradicts the condition of the theorem.

Thus, the algorithm of decoding of burst errors consists in the shift of the decoding word and the calculation of the syndrome until the degree of the syndrome becomes less than b. The addition of the shifted word and the syndrome then results in the shifted codeword. The last step is the reverse cyclic shift of the obtained codeword. The algorithm can be formulated as follows:

1. Calculation of the syndromes $S_i(x) = x^{-i} f(x) \bmod g(x)$, $i = 0, 1, \ldots$ until

$$\deg S_i(x) \leq b - 1. \tag{2.72}$$

2. If $S_i(x)$ satisfy (2.72) then calculate

$$a_i(x) = x^{-i}f(x) + S_i(x). \tag{2.73}$$

3. Calculation of the decoding result $\hat{a}(x)$ in form

$$\hat{a}(x) = x^i a_i(x) \bmod (x^n - 1). \tag{2.74}$$

4. If $S_i(x)$ does not satisfy (2.72) for none of $i = 0, 1, \ldots, n-1$, then it is assumed that the undecodable error occurs.

To calculate the syndromes $S_i(x)$ we need the following result.

Theorem 2.5 The syndrome of the i-fold cyclic shift $(i = \pm0, \pm1, \ldots)$ of the received word is equal to i-fold cyclic shift of the syndrome of the received word

$$(x^i f(x)) \bmod g(x) = x^i (f(x)) \bmod g(x). \tag{2.75}$$

Proof In accordance with the definition of the syndrome

$$f(x) = m(x) \cdot g(x) + S(x).$$

Then $(x^i f(x)) \bmod g(x) = (x^i m(x) \cdot g(x) + x^i S(x)) \bmod g(x) = x^i S(x) \bmod g(x)$

The described algorithm of burst errors correcting can be implemented with the help of the device shown in Figure 2.13. The logic element OR NOT in this decoder has $n - k - b$

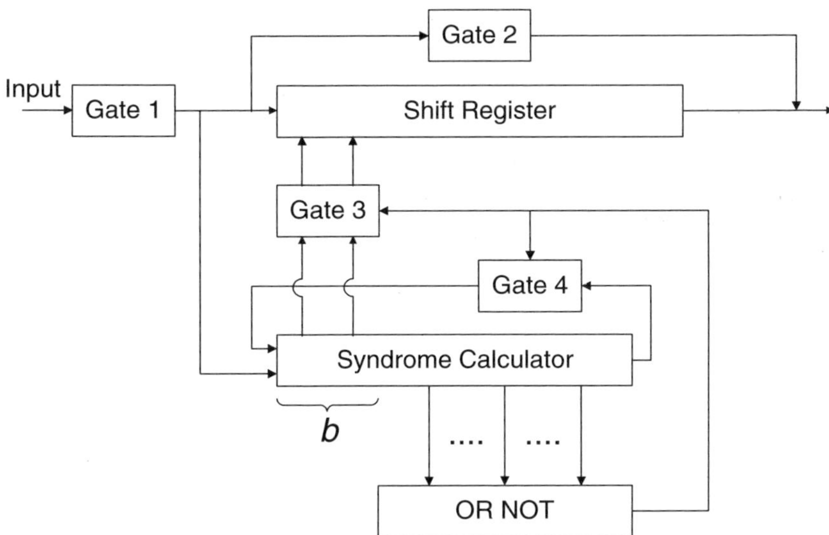

Figure 2.13 The decoder for burst error correction

inputs that are fed by the output of the syndrome calculator. If all these inputs are equal to zero it means that b less significant digits of the syndrome form the error burst of length b or less, which can be corrected. In this case the feedback of the syndrome calculator is broken with the help of Gate 4 and the shifted word is added with the syndrome. After this the corrected word is shifted $n - i$ times (which is equivalent to i reverse shifts).

Example 2.16 The binary cyclic (15,9) code with the generator polynomial $g(x) = x^6 + x^5 + x^4 + x^3 + 1$ is capable of correcting burst-errors of length 3. Let $e(x) = x^7 + x^6 + x^5$ be the burst-error of length 3. The syndrome of word $e(x)$ is

$$S(x) = e(x) \bmod g(x) = x^4 + x.$$

If we shift this syndrome to the less significant digits, then the contents of the syndrome register will be as follows:

Shift	Content of the syndrome register
0	$x^4 + x$
1	$x^3 + 1$
2	$x^5 + x^4 + x^3$
3	$x^4 + x^3 + x^2$
4	$x^3 + x^2 + x$
5	$x^2 + x + 1$

After 5 shifts $n - k - b = 15 - 9 - 3 = 3$ most significant digits of the syndrome are equal to zero; and less significant digits form the burst-error. It is possible to correct the error by shifting the decoding word 5 times to the less significant digits and adding it with the content of the syndrome register on modulo 2. The decoder is shown in Figure 2.14.

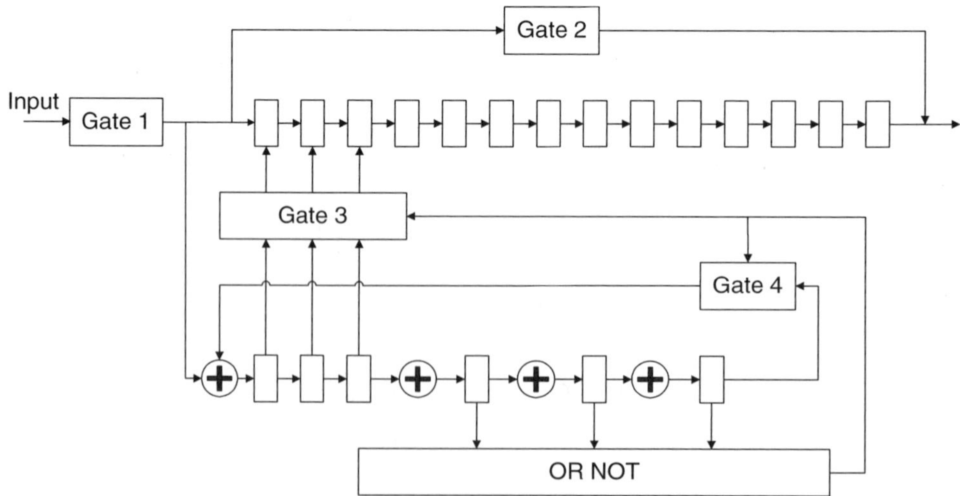

Figure 2.14   The decoder of (15,9) code

2.5 BOUNDS ON MINIMUM DISTANCE

In this section we recall some well-known bounds on the minimum code distance. Let V be a binary code of length n, consisting of M codewords and capable of correcting t errors. If a code can correct t errors then the minimum distance of this code d satisfies the inequality $d \geq 2t + 1$ or, in other words, the spheres of radius t surrounding all codewords are disjoint. The number of binary vectors of length n at the exact distance i from a given binary vector (of the same length n) equals $\binom{n}{i}$. Then each of these spheres of radius t contains $\sum_{i=0}^{t}\binom{n}{i}$ vectors (assuming that $\binom{n}{0} = 1$). On the other hand, the number of all binary vectors of length n equals to 2^n. This gives the *sphere-packing bound* known also as the *Hamming bound* [5]:

Theorem 2.6 For any binary code of length n, consisting of M code words and capable of correcting t-fold errors,

$$M \cdot \sum_{i=0}^{t} \binom{n}{i} \leq 2^n. \tag{2.76}$$

For a q-ary code (2.76) can be written as follows:

$$M \cdot \sum_{i=0}^{t} (q-1)^i \cdot \binom{n}{i} \leq q^n. \tag{2.77}$$

For linear q-ary (n,k) code $M = q^k$ and its code rate $R = \dfrac{k}{n} = \dfrac{\log_q M}{n}$. Therefore, for an arbitrary code its rate is defined as $R = \dfrac{\log_q M}{n}$. Taking this into account (2.77) can be represented in the following form:

$$n - k \geq \log_q \sum_{i=0}^{t} (q-1)^i \cdot \binom{n}{i}. \tag{2.78}$$

For $t < \frac{(q-1)n}{q}$ the sum in (2.78) can be upper estimated as $(q-1)^t \cdot \binom{n}{t}$ since

$$(q-1)^t \cdot \binom{n}{t} > \sum_{i=0}^{t-1} (q-1)^i \cdot \binom{n}{i}. \tag{2.79}$$

Applying Stirling's formula to the binomial coefficient $\binom{n}{t}$ obtains the following approximation:

$$\binom{n}{t} \approx \binom{n}{n\delta/2} \approx q^{nH_q(\delta/2)}, \tag{2.80}$$

where $\delta = \dfrac{d}{n}$ is the *relative minimum distance*, $0 < \delta < 1$, $H_q(x) = -x\log_q x - (1-x)\log_q (1-x)$ is the q-ary entropy function. Most of the results in this section are given in asymptotic, as $n \to \infty$. In this case it is more convenient to use the *asymptotic code rate* $R(\delta)$ as a function of relative minimum distance, i.e., the code rate of maximal cardinality as a function of $\delta = \dfrac{d}{n}$ with $n \to \infty$. It follows that for any q-ary code its asymptotic rate $R(\delta)$:

$$R(\delta) \leq 1 - \frac{\delta}{2}\log_q(q-1) - H_q(\delta/2), \quad 0 < \delta < 1, \tag{2.81}$$

Binary case (2.81) simplifies to the following form known as *asymptotic Hamming bound*:

$$R(\delta) \leq 1 - H_2\left(\frac{\delta}{2}\right), \quad 0 < \delta < 1. \tag{2.82}$$

The binary entropy function is presented in Figure 2.15.

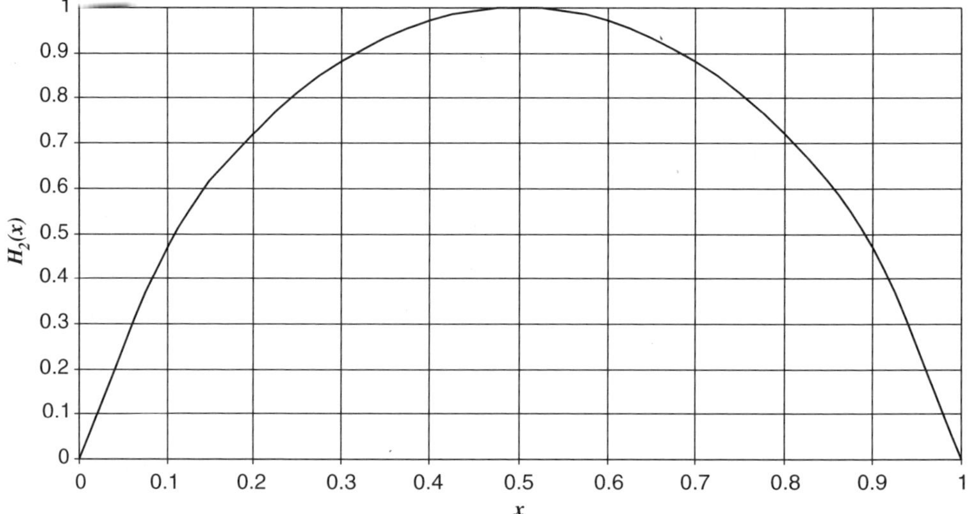

Figure 2.15 Binary entropy function $H_2(x)$

The codes achieving the Hamming bound are called *perfect codes*. Hence spheres of radius t surrounding codewords of a perfect code contain all the vectors of the corresponding vector space. It means that a perfect code can correct any error pattern of weight no more than t, and cannot correct any error pattern of weight more than t. For example, binary Hamming codes are perfect codes.

For a code capable of correcting t errors, the ratio of the overall number of vectors in all the spheres of radius t surrounding the code words to the number of vectors of the whole vector space is called the *packing density* of a code. That is, the packing density is the ratio of 'volume' of all the spheres of radius t surrounding the code words to the 'volume' of vector space, and the Hamming bound means that the *packing density* of any code is at most 1. For perfect codes the packing density is equal to 1, i.e., a perfect code correcting t errors is capable of packing the whole vector space by spheres of radius t.

The Hamming bound is an upper bound on the code rate of the code with a given (relative) minimum distance. Now we consider the Gilbert-Varshamov bound, which is a lower bound that shows an existence of good codes [5], in particular, among linear codes.

Theorem 2.7 If the following inequality holds true

$$q^{n-k} > \sum_{i=0}^{d-2} (q-1)^i \cdot \binom{n-1}{i}$$ (2.83)

then there exists a linear block q-ary code of length n with number of information symbols k that has the minimum distance no less than d.

The proof of the theorem can be found in [5]. The asymptotic Gilbert-Varshamov bound has form

$$R(\delta) \geq 1 - \delta \log_q(q-1) - H_q(\delta), \quad 0 < \delta < \frac{q-1}{q}$$ (2.84)

where $\delta = \dfrac{d}{n} < \dfrac{q-1}{q}$. It is known that codes satisfying Gilbert-Varshamov bound not only exist, but almost all linear coders satisfy it asymptotically tightly. Therefore the construction of *algebraic-geometry codes* [7] was a very surprising discovery such as

$$R(\delta) \geq 1 - \delta - \frac{1}{\sqrt{q}-1}$$ (2.85)

These codes are even asymptotically better than Gilbert-Varshamov bound for $q \geq 49$. Recently very effective algorithms of generation of such codes [8] as well as their decoding (and even beyond of $d/2$ [9]) were discovered which make this class of codes very attractive for practical applications.

For $q = 2$ (2.84) can be written as

$$R(\delta) \geq 1 - H_2(\delta), \quad 0 \leq \delta < \frac{1}{2}.$$ (2.86)

On the other hand, there is a very popular conjecture that for binary codes the Gilbert-Varshamov is asymptotically tight, i.e.

$$R(\delta) \overset{?}{=} 1 - H_2(\delta), \quad 0 \leq \delta < \frac{1}{2}.$$

Now we have the upper Hamming bound (2.81) and the lower Gilbert-Varshamov bound (2.84). Next theorems address the largest gap between these bounds.

Theorem 2.8 (Plotkin bound). The minimum distance d of any q-ary code block code of length n containing M words satisfies the following inequality:

$$d \leq \frac{(q-1)nM}{q(M-1)}.$$ (2.87)

The proof of Theorem 2.8 can be found in [4]. There is a simple recursion: $A_q(n,d) \leq q^{n-n'}A_q(n',d)$, where $A_q(n,d)$ is the maximal cardinality of a q-ary code of length n and distance d. This recursion together with Theorem 2.8 leads to a more general form of the Plotkin bound: $A_q(n, d) \leq d \cdot q^{n - \frac{q(d-1)}{q-1}}$ or, in asymptotic form

$$R(\delta) \leq 1 - \frac{q}{q-1}\delta. \tag{2.88}$$

as $n \to \infty$. This bound is more tight than Hamming bound for low code rates.

The next step was done by P.Elias and L.Bassalygo [10], [11]:

Theorem 2.9 (Elias-Bassalygo bound) For any q-ary (n, M, d) code the following inequality holds true:

$$R(\delta) \leq 1 - H_q\left(\left(1 - \frac{1}{q}\right) \cdot \left(1 - \sqrt{1 - \frac{q}{q-1}\delta}\right)\right), \quad 0 < \delta < \frac{q-1}{q}. \tag{2.89}$$

The proof of Theorem 2.9 can be found in [5].

The Elias-Bassalygo bound is better than Hamming and Plotkin bounds for any $0 < \delta < \frac{q-1}{q}$. At the present time the latest step towards tightening the upper bounds is the McEliece-Rodemich-Rumsey-Welch bound derived with the help of the Linear Programming method (see [5]). It consists of two parts. Let us consider them as two theorems:

Theorem 2.10 [5, 12] For any binary block code of length n with the minimum distance d the code rate R satisfies the following inequality as $n \to \infty$:

$$R(\delta) \leq H_2\left(\frac{1}{2} - \sqrt{\delta(1 - \delta)}\right). \tag{2.90}$$

The discussed bounds for binary block codes are represented in Figure 2.16.

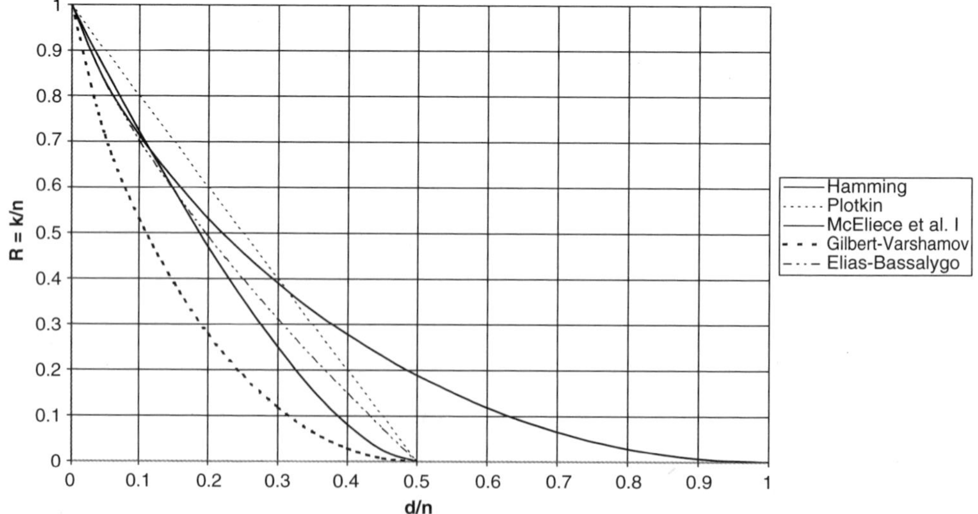

Figure 2.16 Bounds for binary block codes

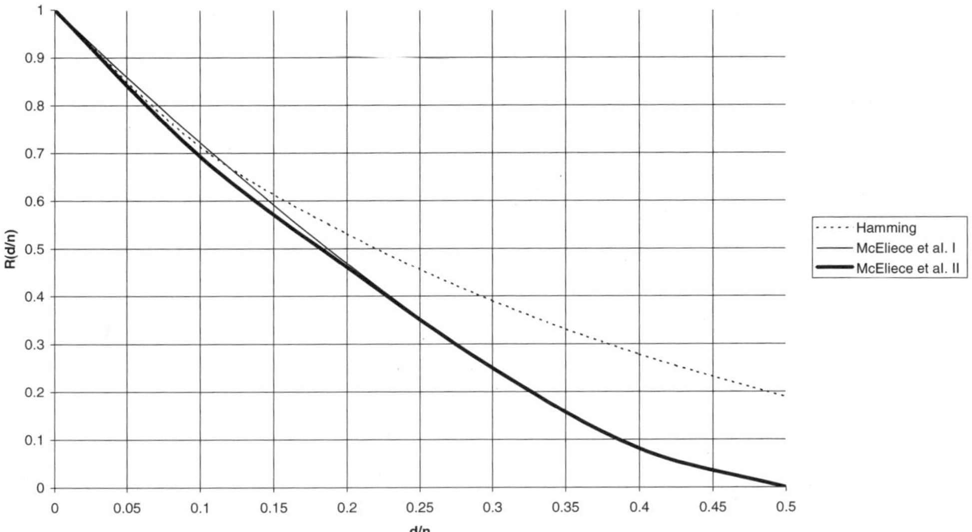

Figure 2.17. First and second McEliece-Rodemich-Rumsey-Welch Bounds for binary block codes.

As can be seen from plots in Figure 2.16, for values of $\delta > 0.1$ all codes are between Gilbert-Varshamov and McEliece-Rodemich-Rumsey-Welch I bound. For low values of δ the Hamming bound is less optimistic than the McEliece-Rodemich-Rumsey-Welch I bound. In this region of δ values the second part of the McEliece-Rodemich-Rumsey-Welch bound (or McEliece-Rodemich-Rumsey-Welch II bound) can be used:

Theorem 2.11. [5, 12] For any binary block code of length n with the minimum distance d the code rate R satisfies the following inequality as $n \to \infty$:

$$R(\delta) \le B(\delta), \tag{2.91}$$

where $B(\delta) = \min B(u, \delta), 0 < u \le 1 - 2\delta,$
$$B(u, \delta) = 1 + h(u^2) - h(u^2 + 2\delta u + 2\delta)$$
$$h(x) = H_2\left(\frac{1}{2} - \frac{1}{2}\sqrt{1 - x}\right)$$

The bound given by (2.91) improves the first McEliece-Rodemich-Rumsey-Welch bound in the region of low values of δ (it coincides with the first bound for $\delta \ge 0.273$). The comparison of the first and second McEliece-Rodemich-Rumsey-Welch bound with the Hamming bound is depicted in Figure 2.17.

REFERENCES

1. Shannon, C. E. (1948). *A Mathematical Theory of Communication*. Bell System Tech.
2. Peterson, W. W. and Weldon, E. J. (1972). *Error -Correcting codes*. MIT Press, Cambridge, MA.
3. Birkhoff, G. and Mac Lane, S. (1941). *A Survey of Modern Algebra*, Macmillan, New York.

4. Gallager, R. G. (1968). *Information Theory and Reliable Communication*. John Wiley & Sons, Chichester, UK.

5. MacWilliams, F. J. and Sloan, J. J. (1977). *The Theory of Error-Correcting Codes*, North-Holland, Amsterdam, The Netherlands.

6. Evseev, G. S. (1983). Complexity of decoding for linear codes, *Probl. Inform. Transm.*, **19**(1), 3–8 (in Russian) and 1–6 (English translation), 1983.

7. Tsfasman, M. A. and Vladuts, S. G. (1991). *Algebraic-Geometry Codes*, Kluwer, Dordrecht, The Netherlands.

8. Tsfasman, M. A., Vladuts, S. G., and Zink, T. (1982). Modular curves, Shimura curves, and Goppa codes better than Varshamov-Gilbert Bound, *Math. nachr.*, **109**, 21–28.

9. Guruswami, V. and Sudan, M. (1999). Improved Decoding of Reed–Solomon Codes and Algebraic Geometry Codes, *IEEE Trans. Inform. Theory*, **45**(6), 1757–1767.

10. Bassalygo, L. A. (1965). New upper bounds for error-correcting codes, *Problems of Information Transmission*, **4**, 41–44.

11 Shannon, C. E., Gallager R. G., and Berlekamp E. R. (1967). Lower bounds to error probability for coding on discrete memoryless channels, *Info. and Control*, **10**, 65–103 and 522–552.

12. R. J. McEliece, E. R. Rodemich, H. C. Rumsey, Jr. and L. R. Welch, New upper bounds on the rate of a code via Delsarte - MacWilliams inequalities, *IEEE Trans. Inform. Theory*, vol. 23, no. 2, pp. 157–166, March 1977.

3

General Methods of Decoding of Linear Codes

In this chapter we consider the general methods of linear codes decoding, i.e. the decoding methods that do not need any special properties of codes, except linearity. We start with minimum distance (Hamming distance) decoding, which coincides with the maximum likelihood decoding for channels with independent errors, then we consider more comprehensive methods of decoding linear codes.

3.1 MINIMUM DISTANCE DECODING

Usually we take the words minimum distance decoding to mean the procedure of searching for the codeword that is the closest one to the received word. Such procedures can be realised by an exhaustive search on the set of codewords or on the set of syndromes of probable error vectors.

The exhaustive search algorithm on the set of codewords consists of comparing the received word with all codewords and choosing of the closest codeword.

The exhaustive search algorithm on the set of syndromes of error vectors can be realised if we have table T, in which syndromes correspond to coset leaders. The algorithm is as follows:

1. Calculation of the syndrome $\mathbf{s}$ of the received word $\mathbf{b}$;

2. The search of the syndrome $\mathbf{s}$ and the corresponding coset leader $\mathbf{e}$ in the table T;

3. Calculation of decoded vector $\hat{\mathbf{a}}$:

$$\hat{\mathbf{a}} = \mathbf{b} + \mathbf{e}.$$

The first algorithm usually needs more operations, which is related to the necessity to generate all codewords (to execute q^k encoding procedures) and to compare them with the received word (q^k comparisons). The decoding on the syndromes requires keeping in memory the table with q^r words of length n.

Error Correcting Coding and Security for Data Networks   G. Kabatiansky, E. Krouk and S. Semenov
© 2005 John Wiley & Sons, Ltd   ISBN: 0-470-86754-X

It is usually difficult to implement these algorithms in practice due to their high complexity. Now we will consider less complex algorithms which, however, cannot provide full decoding, i.e. the decoding of all errors, which can be corrected with the help of given code (all coset leaders).

3.2 INFORMATION SET DECODING

The key concept for many general methods of decoding is the concept of the *information set* of the code. It is known that for systematic code first k symbols $a_0, \ldots, a_{k-1}$ fully define the word $a_0, \ldots, a_{k-1}, a_k, \ldots, a_{n-1}$ of the (n, k) code. That means there is only one word in the code where the symbols $a_0, \ldots, a_{k-1}$ occupy the positions $\{0, \ldots, k-1\}$.

However, the set of positions $\{0, \ldots, k-1\}$ is not a unique one. For every code there exists many other sets $\{j_1, \ldots, j_k\}$, $(0 \leq j_1 < \ldots < j_k \leq n-1)$, such that for any $a_{j_1}, \ldots, a_{j_k}$ there exists only one word in the code where the symbols $a_{j_1}, \ldots, a_{j_k}$ occupy the positions $\{j_1, \ldots, j_k\}$. Hence, the symbols on the positions $\{j_1, \ldots, j_k\}$ also fully define the codeword.

Definition 3.1 The set of positions $\{j_1, \ldots, j_k\}$, $(0 \leq j_1 < \cdots < j_k \leq n-1)$ is called the information set of the code V if the symbols $a_{j_1}, \ldots, a_{j_k}$ uniquely define the codeword from V.

Let $\mathbf{G}$ be the generator matrix of the code V. Let us denote by $\mathbf{G}(\gamma)$ the matrix constructed from the columns of $\mathbf{G}$ enumerated by the elements of set γ. It is obvious that set γ is the information set if, and only if, the mapping $f_\gamma : V \to A^k$, which put in correspondence to the codeword it coordinates with numbers from γ, is one-to-one mapping. This fact as was shown in chapter 2, is equivalent to the nonsingularity of matrix of mapping f_γ, which is the matrix $\mathbf{G}(\gamma)$ in the basis $\{\mathbf{v}_1, \ldots, \mathbf{v}_k\}$. Thus, the following statement is correct:

Lemma 3.1 The set of positions $\gamma = \{j_1, \ldots, j_k\}$ is the information set if, and only if, the matrix $\mathbf{G}(\gamma)$ is nonsingular.

If in the received erroneous word there is at least one information set without erroneous symbols (i.e. symbols with indexes from this information set do not contain errors), then the transmitted word can be restored on the basis of this information set. In this case the decoding procedure can be regarded as the search of an information set that is free of errors. As this takes place, the issue of the 'stop rule' is very important, i.e. we should choose the rule according to whether it is possible to identify that the information set free of errors is found. Hereafter we will consider the decoding of t-fold errors.

Let us describe now the decoding algorithm based on the information sets. Let $\gamma = \{j_1, \ldots, j_k\}$ be the information set of the code V, and let $\mathbf{G}$ and $\mathbf{H}$ be the generator and the parity matrix of this code. Let us denote by $\mathbf{G}_\gamma$ the matrix

$$\mathbf{G}_\gamma = (\mathbf{G}(\gamma))^{-1} \cdot \mathbf{G}. \tag{3.1}$$

It is obvious that the columns of matrix $\mathbf{G}_\gamma$ with numbers $\{j_1, \ldots, j_k\}$ form the identity $(k \times k)$-matrix. Multiplying the vector $(a_{j_1}, \ldots, a_{j_k})$ by matrix $\mathbf{G}_\gamma$ results in the codeword with symbols $a_{j_1}, \ldots, a_{j_k}$ on the positions $\{j_1, \ldots, j_k\}$:

$$
(a_{j_1}, \ldots, a_{j_k}) \cdot \mathbf{G}_\gamma = (a_{j_1}, \ldots, a_{j_k}) \cdot
\begin{bmatrix}
.. & 1 & ... & 0 & ... & 0 & .. \\
.. & 0 & ... & 1 & ... & 0 & .. \\
.. & .. & ... & .. & ... & .. & .. \\
.. & 0 & ... & 0 & ... & 1 & ..
\end{bmatrix}
$$

$$
= (\ \ .. \ \ a_{j_1}, \ \ .., \ \ a_{j_2}, \ \ .., \ \ a_{j_k}, \ \ .. \ \ )
$$

Let us put matrix $\mathbf{G}_\gamma$ in correspondence to parity matrix

$$
\mathbf{H}_\gamma = (\mathbf{H}(\bar{\gamma}))^{-1} \cdot \mathbf{H}, \tag{3.2}
$$

where $\bar{\gamma} = \{1, 2, \ldots, n\} \backslash \gamma$, i.e. the set of positions that are not included in γ, and $\mathbf{H}(\bar{\gamma})$ is the matrix formed by the columns of matrix $\mathbf{H}$ with indexes from $\bar{\gamma}$. The columns of matrix $\mathbf{H}_\gamma$ with indexes not included in γ form the identity $(r \times r)$-matrix. Hence, if all nonzero elements of error vector $\mathbf{e}$ are located in the set $\bar{\gamma}$ (i.e. γ is free of errors), then the weight of the syndrome

$$
\mathbf{s}_\gamma(\mathbf{e}) = \mathbf{e} \cdot \mathbf{H}_\gamma^T, \tag{3.3}
$$

is equal to weight of error vector. Therefore, the algorithm of the information set decoding of ν-fold errors, $\nu \leq t = \left\lceil \dfrac{d-1}{2} \right\rceil$, can be formulated as follows:

Let $\Gamma = \{\gamma_1, \gamma_2, \ldots, \gamma_l\}$ be the set of information sets of the code. Let us assume that the set Γ contains a reasonable number of information sets to correct of ν-fold errors.

1. Calculation of the syndromes $\mathbf{s}_{\gamma_i}(\mathbf{b})$, where $\mathbf{b} = (b_0, b_1, \ldots, b_{n-1})$ is the received vector, $\mathbf{b} = \mathbf{a} + \mathbf{e}$, $\mathbf{a}$ is the transmitted vector and $\mathbf{e}$ is the error vector, until information set $\gamma = \{j_1, j_2, \ldots, j_k\}$ is found such that the weight of the corresponding syndrome

$$
w(\mathbf{s}_\gamma(\mathbf{b})) \leq \nu. \tag{3.4}
$$

2. If the condition (3.4) is satisfied the codeword $\hat{\mathbf{a}}$ is regarded as the decoded word.

$$
\hat{\mathbf{a}} = \mathbf{b}(\gamma) \cdot \mathbf{G}_\gamma = (b_{j_1}, b_{j_2}, \ldots, b_{j_k}) \cdot \mathbf{G}_\gamma. \tag{3.5}
$$

3. If none of the information sets γ_i satisfy the condition (3.4) the calculation (3.5) is not executed and it is assumed that an uncorrectable error is detected.

Example 3.1 Consider the decoding of the binary $(7, 4)$ code with $d = 3$. Let the generator matrix $\mathbf{G}$ and the parity matrix $\mathbf{H}$ be as follows

$$\mathbf{G} = \begin{bmatrix} 1 & 0 & 0 & 0 & 1 & 0 & 1 \\ 0 & 1 & 0 & 0 & 1 & 1 & 1 \\ 0 & 0 & 1 & 0 & 1 & 1 & 0 \\ 0 & 0 & 0 & 1 & 0 & 1 & 1 \end{bmatrix}; \quad \mathbf{H} = \begin{bmatrix} 1 & 1 & 1 & 0 & 1 & 0 & 0 \\ 0 & 1 & 1 & 1 & 0 & 1 & 0 \\ 1 & 1 & 0 & 1 & 0 & 0 & 1 \end{bmatrix}. \tag{3.6}$$

The set of symbols $\gamma' = \{0, 1, 2, 6\}$ is the information set of the code and the set $\gamma'' = \{0, 1, 2, 4\}$ is not the information set of the code because the determinant of the matrix $\mathbf{G}(\gamma')$

$$|\mathbf{G}(\gamma')| = \begin{vmatrix} 1 & 0 & 0 & 1 \\ 0 & 1 & 0 & 1 \\ 0 & 0 & 1 & 0 \\ 0 & 0 & 0 & 1 \end{vmatrix} \neq 0$$

and the determinant of the matrix $\mathbf{G}(\gamma'')$

$$|\mathbf{G}(\gamma'')| = \begin{vmatrix} 1 & 0 & 0 & 1 \\ 0 & 1 & 0 & 1 \\ 0 & 0 & 1 & 1 \\ 0 & 0 & 0 & 0 \end{vmatrix} = 0.$$

In the same way we can verify that $\gamma^{(0)} = \{0, 1, 2, 3\}$, $\gamma^{(1)} = \{3, 4, 5, 6\}$, $\gamma^{(2)} = \{0, 1, 2, 6\}$ are the information sets. For these information sets $\mathbf{H}_{\gamma^{(0)}} = \mathbf{H}$, $\mathbf{G}_{\gamma^{(0)}} = \mathbf{G}$;

$$\mathbf{H}_{\gamma^{(1)}} = (\mathbf{H}(0, 1, 2))^{-1} \cdot \mathbf{H} = \begin{bmatrix} 1 & 1 & 1 \\ 0 & 1 & 1 \\ 1 & 1 & 0 \end{bmatrix}^{-1} \cdot \mathbf{H} = \begin{bmatrix} 1 & 1 & 0 \\ 1 & 1 & 1 \\ 1 & 0 & 1 \end{bmatrix} \cdot \begin{bmatrix} 1 & 1 & 1 & 0 & 1 & 0 & 0 \\ 0 & 1 & 1 & 1 & 0 & 1 & 0 \\ 1 & 1 & 0 & 1 & 0 & 0 & 1 \end{bmatrix}$$

$$= \begin{bmatrix} 1 & 0 & 0 & 1 & 1 & 1 & 0 \\ 0 & 1 & 0 & 0 & 1 & 1 & 1 \\ 0 & 0 & 1 & 1 & 1 & 0 & 1 \end{bmatrix},$$

$$\mathbf{G}_{\gamma^{(1)}} = (\mathbf{G}(3, 4, 5, 6))^{-1} \cdot \mathbf{G} = \begin{bmatrix} 1 & 0 & 1 & 1 & 0 & 0 & 0 \\ 1 & 1 & 1 & 0 & 1 & 0 & 0 \\ 1 & 1 & 0 & 0 & 0 & 1 & 0 \\ 0 & 1 & 1 & 0 & 0 & 0 & 1 \end{bmatrix};$$

$$\mathbf{H}_{\gamma^{(2)}} = (\mathbf{H}(3, 4, 5))^{-1} \cdot \mathbf{H} = \begin{bmatrix} 1 & 1 & 0 & 1 & 0 & 0 & 1 \\ 1 & 1 & 1 & 0 & 1 & 0 & 0 \\ 1 & 0 & 1 & 0 & 0 & 1 & 1 \end{bmatrix};$$

$$\mathbf{G}_{\gamma^{(2)}} = (\mathbf{G}(0, 1, 2, 6))^{-1} \cdot \mathbf{G} = \begin{bmatrix} 1 & 0 & 0 & 0 & 1 & 0 & 1 \\ 0 & 1 & 0 & 0 & 1 & 1 & 1 \\ 0 & 0 & 1 & 0 & 1 & 1 & 0 \\ 0 & 1 & 1 & 0 & 0 & 0 & 1 \end{bmatrix}.$$

The set of information sets $\Gamma = \{\gamma^{(0)}, \gamma^{(1)}, \gamma^{(2)}\}$ allows decoding any 1-fold errors in (7,4) code. Let $\mathbf{a} = (0, 0, 0, 0, 0, 0, 0)$ be the transmitted word and $\mathbf{b} = (0, 0, 0, 1, 0, 0, 0)$ be the received word. The decoding procedure in accordance with the algorithm described above is as follows:

$$\mathbf{s}_{\gamma^{(0)}}(\mathbf{b}) = \mathbf{b} \cdot \mathbf{H}_{\gamma^{(0)}}^{\mathrm{T}} = (0, 1, 1) \Rightarrow w(\mathbf{s}_{\gamma^{(0)}}(\mathbf{b})) = 2 > 1 = \frac{d-1}{2};$$

$$\mathbf{s}_{\gamma^{(1)}}(\mathbf{b}) = \mathbf{b} \cdot \mathbf{H}_{\gamma^{(1)}}^{\mathrm{T}} = (1, 0, 1) \Rightarrow w(\mathbf{s}_{\gamma^{(1)}}(\mathbf{b})) = 2 > 1 = \frac{d-1}{2};$$

$$\mathbf{s}_{\gamma^{(2)}}(\mathbf{b}) = \mathbf{b} \cdot \mathbf{H}_{\gamma^{(2)}}^{\mathrm{T}} = (1, 0, 0) \Rightarrow w(\mathbf{s}_{\gamma^{(2)}}(\mathbf{b})) = 1 = \frac{d-1}{2}.$$

With the help of the information set $\gamma^{(2)}$ we can calculate the decoded word

$$\hat{\mathbf{a}} = (b_0, b_1, b_2, b_6) \cdot \mathbf{G}_{\gamma^{(2)}} = (0\,0\,0\,0) \cdot \begin{bmatrix} 1 & 0 & 0 & 0 & 1 & 0 & 1 \\ 0 & 1 & 0 & 0 & 1 & 1 & 1 \\ 0 & 0 & 1 & 0 & 1 & 1 & 0 \\ 0 & 0 & 0 & 1 & 0 & 0 & 1 \end{bmatrix} = (0\,0\,0\,0\,0\,0\,0) = \mathbf{a}.$$

The cardinal number of set Γ, i.e. the number of information sets, which is required for decoding generally increases very quickly with the increasing of the code length and the number of correctable errors. Since each information set γ from Γ needs in keeping or calculating matrices $\mathbf{G}_\gamma$ and (or) $\mathbf{H}_\gamma$, then the complexity of the algorithm increases. The simplification of the information set decoding is associated with two modifications of this algorithm - *permutation decoding* and decoding with the help of *covering polynomials* (or *covering-set decoding*).

As was defined above, the permutation π ($\pi \in S_n$) is the one-to-one self-mapping of set $\{1, 2, \ldots, n\}$. Each permutation $\pi \in S_n$ corresponds to the linear operator on space A^n, i.e. $\pi(a_1, \ldots, a_n) = (a_{\pi(1)}, \ldots, a_{\pi(n)})$. Arbitrary permutation transfers the word of the code V to some other word (normally this word does not belong to the code V). However, there exist some permutations that transfer any codeword to the codeword of the same code. Such permutations are said to be preserving the code permutation, and the code is said to be invariant relative to this permutation. It is easy to verify that the set of permutations preserving the code V forms the subgroup in the group S_n of all permutations. This subgroup is denoted as Aut V.

Example 3.2   Consider the binary linear (3, 2)-code consisting of 4 words (000), (110), (100), (010). The permutation of the first and the second symbol of any codeword transfers it to the codeword, but the permutation of the second and the third symbol transfers the codeword (110) to the word (101), which does not belong to the code.

Let $\mathbf{G}$ be the generator matrix and $\mathbf{H}$ be the parity matrix of the code V

$$\mathbf{G} = \begin{bmatrix} \mathbf{g}_1 \\ \cdot \\ \cdot \\ \cdot \\ \mathbf{g}_k \end{bmatrix}; \qquad \mathbf{H} = \begin{bmatrix} \mathbf{h}_1 \\ \cdot \\ \cdot \\ \cdot \\ \mathbf{h}_r \end{bmatrix}.$$

It is a necessary condition for some permutation π to preserve the code V if matrix

$$\pi(\mathbf{G}) = \begin{bmatrix} \pi(\mathbf{g}_1) \\ \cdot \\ \cdot \\ \cdot \\ \pi(\mathbf{g}_k) \end{bmatrix}$$

satisfies the following equation

$$\pi(\mathbf{G}) \cdot \mathbf{H}^{\mathrm{T}} = \mathbf{0}.$$

Let π be the permutation preserving the code V. Then for any vector $\mathbf{b} = \mathbf{a} + \mathbf{e}, \quad \mathbf{a} \in V,$

$$\pi(\mathbf{b}) = \pi(\mathbf{a}) + \pi(\mathbf{e}) = \mathbf{a}' + \mathbf{e}',$$

where $\mathbf{a}'$ is some codeword and $\mathbf{e}'$ is the error vector, which has the same weight as vector $\mathbf{e}$. If the weight of vector $\mathbf{e}$ is no more than t, then the weight of vector $\mathbf{e}'$ also is no more than t.

With the help of permutations preserving the code, information set decoding can be realised as follows. Let $\mathbf{G}$ be the generator matrix and $\mathbf{H}$ be the parity matrix of the code V, and let both these matrices be in systematic form, i.e. corresponding to the information set $\gamma^{(0)} = \{0, 1, \ldots, k - 1\}$. Let Aut $V = \{\pi_1, \ldots, \pi_l\}$ be the set of permutations preserving the code. Let us calculate the syndromes of vectors $\pi_i(\mathbf{b})$ with the help of information set $\gamma^{(0)}$

$$\mathbf{s}(\pi_i(\mathbf{b})) = \pi_i(\mathbf{b}) \cdot \mathbf{H}^{\mathrm{T}}, \tag{3.7}$$

then we calculate the weight of syndromes (3.7). If some permutation π transfers vector $\mathbf{e}$ to vector $\mathbf{e}'$, where all nonzero components are located on the positions $\{k, \ldots, n - 1\}$, then the information set $\gamma^{(0)}$ is free of errors, and the weight of the corresponding syndrome is no more than t. In this case it is sufficient to use the permutation π^{-1} to vector $\mathbf{a}'$ to restore the transmitted codeword.

Example 3.3 Consider the permutation decoding of $(7, 4)$-code with $d = 3$, which was defined in the Example 3.1. This code is invariant relative to the cyclic permutation T, $T(i) = (i + 1) \bmod 7$. Really,

$$T(\mathbf{G}) = \begin{bmatrix} 1 & 1 & 0 & 0 & 0 & 1 & 0 \\ 1 & 0 & 1 & 0 & 0 & 1 & 1 \\ 0 & 0 & 0 & 1 & 0 & 1 & 1 \\ 1 & 0 & 0 & 0 & 1 & 0 & 1 \end{bmatrix},$$

$$T(\mathbf{G}) \cdot \mathbf{H}^{\mathrm{T}} = \mathbf{0}.$$

Let us decode the word $\mathbf{b} = \mathbf{a} + \mathbf{e} = (1011000) + (0100000) = (1111000)$. Let us calculate

$$\mathbf{s}_{\gamma^{(0)}}(T^i(\mathbf{b})) \text{ for } i = 0, 1, \ldots, 6 \text{ since } T^7 = I = T^0.$$

$$\mathbf{s}_{\gamma^{(0)}}(\mathbf{b}) = \mathbf{b} \cdot \mathbf{H}^{\mathrm{T}} = (111) \Rightarrow w(\mathbf{s}_{\gamma^{(0)}}(\mathbf{b})) = 3 > 1 = \frac{d-1}{2};$$

$$\mathbf{s}_{\gamma^{(0)}}(T(\mathbf{b})) = T(\mathbf{b}) \cdot \mathbf{H}^{\mathrm{T}} = (110) \Rightarrow w(\mathbf{s}_{\gamma^{(0)}}(T(\mathbf{b}))) = 2 > 1 = \frac{d-1}{2};$$

$$\mathbf{s}_{\gamma^{(0)}}(T^2(\mathbf{b})) = T^2(\mathbf{b}) \cdot \mathbf{H}^{\mathrm{T}} = (011) \Rightarrow w(\mathbf{s}_{\gamma^{(0)}}(T^2(\mathbf{b}))) = 2 > 1 = \frac{d-1}{2};$$

$$\mathbf{s}_{\gamma^{(0)}}(T^3(\mathbf{b})) = T^3(\mathbf{b}) \cdot \mathbf{H}^{\mathrm{T}} = (100) \Rightarrow w(\mathbf{s}_{\gamma^{(0)}}(T^3(\mathbf{b}))) = 1 = \frac{d-1}{2},$$ and the condition (3.4) is satisfied. With the help of information set $\gamma^{(0)}$ calculate the word $\mathbf{a}' = T^3(\hat{\mathbf{a}})$, $\hat{\mathbf{a}}$ is the decoded word

$$\mathbf{a}' = T^3(\hat{\mathbf{a}}) = (0001) \cdot \mathbf{G} = (0001011),$$

and

$$\hat{\mathbf{a}} = T^{-3}(\mathbf{a}') = (1011000) = \mathbf{a}.$$

Another way of 'clearing' the information set from errors is the covering of errors in the information set. This method is called decoding with the help of covering polynomials[1].

Let θ be the vector (covering polynomial), which coincides with error vector $\mathbf{e}$ on the positions of the information set γ and with zeroes on the other positions. Then for vector $(\mathbf{b} - \theta)$ the information set γ is free of errors, and the weight of the corresponding syndrome

$$\mathbf{s}_{\gamma}(\mathbf{b} - \theta) = (\mathbf{e} - \theta) \cdot \mathbf{H}^{\mathrm{T}}$$

is

$$w(\mathbf{s}_{\gamma}(\mathbf{b} - \theta)) \leq t - w(\theta). \tag{3.8}$$

If we search vectors θ in increasing order of their weights (starting with $\theta_0 = (0\ldots0)$) until some θ^* satisfy (3.8), then it will be possible to restore the transmitted vector with the help of vector $\mathbf{b}^* = (\mathbf{b} - \theta^*)$ and the information set γ (of course, if the weight of error vector does not exceed t).

Example 3.4 Let us use decoding with the help of covering polynomials for the case considered in Example 3.1. Let the set of covering polynomials be $\theta_0 = (0000000)$, $\theta_1 = (1000000)$, $\theta_2 = (0100000)$, $\theta_3 = (0010000)$, $\theta_4 = (0001000)$. Consider the

---

[1]The term 'covering polynomial' is well-established in coding theory but is not exactly correct. The proper term is 'covering vector' 'covering word'.

information set $\gamma^{(0)}$. Let the received vector be $\mathbf{b} = (0001000) = \mathbf{a} + \mathbf{e} = (0000000) + (0001000)$. Calculate the syndromes

$$\mathbf{s}_{\gamma^{(0)}}(\mathbf{b} - \theta_0) = (0001000) \cdot \mathbf{H}^{\mathrm{T}} = (011) \Rightarrow w(\mathbf{s}_{\gamma^{(0)}}(\mathbf{b} - \theta_0)) = 2 > 1 = \frac{d-1}{2};$$

$$\mathbf{s}_{\gamma^{(0)}}(\mathbf{b} - \theta_1) = (1001000) \cdot \mathbf{H}^{\mathrm{T}} = (110) \Rightarrow w(\mathbf{s}_{\gamma^{(0)}}(\mathbf{b} - \theta_1)) = 2 > 1 = \frac{d-1}{2} - w(\theta_1);$$

$$\mathbf{s}_{\gamma^{(0)}}(\mathbf{b} - \theta_2) = (0101000) \cdot \mathbf{H}^{\mathrm{T}} = (100) \Rightarrow w(\mathbf{s}_{\gamma^{(0)}}(\mathbf{b} - \theta_2)) = 1 > 0 = \frac{d-1}{2} - w(\theta_2);$$

$$\mathbf{s}_{\gamma^{(0)}}(\mathbf{b} - \theta_3) = (0011000) \cdot \mathbf{H}^{\mathrm{T}} = (101) \Rightarrow w(\mathbf{s}_{\gamma^{(0)}}(\mathbf{b} - \theta_3)) = 2 > 0 = \frac{d-1}{2} - w(\theta_3);$$

$$\mathbf{s}_{\gamma^{(0)}}(\mathbf{b} - \theta_4) = (0000000) \cdot \mathbf{H}^{\mathrm{T}} = (000) \Rightarrow w(\mathbf{s}_{\gamma^{(0)}}(\mathbf{b} - \theta_4)) = 0 = \frac{d-1}{2} - w(\theta_4).$$

Then $\theta^* = \theta_4$, $\mathbf{b}^* = (\mathbf{b} - \theta^*) = (0000000)$, and the transmitted word can be restored with the help of $\gamma^{(0)}$ and vector $\mathbf{b}^*$:

$$\hat{\mathbf{a}} = (b_0^*, b_1^*, \ldots, b_{k-1}^*) \cdot \mathbf{G}_{\gamma^{(0)}} = (0000000) = \mathbf{a}.$$

The best results can be obtained with the joint use of the algorithms considered above. Let $\Gamma = \{\gamma^{(0)}, \ldots, \gamma^{(m)}\}$ be the set of the information sets of code V and $\theta^{(0)}, \ldots, \theta^{(m)}$ be the sets of covering polynomials corresponding to information sets $\gamma^{(0)}, \ldots, \gamma^{(m)}$: $\theta^{(0)} = \{\theta_{00}, \ldots, \theta_{0l_0}\}, \ldots, \theta^{(m)} = \{\theta_{m0}, \ldots, \theta_{ml_m}\}$ with $\theta_{j0} = \mathbf{0}$ and $w(\theta_{j0}) < w(\theta_{j1}) \leq \ldots \leq w(\theta_{jl_j})$, $j = 0, \ldots, m$. The decoding algorithm based on the joint use of information sets and covering polynomials is as follows:

Covering-Set Decoding:

1. Calculate the vector $\tilde{\mathbf{b}}_{ij} = \mathbf{b} - \theta_{ij}$ and the syndrome $\mathbf{s}_{\gamma^{(i)}}(\tilde{\mathbf{b}}_{ij}) = \tilde{\mathbf{b}}_{ij} \cdot \mathbf{H}^{\mathrm{T}}$ for each pair $\gamma^{(i)}$, θ_{ij} ($i = 0, \ldots, m; j = 0, \ldots, l_j$) until the pair i^*, j^* will be found, such that the following condition is satisfied

$$w(\mathbf{s}_{\gamma^{(i^*)}}(\tilde{\mathbf{b}}_{i^*j^*} - \theta)) \leq t - w(\theta_{i^*j^*}). \tag{3.9}$$

2. If the condition (3.9) is satisfied, calculate

$$\hat{\mathbf{a}} = \tilde{\mathbf{b}}(\gamma^{(i^*)}) \cdot \mathbf{G}_{\gamma^{(i^*)}},$$

where $\tilde{\mathbf{b}}(\gamma^{(i^*)})$ is the subvector of vector $\tilde{\mathbf{b}}_{i^*j^*}$ combined from the elements of vector $\tilde{\mathbf{b}}_{i^*j^*}$, which belong to the information set $\gamma^{(i^*)}$

3. If any pair $\gamma^{(i)}$, θ_{ij} does not satisfy the condition (3.9), then it is assumed that the transmitted word was corrupted by the uncorrectable error.

The set of set Γ and sets of the covering polynomials $\Theta = \{\theta^{(0)}, \ldots, \theta^{(m)}\}$ is called the decoding set and is denoted as $DS = \{\Gamma, \Theta\}$.

It is often convenient not to use all covering polynomials for decoding but only those with weight no more than ν, i.e. vectors with nonzero elements located on the positions of set γ, and the number of these nonzero elements does not exceed ν. Such kinds of vector we denote as $\theta_\gamma(\nu)$.

Example 3.5   Consider (7,4) code from the Example 3.1. Consider $\gamma_0 = \{0, 1, 2, 3\}$. The set of polynomials $\theta_{\gamma^{(0)}}(1)$ consists of five polynomials $\theta_{00} = (0000000)$, $\theta_{01} = (1000000)$, $\theta_{02} = (0100000)$, $\theta_{03} = (0010000)$, $\theta_{04} = (0001000)$.

It is easy to verify that if γ is the information set of the code V and π is the permutation preserving the code, then the set $\pi(\gamma)$ is also the information set of code V.

Example 3.6   Consider the decoding of the (15, 5) Hamming code with distance $d = 7$. It is possible to use the decoding set $DS = \{\Gamma, \Theta(1)\}$ to decode this code, where $\Gamma = \{\gamma^{(0)}, \gamma^{(1)} = T^5(\gamma^{(0)})\}$, $\gamma^{(0)} = \{0, 1, 2, 3, 4\}$, and $T^5(\gamma^{(0)}) = \{5, 6, 7, 8, 9\}$ is the cyclic shift of the information set $\gamma^{(0)}$ by 5 positions. Let the received word $\mathbf{b}$ be $\mathbf{b} = \mathbf{a} + \mathbf{e} = (001110110010100) + (010100010000000)$, i.e. the received word is corrupted in the first, third and seventh position. Calculate $\mathbf{s}_{0j}(\mathbf{b}) = \tilde{\mathbf{b}}_{0j} \cdot \mathbf{H}_{\gamma^{(0)}}^{\mathrm{T}}$ with the help of information set $\gamma^{(0)}$:

$$\tilde{\mathbf{b}}_{00} = \mathbf{b} + \theta_{00} = \mathbf{b} + (000000000000000),$$

$$\mathbf{s}_{00}(\mathbf{b}) = \tilde{\mathbf{b}}_{00} \cdot \mathbf{H}_{\gamma^{(0)}}^{\mathrm{T}} = (0011110110) \Rightarrow w(\mathbf{s}_{00}(\mathbf{b})) = 6 > 3 = \frac{d-1}{2} - w(\theta_{00});$$

$$\tilde{\mathbf{b}}_{01} = \mathbf{b} + \theta_{01} = \mathbf{b} + (100000000000000),$$

$$\mathbf{s}_{01}(\mathbf{b}) = \tilde{\mathbf{b}}_{01} \cdot \mathbf{H}_{\gamma^{(0)}}^{\mathrm{T}} = (1101000100) \Rightarrow w(\mathbf{s}_{01}(\mathbf{b})) = 4 > 2 = \frac{d-1}{2} - w(\theta_{01});$$

$$\cdots\cdots\cdots\cdots\cdots\cdots\cdots\cdots\cdots\cdots\cdots\cdots\cdots\cdots$$

$$\tilde{\mathbf{b}}_{05} = \mathbf{b} + \theta_{05} = \mathbf{b} + (000010000000000),$$

$$\mathbf{s}_{05}(\mathbf{b}) = \tilde{\mathbf{b}}_{05} \cdot \mathbf{H}_{\gamma^{(0)}}^{\mathrm{T}} = (1110010011) \Rightarrow w(\mathbf{s}_{05}(\mathbf{b})) = 6 > 2 = \frac{d-1}{2} - w(\theta_{05});$$

Now let us use the information set $\gamma^{(1)} = T^5(\gamma^{(0)})$, and the corresponding matrices $\mathbf{G}_{\gamma^{(1)}} = T^5(\mathbf{G}_{\gamma^{(0)}})$ and $\mathbf{H}_{\gamma^{(1)}} = T^5(\mathbf{H}_{\gamma^{(0)}})$ are

$$\mathbf{G}_{\gamma^{(1)}} = \begin{bmatrix} 1 & 0 & 0 & 1 & 0 & 1 & 0 & 0 & 0 & 0 & 1 & 1 & 1 & 0 & 1 \\ 1 & 1 & 0 & 0 & 1 & 0 & 1 & 0 & 0 & 0 & 0 & 1 & 1 & 1 & 0 \\ 1 & 1 & 1 & 1 & 0 & 0 & 0 & 1 & 0 & 0 & 1 & 1 & 0 & 1 & 0 \\ 0 & 1 & 1 & 1 & 1 & 0 & 0 & 0 & 1 & 0 & 0 & 1 & 1 & 0 & 1 \\ 0 & 0 & 1 & 0 & 1 & 0 & 0 & 0 & 0 & 1 & 1 & 1 & 0 & 1 & 1 \end{bmatrix},$$

$$\mathbf{H}_{\gamma^{(1)}} = \begin{bmatrix} 0 & 0 & 0 & 0 & 0 & 1 & 0 & 1 & 0 & 1 & 1 & 0 & 0 & 0 & 0 \\ 0 & 0 & 0 & 0 & 0 & 1 & 1 & 1 & 1 & 1 & 0 & 1 & 0 & 0 & 0 \\ 0 & 0 & 0 & 0 & 0 & 1 & 1 & 0 & 1 & 0 & 0 & 0 & 1 & 0 & 0 \\ 0 & 0 & 0 & 0 & 0 & 0 & 1 & 1 & 0 & 1 & 0 & 0 & 0 & 1 & 0 \\ 0 & 0 & 0 & 0 & 0 & 1 & 0 & 0 & 1 & 1 & 0 & 0 & 0 & 0 & 1 \\ 1 & 0 & 0 & 0 & 0 & 1 & 1 & 1 & 0 & 0 & 0 & 0 & 0 & 0 & 0 \\ 0 & 1 & 0 & 0 & 0 & 0 & 1 & 1 & 1 & 0 & 0 & 0 & 0 & 0 & 0 \\ 0 & 0 & 1 & 0 & 0 & 0 & 0 & 1 & 1 & 1 & 0 & 0 & 0 & 0 & 0 \\ 0 & 0 & 0 & 1 & 0 & 1 & 0 & 1 & 1 & 0 & 0 & 0 & 0 & 0 & 0 \\ 0 & 0 & 0 & 0 & 1 & 0 & 1 & 0 & 1 & 1 & 0 & 0 & 0 & 0 & 0 \end{bmatrix}.$$

Calculate $\mathbf{s}_{1j}(\mathbf{b}) = \tilde{\mathbf{b}}_{1j} \cdot \mathbf{H}^{T}_{\gamma^{(1)}}$ with the help of information set $\gamma^{(1)}$:

$$\tilde{\mathbf{b}}_{10} = \mathbf{b} + \theta_{10} = \mathbf{b} + (000000000000000),$$

$$\mathbf{s}_{10}(\mathbf{b}) = \tilde{\mathbf{b}}_{10} \cdot \mathbf{H}^{T}_{\gamma^{(1)}} = (1101010100) \Rightarrow w(\mathbf{s}_{10}(\mathbf{b})) = 5 > 3 = \frac{d-1}{2} - w(\theta_{10});$$

$$\tilde{\mathbf{b}}_{11} = \mathbf{b} + \theta_{11} = \mathbf{b} + (0000100000000000),$$

$$\mathbf{s}_{11}(\mathbf{b}) = \tilde{\mathbf{b}}_{11} \cdot \mathbf{H}^{T}_{\gamma^{(1)}} = (0011100110) \Rightarrow w(\mathbf{s}_{11}(\mathbf{b})) = 5 > 2 = \frac{d-1}{2} - w(\theta_{11});$$

$$\cdots\cdots\cdots\cdots\cdots\cdots\cdots\cdots\cdots\cdots\cdots\cdots\cdots\cdots\cdots\cdots\cdots\cdots\cdots\cdots\cdots$$

$$\tilde{\mathbf{b}}_{13} = \mathbf{b} + \theta_{13} = \mathbf{b} + (0000000100000000),$$

$$\mathbf{s}_{13}(\mathbf{b}) = \tilde{\mathbf{b}}_{13} \cdot \mathbf{H}^{T}_{\gamma^{(1)}} = (0000001010) \Rightarrow w(\mathbf{s}_{13}(\mathbf{b})) = 2 = \frac{d-1}{2} - w(\theta_{13}).$$

With the help of the word $\tilde{\mathbf{b}}_{13}$ and the information set $\gamma^{(1)}$ we can restore the word $\hat{\mathbf{a}}$:

$$\hat{\mathbf{a}} = \tilde{\mathbf{b}}_{13}(\gamma^{(1)}) \cdot \mathbf{G}_{\gamma^{(1)}} = (01100) \cdot \mathbf{G}_{\gamma^{(1)}} = (001110110010100) = \mathbf{a}.$$

In this example the decoding set can be constructed with the help of the permutation T^5 rather than the usage of the information set $\gamma^{(1)}$. In this case the decoding algorithm consists of calculation of $\mathbf{s}_{0j}(\mathbf{b})$ and then the calculation of $\mathbf{s}_{0j}(T^5(\mathbf{b}))$.

Actually finding set Γ is very difficult. A few nontrivial examples are found in [1], [2], and [3]; see also [4]. Therefore, to implement the general information-set decoding algorithm, we have to specify a way of choosing information sets. One obvious suggestion is to take random uniformly distributed k-subsets of set $\{0, 1, \ldots, n-1\}$. We call the following algorithm *generalised covering-set decoding* because in essence, it produces a random covering design and can be regarded as a generalisation of the algorithms considered above.

Generalised Covering-Set Decoding:

1. Set $\hat{\mathbf{a}} = \mathbf{0}$.

2. Choose randomly a k-subset γ. Form a list of codewords $M(\gamma) = \{\mathbf{c} \in V | \mathbf{c}(\gamma) = \mathbf{b}(\gamma)\}$.

3. If there is a $\mathbf{c} \in M(\gamma)$ such that

$$dist(\mathbf{c}, \mathbf{b}) < dist(\hat{\mathbf{a}}, \mathbf{b})$$

assign $\hat{\mathbf{a}} \leftarrow \mathbf{c}$.

4. Repeat the last two steps $L_n(k)$ times. Output $\hat{\mathbf{a}}$.

The number of steps $L_n(k)$ needed to execute the algorithm will be discussed later.

An improvement of this algorithm was achieved in two steps in [5], [6]. The idea in [5] is to organise the syndrome table more economically by computing the syndrome separately for the 'left' and 'right' parts of the received vector $\mathbf{b}$.

Suppose that the actual number of errors is t. Let us split the set $\{0, 1, \ldots, n-1\}$ into two parts, $l = \{0, 1, \ldots, m-1\}$ and $r = \{m, m+1, \ldots, n-1\}$, and let $[\mathbf{H}_l | \mathbf{H}_r]$ be the corresponding partition of the parity-check matrix $\mathbf{H}$. Any error vector $\mathbf{e} = (\mathbf{e}_l|\mathbf{e}_r)$ with $\mathbf{e} \cdot \mathbf{H}^T = \mathbf{e}_l \cdot \mathbf{H}_l^T + \mathbf{e}_r \cdot \mathbf{H}_r^T = \mathbf{s}$ is a plausible candidate for the decoding output. Assume, in addition, that the number of errors within the subset l equals u, where the numbers u and m are chosen in accordance with the natural restrictions $u \leq m$, $t - u \leq n - m$. For every possible m-vector $\mathbf{e}_l$, compute the product $\mathbf{s}_l = \mathbf{e}_l \cdot \mathbf{H}_l^T$ and store it, together with the vector $\mathbf{e}_l$, as an entry of the table X_l. Likewise, form the table X_r and look for a pair of entries $(\mathbf{s}_l, \mathbf{s}_r)$ that add up to the received syndrome $\mathbf{s}$. Therefore, for every given $\mathbf{s}_r$ occurring in X_r, we should inspect X_l for the occurrence of $\mathbf{s} - \mathbf{s}_r$. One practical way to do this is to order X_l with respect to the entries $\mathbf{s}_l$.

However, in reality we know neither the number of errors nor their distribution. Therefore, we have to repeat the described procedure for several choices of m and u. In doing so, we may optimise the choice in order to reduce the total size of memory used for the tables X_l and X_r. For every choice of m there are no more than t different options for the choice of u. Hence, by repeatedly building the tables, though not more than nt times, we shall capture any distribution of t errors. Finally, the entire procedure should be repeated for all $t = 1, 2, \ldots, d$ until we find the error vector that has the 'received' syndrome $\mathbf{s}$. Let us give a more formal description of the algorithm.

Split Syndrome Decoding:

Precomputation stage: For every weight t, $1 \leq t \leq d$ find the point m such that the tables X_l and X_r have an (almost) equal size. Store the pair (m, u) in the set $E(t)$.

1. Compute $\mathbf{s} = \mathbf{b} \cdot \mathbf{H}^T$ and set $t = 1$.

2. For every entry of $E(t)$, form the tables X_l and X_r as described.

3. Order X_l with respect to the entries $\mathbf{s}_l$.

4. For every entry of X_r check whether X_l contains the vector $\mathbf{s}_l = \mathbf{s} - \mathbf{s}_r$. If this is found, then output $\hat{\mathbf{a}} = \mathbf{b} - (\mathbf{e}_l|\mathbf{e}_r)$ and STOP.

5. Otherwise, set $t = t + 1$ and repeat Steps 2–5 while $t < d$.

3.3 A SUPERCODE DECODING ALGORITHM

A supercode decoding algorithm is based on the ideas illustrated in the previous section. A more detailed description of this algorithm can be found in [7]. The basic idea of the supercode algorithm is to combine lists of candidates obtained after decoding of several 'supercodes' of V, i.e., linear codes V' such that $V \subset V'$. We begin with an example that illustrates some of the ideas of the algorithm.

Example 3.7 Consider the (48, 24, 12) *Binary Extended QR Code*. The aim is to construct a decoder that corrects five errors. Suppose the first 24 coordinates form an information set of the code. Since the code is self-dual, the last 24 coordinates also form an information set. The decoding algorithm consists of two stages, one for each of the two choices of information sets. We only explain one of them; the other is symmetric. Suppose the parity-check matrix $\mathbf{H}$ is reduced to the form $[\mathbf{I}_{24}|\mathbf{A}]$. Let

$\mathbf{e} = (\mathbf{e}_l|\mathbf{e}_r)$ be an error vector of weight ≤ 5, where $\mathbf{e}_l$ and $\mathbf{e}_r$ are the first and the second halves, respectively. At this stage we aim at correcting error vectors satisfying $wt(\mathbf{e}_l|\mathbf{e}_r) = (5, 0)$ or $(4, 1)$ or $(3, 2)$; the remaining possibilities will be covered by the second stage.

Let $\mathbf{b} = (\mathbf{b}_l|\mathbf{b}_r)$ be the received vector. First, the decoder assumes that $wt(\mathbf{e}_r) \leq 1$. There are 25 such error vectors. Each of them is subtracted from $\mathbf{b}_r$. The obtained vector is then encoded with the code and compared to $\mathbf{b}$. If the distance between them is less than or equal to 5, the decoding stops.

Otherwise, let $\mathbf{s} = \mathbf{b} \cdot \mathbf{H}^T$ be the received syndrome. Let $\mathbf{H}_i$, $1 \leq i \leq 4$, be the submatrix of $\mathbf{H}$ formed by rows $6(i - 1) + j$, $1 \leq j \leq 6$, and let $\mathbf{s}_i = \mathbf{b} \cdot \mathbf{H}_i^T$ be the corresponding part of the syndrome. Denote by $\mathbf{A}_i$ the corresponding six rows of the matrix $\mathbf{A}$. The partition into submatrices defines a partition of the first 24 coordinates of the code into four parts

$$N_i = [6(i - 1) + 1, 6(i - 1) + 2, \ldots, 6(i - 1) + 6], \quad 1 \leq i \leq 4.$$

The syndrome $\mathbf{s}$ is divided into four parts $\mathbf{s}_1, \mathbf{s}_2, \mathbf{s}_3, \mathbf{s}_4$ where part $\mathbf{s}_i$ is formed by the sum of the columns in $\mathbf{H}_i$ that correspond to the positions of errors. If a part, say N_1, is error-free, then there is an error pattern $\mathbf{e}$ with $wt(\mathbf{e}_r) \leq 2$ such that $\mathbf{e}_r \cdot \mathbf{A}_1^T = \mathbf{s}_1$. Any single error in N_1 affects one coordinate in the syndrome. Therefore, if N_1 contains one error, by inspecting all error patterns $\mathbf{e}_r$ of weight ≤ 2 in the information part we shall find a syndrome $\mathbf{s}' = \mathbf{e}_r \cdot \mathbf{A}_1^T$ at a distance one from $\mathbf{s}_1$. Therefore, this step can be accomplished as follows. For each i, $1 \leq i \leq 4$, we make a list of error patterns $\mathbf{e}_r$ with $wt(\mathbf{e}_r) \leq 2$ that yield a syndrome $\mathbf{s}'$ at a distance ≤ 1 from $\mathbf{s}_i$. An error pattern is a plausible candidate if it appears in three out of four lists.

Since we do not know the actual error distribution, we need to store four tables of error patterns for each of the two message sets. Each table consists of 64 records, one for each possible value of $\mathbf{s}_i$. Finally, each record is formed by all error patterns $\mathbf{e}_r$ of weight 2 or less for which $dist(\mathbf{s}_i, \mathbf{e}_r \cdot \mathbf{A}_i^T) \leq 1$.

The decoding is repeated for each of the two information sets. For a given information set, we compile a list of error patterns that appear in three out of four tables T_i in the record corresponding to the received syndrome $\mathbf{s}_i$. Each error pattern is subtracted from the 24 coordinates of $\mathbf{b}$ that correspond to the message part. The obtained message set is then encoded with the code. The decoding stops when we find a vector at a distance of at most 5 from $\mathbf{b}$.

The total size of memory used by tables T_i is 8 kbytes. The decoding requires about 3000 operations with binary vectors of length 24 and seems to be the simplest known for this code. Note that the straightforward search over all error patterns of weight ≤ 2 in the message part would require about twice as many operations.

Let us now pass to the general case. The algorithm involves an (exponential) number of iterations. Each iteration is performed for a given information set with respect to V and consists of $O(n - k)$ steps of decoding different codes V'. Let $\hat{\mathbf{a}}$ be the current decision, which is updated in the course of the decoding. The initial value is set to 0.

First, we describe what happens after we fix an information set $\gamma \subset \{0, 1, \ldots, n - 1\}$ with respect to the code V. Let $\mathbf{H}$ be an $((n - k) \times n)$ parity-check matrix of V. Choose the basis of V in such a way that $\mathbf{H}$ is diagonal on $\{0, 1, \ldots, n - 1\} \backslash \gamma$, i.e., $\mathbf{H} = [\mathbf{A}|\mathbf{I}_{n-k}]$. The idea is to look for the part of the syndrome least affected by errors. Let y, $0 \leq y \leq n - k$ be an integer parameter whose value will be chosen later. Represent the matrix $\mathbf{H}$ in the form

$$\mathbf{H} = \begin{bmatrix} \mathbf{A}_1 & \mathbf{I}_y & 0 \\ \mathbf{A}_2 & 0 & \mathbf{I}_{n-k-y} \end{bmatrix}.$$

Let $\mathbf{b}$ be a received vector, i.e., $\mathbf{b} = \mathbf{a} + \mathbf{e}$, where $\mathbf{a} \in V$ is the closest codeword to $\mathbf{b}$. Isolate the first $k + y$ coordinates of V and denote by $V(y)$ the linear code orthogonal to $\mathbf{H}_y = [\mathbf{A}_1 | \mathbf{I}_y]$. Let $\mathbf{s}_y$ be the syndrome of $\mathbf{b}$ with respect to $\mathbf{H}_y$. Decoding in $V(y)$ amounts to solving the equation

$$\mathbf{u} \cdot \mathbf{H}_y^T = \mathbf{s}_y \tag{3.10}$$

with respect to the unknown vector $\mathbf{u}$. Suppose $\mathbf{u}$ is represented as $\mathbf{u} = (\mathbf{u}_1 | \mathbf{u}_2)$, where $\mathbf{u}_1$ is a k-vector and $\mathbf{u}_2$ is a y-vector. Then in (3.10) we are looking for vectors satisfying

$$(\mathbf{u}_1 | \mathbf{u}_2) \cdot [\mathbf{A}_1 | \mathbf{I}_y]^T = \mathbf{s}_y \tag{3.11}$$

To build the list of solutions to this equation, we again use the split-syndrome algorithm. Suppose that we also know that $wt(\mathbf{u}_1) \leq e_1$ and $wt(\mathbf{u}_2) \leq e_2$ (below we abbreviate this as $wt(\mathbf{u}_1 | \mathbf{u}_2) \leq (e_1 | e_2)$). This restriction allows us to reduce the size of the list of solutions to (3.11). Here e_1 and e_2 again are integer parameters whose values are chosen below. Partition the subset $\{0, 1, \ldots, n - 1\} \backslash \gamma$ of size n-k into $s = (n - k)/y$ consecutive segments of length y (we assume that s is integer). Repeat the decoding for all s placements of the y-segment within the check part of V. The ith placement supplies us with a list $K_i = \{\mathbf{u} = (\mathbf{u}_1 | \mathbf{u}_2)\}$, where every vector $\mathbf{u}$ satisfies (3.11). We are only going to test those error vectors $\mathbf{u}_1$ that appear as first parts of $\mathbf{u}$ for at least l lists K_i, where l is another parameter of the procedure. Form a list

$$K = K(\gamma) = \left\{ \mathbf{u}_1 | \mathbf{u} = (\mathbf{u}_1 | \mathbf{u}_2) \in \bigcap_{j=1}^{l} K_{i_j} \quad \text{for some } 1 \leq i_1 < i_2 \cdots < i_l \leq s \right\} \tag{3.12}$$

Entries of this list are possible error vectors in the coordinates of γ. Therefore, we subtract them from $\mathbf{b}(\gamma)$ to form a list $J(\gamma) = \{\mathbf{b}(\gamma) - \mathbf{u}_1 | \mathbf{u}_1 \in K(\gamma)\}$. For every vector $\mathbf{z}$ in this list we examine all code vectors $\mathbf{c}'$ with $\mathbf{c}'(\gamma) = \mathbf{z}$. We update the current decision $\hat{\mathbf{a}}$ by setting $\hat{\mathbf{a}} = \mathbf{c}'$ if this procedure finds a vector with $dist(\mathbf{b}, \mathbf{c}') < dist(\mathbf{b}, \hat{\mathbf{a}})$. Now let us give the formal description of the algorithm [7].

Supercode Decoding:

1. Compute the syndrome $\mathbf{s} = \mathbf{b} \cdot \mathbf{H}^T$. Set $\hat{\mathbf{a}} = \mathbf{0}$.

2. Choose a random subset $\gamma \in \{0, 1, \ldots, n - 1\}$, $|\gamma| = k$. Bring the matrix $\mathbf{H}$ to the form $\mathbf{H}' = [\mathbf{A} | \mathbf{I}_{n-k}]$, where the columns of $\mathbf{A}$ have their numbers in γ.

3. Split the subset $n - k$ into $s = (n - k)/y$ segments of length y. For every i, $1 \leq i \leq s$, do the following two steps:

   3.1. Form the $(y \times (k + y))$ matrix $\mathbf{H}_i = [\mathbf{A}_i | \mathbf{I}_y]$, isolating rows $y(i - 1) + j$, $1 \leq j \leq y$, of the parity-check matrix $\mathbf{H}$. Form the vector $\mathbf{s}_i = (\mathbf{s}_{y(i-1)+j})$, $1 \leq j \leq y$.

   3.2. Apply the split-syndrome algorithm to form a list K_i of vectors $\{\mathbf{u} = (\mathbf{u}_1 | \mathbf{u}_2)\}$ with $wt(\mathbf{u}_1 | \mathbf{u}_2) \leq (e_1 | e_2)$ that satisfy the equation $\mathbf{s}_i = \mathbf{u} \cdot \mathbf{H}_i^T$.

4. Form the list $K(\gamma)$ of those vectors $\mathbf{u}_1$ that appear in at least l lists K_i, see (3.12).

5. For every k-vector $\mathbf{m} = \mathbf{b}(\gamma) - \mathbf{u}_1$, $\mathbf{u}_1 \in K(\gamma)$ generate successively all code vectors $\mathbf{c}' \in V$ whose projection on the chosen k-subset γ equals $\mathbf{m}$. If $dist(\mathbf{b}, \mathbf{c}') < dist(\mathbf{b}, \hat{\mathbf{a}})$, assign $\hat{\mathbf{a}} = \mathbf{c}'$.

6. Output $\hat{\mathbf{a}}$.

The steps 2–5 are performed $L_n(k, e_1)$ times. The value of $L_n(k, e_1)$ will be discussed later.

This algorithm rests on a number of assumptions. First, we have assumed that γ is an information set. In reality, however, γ is not to be found immediately; therefore, we may not be able to diagonalise the parity-check and generator matrices of V. Next we assume that we are able to control the weight of errors on different parts of the received vector. The detailed explanation why these assumptions hold true can be found in [7].

We conclude this section by estimating the performance of our algorithm.

Theorem 3.1 [7]. For almost all long linear codes the supercode algorithm performs complete minimum-distance decoding.

Proof If all our assumptions about the code and the number of errors hold true, the 'true' error vector will appear in at least l lists K_i for a certain choice of γ. Then it will be included in the list $J(\gamma)$ and will be encoded on Step 5 into a code vector (a list of code vectors). Obviously, one of these code vectors is the transmitted one. It will be chosen as the decoding result if it is the closest to the received vector $\mathbf{b}$. Therefore, decoding with the supercode algorithm can result in a wrong codeword, i.e., a codeword other than the transmitted one if one of the following events takes place.

1. The weight of the error is greater than d;

2. The correct code vector appears in one of the lists $K(\gamma)$ but is not the closest to the transmitted vector $\mathbf{b}$;

3. Repeated random choice fails to produce a partition with the desired error distribution.

For almost all codes, the first and second events form the part of the inherent error of any decoding algorithm (even an exhaustive search would yield an error). The error probability of the complete maximum-likelihood decoding for most codes is known to behave as $p_c = q^{-O(n)}$ [8]. The third event occurs only with probability n^{-n}. We conclude that for almost all codes, except for a fraction of codes that decays exponentially in the code length, the decoding error probability up to $o(1)$ terms behaves as the probability p_c of the complete maximum-likelihood decoding.

3.4 THE COMPLEXITY OF DECODING IN THE CHANNEL WITH INDEPENDENT ERRORS

The complexity of decoding algorithms is currently a subject of extensive study in coding theory. However, the way of measuring complexity itself is seldom discussed or specified, which sometimes results in algorithms that are performed under different computation models being listed as comparable. Here we work with the following two models: random-access machines (RAMs) [9] and Boolean circuits. The RAM is a computing device that has an unrestricted amount of memory with direct access and performs basic operations similar to those performed by Turing machines. This computational model corresponds to 'real-life' computers. Most algorithms in coding theory that are currently being studied are formulated under the implicit assumption of this model. The complexity is measured by the number of operations (*time complexity*) and the amount of memory used by the algorithm (*space complexity*). Implementation of decoders by Boolean circuits allows basic operations to be performed in parallel. This approach has a long history in coding theory [10]. The complexity is measured by the number of gates in the circuit (*size*) and the length of the longest path from an input gate to an output gate (*depth*). We discuss implementation of decoders under both models.

The analysis of the complexity of the maximum likelihood (ML) decoding is based on Lemma 2.2, which reduces the ML decoding to the combinatorial problem of the correction of the given number of errors. This lemma claims that in a channel with independent errors it is possible to provide an error probability of not more than two times worse than that provided by ML decoding if decoded in the sphere of radius d_{GV} (see Section 2.5), i.e. to correct t-fold errors, where $t < d_{GV}$, and d_{GV} is the maximal number such that

$$\sum_{i=1}^{d_{GV}-1} \binom{n}{i} \cdot (q-1)^i \leq q^{n-k},$$

where q is the number of symbols in the alphabet, n is the code length, and k is the number of information symbols of the code. The *relative* GV distance δ_{GV} is the limit value of δ_{GV}/n as $n \to \infty$. Let $R = k/n$ be the rate of the code, then $\delta_{GV} = \delta_{GV}(R)$ is the smallest positive root of the equation $R = 1 - H_q(\delta)$, where $H_q(x) = x \cdot \log_q(q-1) - \cdot \log_q x - (1-x) \cdot x \log_q(1-x)$ is the entropy function.

By their nature, decoding algorithms allow a certain error rate due to the occasional high number of errors in the channel. The idea of reducing the decoding complexity is to allow an algorithmic error whose rate has, at most, the same order as that of inherent error events. The overall decoding error rate for long codes is then essentially the same as that for minimum-distance decoding.

This line of research was initiated by the work of Evseev [11]. He has studied decoding in discrete additive channels. Let X be a finite input alphabet and $Y \supseteq X$ a finite output alphabet of the channel (say, an additive group). A channel is called *additive* if $\Pr(\mathbf{y}|\mathbf{x}) = \Pr(\mathbf{y} - \mathbf{x})$, i.e., the error process does not depend on the message transmitted. Let p_{ML} be the error probability of maximum-likelihood decoding. Evseev has proved that any decoding algorithm that examines q^{n-k} most probable error patterns has error probability $p \leq 2p_{ML}$. Specialising this for the q-ary symmetric channel and using the definition of d_{GV}, we observe that given a (n, k, d) linear code, inspecting all possible errors in the sphere of radius

$d_{GV}(n,k)$ around the received word rather than all error patterns, at most doubles the decoding error probability. Based on this, Evseev [11] proposed a general decoding algorithm whose asymptotic complexity $q^{k(1-R)(1+o(1))}$ improved all the methods known at that time.

This work opened a new page in the study of decoding algorithms of general linear codes. Papers [3], [5], [6], [12], [13] and [14] introduced new decoding methods and provided theoretical justification of those already known.

In the asymptotic setting this approach was later supplemented by the following important result.

Theorem 3.2 [15]:   The covering radius of almost all (n, k) random linear codes equals $d_{GV} \cdot (1 + o(1))$.

Remark Most of the results hereafter are formulated for long linear codes. Strictly speaking, this means that we study families of (n, k_n, d_n) codes of growing length. Let V_n be such a family. Suppose there exists the limit $R = \lim_{n\to\infty} k_n/n$ called the rate of the family. The statement of the theorem means that if ρ_n is the covering radius of V_n, then with probability $\to \infty$ the quotient $\rho_n/n \to \delta_{GV}(R)$.

Theorem 3.2 implies that for long linear codes, correcting a little more than d_{GV} errors ensures the same output error rate as complete minimum-distance decoding. For these reasons, the algorithms considered below restrict themselves to decoding in the sphere of radius $d_{GV} = n \cdot \delta_{GV}(R)$. In the rare case that the algorithms find no codeword at all, we can take any codeword as the decoding result. Asymptotically this will not affect the decoding error rate. We wish to underline that these algorithms, in contrast to gradient-like methods [16], [17], perform complete maximum-likelihood decoding only in the limit as $n \to \infty$. Their error probability as a function approaches the error probability of complete maximum likelihood decoding (the limit of their quotient is one).

Recently, Dumer [18], [19] extended both results to the case of much more general channels. Namely, he has proved [18] that for a symmetric channel with finite input X and arbitrary output $Y \supseteq X$, a decoding algorithm that examines $N > q^{n-k}$ most probable vectors of X^n has error probability $p \leq p_{ML} \cdot (1 + (q^{n-k}/N - q^{n-k}))$. This enabled him to construct general maximum-likelihood soft-decision decoding algorithms with reduced complexity similar in spirit to the known hard-decision decoding methods.

Algorithms that we discuss examine a list of plausible candidates for the decoder output. Our sole concern will be that the transmitted codeword appears in this list. If it later fails to be chosen, this is a part of the inherent error event rather than the fault of a specific decoder. Note that in practice we often do not need to store the whole list, keeping only the most plausible candidates obtained so far.

Minimum-distance decoding can be accomplished either by inspecting all codewords of V (time complexity $O(nq^k)$) or by storing the table of syndromes and coset leaders (space complexity $O(nq^{n-k})$). The last method is called *syndrome decoding*. The only known algorithm that yields an asymptotic improvement of these methods based on geometric properties of the code itself, i.e., without introducing an additional algorithmic error rate, is the *zero-neighbours algorithm* [17]. The decoding is accomplished by iterative refinements of the current decision in much the same way as are standard optimisation methods in continuous spaces. However, since our space is discrete, in order to determine the direction that reduces the value of the objective function (the distance between the received vector $\mathbf{y}$ and the closest codeword found), one has to inspect a certain subset of codewords called zero neighbors. It is shown in [17] that this subset lies entirely inside the sphere of radius $2t + 1$ about 0, where t is the covering radius of the code. Both time and space complexity of this decoding are governed by the size of this set. By Theorem 3.2, the covering radius for almost all codes of rate R grows as $n\delta_{GV}(R)$. Thus the complexity of this algorithm for codes of rate R is dominated by the asymptotic size of the sphere of radius $2n\delta_{GV}(R)$, given in Lemma 3.5 below. This leads to the following result.

Theorem 3.3 [17]:   Let V be an $(n,\ Rn)$ linear code. For any $\mathbf{y} \in E_q^n$, zero-neighbors decoding always finds a closest codeword. For almost all codes it can be implemented by a sequential algorithm with both time and space complexity $q^{n\eta_q(R)(1+o(1))}$, where

$$
\eta_q(R) = \begin{cases}
R, & 0 \le R \le 1 - H_q\left(\dfrac{q-1}{2q}\right) \\[2ex]
H_q(2\delta_{GV}) - (1 - R), & 1 - H_q\left(\dfrac{q-1}{2q}\right) < R < 1.
\end{cases}
$$

A parallel implementation of this decoding requires a Boolean circuit of size $q^{n\eta_q(R)(1+o(1))}$ and depth $O(n^2)$.

For instance, for $q = 2$, the complexity of this decoding is exponentially smaller than that of the exhaustive search for $R \ge 1 - H_2(1/4) = 0.189$. This is also smaller than the time complexity $O(nq^{k(1-R)})$ of the decoding algorithm in [11] for high code rates.

However, as shown in [20], this approach seems to have already reached its limits. Namely, a result in [20] shows that any 'gradient-like' decoding method for binary codes, all of whose codewords have even weight, has to examine all zero neighbors. (See [20] for definitions and exact formulations.) For this reason we turn our attention to information-set decoding.

Let us choose the number of steps $L_n(k)$ in Covering-Set Decoding algorithm as follows

$$
L_n(k) = (n \log n) \cdot \binom{n}{d_{GV}} \Big/ \binom{n-k}{d_{GV}}. \tag{3.13}
$$

A key result in [13] and [21] states that as the length of the code grows, any k coordinates form an 'almost' information set, i.e., that the codimension of the space of code vectors in any k coordinates is small. This means that the number of code vectors that project identically on any k coordinates, i.e., the size of the list $M(\gamma)$ for any γ, is small. The following lemma shows that this size grows as $q^{O(n^{1/2})}$ and, therefore, does not contribute to the main term of the complexity estimate. In [7] a slightly better estimate was proved for the corank of square submatrices of a random matrix than the one in [13] and [21].

Lemma 3.2   Let $\mathbf{A}$ be a random $(k \times n)$ matrix over F_q, $k < n$, $k, n \to \infty$. For almost all matrices, the corank of every square $(k \times k)$ submatrix $\mathbf{B}$ is

$$
corank\ \mathbf{B} \le \sqrt{\log_q \binom{n}{k}}.
$$

Proof Let us estimate the probability $\pi(k, u)$ that a random $k \times k$ matrix has rank u. Every $(k \times k)$-matrix of rank corresponds to a linear mapping that takes E_q^k to a u-dimensional space F that can be viewed as a subspace of E_q^k. The number of image subspaces is

$$\begin{bmatrix} k \\ u \end{bmatrix} = \prod_{j=0}^{u-1} (q^k - q^j)/(q^u - q^j).$$

Likewise, the number of kernels of a rank u mapping is $\begin{bmatrix} k \\ u \end{bmatrix}$. Every choice of the basis in the subspace $F \subset E_q^k$ accounts for a different matrix. The number of bases in F equals $\prod_{j=0}^{u-1} (q^u - q^j)$. Therefore, the number of square matrices of order k and rank u equals

$$\prod_{j=0}^{u-1} (q^k - q^j)^2/(q^u - q^j).$$

Thus

$$\pi(k, u) = q^{-k^2} \prod_{j=0}^{u-1} \frac{(q^k - q^j)^2}{(q^u - q^j)} < q^{-k^2 + ku} \prod_{j=0}^{u-1} \frac{q^{k-j} - 1}{q^{u-j} - 1},$$

where we have estimated $\prod_{j=0}^{u-1} (q^k - q^j) = q^{(1/2)u(u-1)} \prod_{j=0}^{u-1} (q^{k-j} - 1)$ by q^{ku}. Estimating the Gaussian binomial, we obtain

$$\prod_{j=0}^{u-1} \frac{q^{k-j} - 1}{q^{u-j} - 1} < q^{u(k-u)} \prod_{j=0}^{u-1} \frac{q^{u-j}}{q^{u-j} - 1} = q^{u(k-u)} \prod_{j=0}^{u-1} \left(1 + \frac{1}{q^j - 1}\right)$$

$$= q^{u(k-u)} \cdot \frac{q}{q - 1} \cdot \left(1 + \frac{1}{q^2 - 1} + \cdots\right).$$

For $q \geq 2$, the constant factor here is less than 5 since the omitted terms are always less than 1. Thus

$$\pi(k, u) < 5q^{-(k-u)^2}. \tag{3.14}$$

Since there are $\binom{n}{k}$ possibilities for **B**, the probability that there is a submatrix of **A** with corank $> k - l$ is

$$\sum_{u=0}^{l-1} \binom{n}{k} \pi(k, u). \tag{3.15}$$

Substituting (3.14) in (3.15) we obtain that the value of probability (3.15) does not exceed

$$\sum_{u=0}^{l-1}\binom{n}{k}\pi(k,u) \; < \; 5\binom{n}{k}q^{-(k-l)^2}\sum_{u=0}^{l-1}q^{-(k-u)^2+(k-l)^2} \; < \; 5\binom{n}{k}q^{-(k-l)^2}\sum_{i=1}^{l}q^{-(i+1)^2+1}.$$

The last sum is maximal for $q = 2$. It can be checked not to exceed 0.2. Therefore, the required probability is at most $\binom{n}{k}\cdot q^{-(k-l)^2}$, which falls exponentially if

$$corank\; \mathbf{B} > \sqrt{\log_q\binom{n}{k}}. \quad Q.E.D.$$

The time complexity of the generalised covering-set decoding is determined by the quantity $L_n(k)$ (3.13). Let us formulate the properties of the generalised covering-set decoding algorithm as a theorem.

Theorem 3.4 [13], [21]: The covering-set decoding for almost all codes performs minimum distance decoding. The decoding can be implemented by a sequential algorithm with time complexity at most

$$O(n^4(\log n)q^{n\alpha^{(q)}(R)+\sqrt{nH_2(R)/\log_2 q}}) = q^{n\alpha^{(q)}(R)(1+o(1))}, \tag{3.16}$$

where

$$\alpha^{(q)}(R) = (\log_q 2)\left[H_2(\delta_{GV}) - (1-R)\cdot H_2\left(\frac{\delta_{GV}}{1-R}\right)\right] \tag{3.17}$$

and space complexity $O(n^3)$. A parallel implementation of this algorithm requires a circuit of size $q^{n\alpha^{(q)}(R)(1+o(1))}$ and depth $O(n)$.

Now let us consider the complexity of the split syndrome decoding. The total size of tables X_l and X_r used in this algorithm is $O\left(n\cdot\binom{m}{u}\cdot(q-1)^u\right)$ and $O\left(n\cdot\binom{n-m}{t-u}\cdot(q-1)^{t-u}\right)$ correspondingly (see section 3.2). We may optimise on the choice of m and u in order to reduce the total size of memory used for the tables X_l and X_r. Since this size is a sum of two exponential functions, for every error distribution we must choose the point so that both tables are (roughly) equally populated. The size of the memory is then bounded as $O(M(t))$, where

$$M(t) = n\cdot\left(\binom{m}{u}\cdot(q-1)^u\right)^{1/2}n\cdot q^{nH_q(t/n)/2} \tag{3.18}$$

Note that in Step 3 this algorithm makes a call to a sorting subroutine. Suppose we need to order an array of N binary vectors of length n. Sorting can be accomplished by a sequential algorithm (RAM computation) with both time and space complexity $O(nN)$. Alternatively, sorting an array can be implemented by a Boolean circuit of size $O(nN)$ and depth $O(n^2)$

[22], [23]. The complexity of sorting dominates the space complexity of split syndrome decoding and all algorithms dependent on it. The properties of the algorithm can be summarized as follows.

Theorem 3.5 [5].　For most long linear codes the split syndrome decoding algorithm performs minimum-distance decoding. Its sequential implementation has for any code of rate R time complexity $O(n \cdot d_{GV}^2 \cdot M(d_{GV})) = q^{(1/2)n(1-R)(1+o(1))}$ and space complexity $O(n \cdot M(d_{GV}))$. Its parallel implementation requires a Boolean circuit of size $O(n^4 \cdot q^{(1/2)n(1-R)(1+o(1))})$ and depth $O(n^2)$.

The time complexity of the split-syndrome algorithm is smaller than the complexity of the generalised covering-set decoding for high code rates. For instance, for $q = 2$ the complexity exponent of this algorithm is better than $\alpha^{(2)}(R)$ in (3.17) for $R > 0.954$. Therefore, if we puncture the code V (i.e., cast aside some of its parity symbols) so that its rate becomes large, we can gain in complexity by applying split syndrome decoding to the punctured code V'. By Lemma 3.2 we may regard any symbols as parity ones provided that their number is less than $n - k$. This is the basic idea of the second improvement of covering-set decoding, undertaken in [6]. We shall call this algorithm _punctured split syndrome decoding_. The main result of [6] is expressed by the following theorem, which we state for $q = 2$. Let

$$\beta^{(2)}(R) = \min_{u,\alpha} \max\left\{ (1 - u) \cdot \left[1 - H_2\left(\frac{\delta_{GV} - \alpha}{1 - u}\right)\right], \right.$$
$$\left. 1 - R - \frac{1}{2}uH_2\left(\frac{\alpha}{u}\right) - (1 - u) \cdot \left[1 - H_2\left(\frac{\delta_{GV} - \alpha}{1 - u}\right)\right]\right\} \tag{3.19}$$

with $R \leq u \leq 1$ and $\max(0, \delta_{GV} + u - 1) < \alpha < \min(\delta_{GV}, u)$, and let $\alpha^* = \alpha^*(R)$ and $u^* = u^*(R)$ be the values that furnish this minimum for a given rate R.

Theorem 3.6 [6].　For most long binary linear codes the punctured split syndrome algorithm performs complete minimum-distance decoding. Its sequential implementation for a code of rate R has time complexity $2^{n\beta^{(2)}(R)(1+o(1))}$ and space complexity $2^{(1/2)u^* \cdot nH_2(\alpha^*/u^*)(1+o(1))}$.

Setting the parameters α and u to values other than α^* and u^*, we can trade time complexity for space complexity. Taking $\alpha = \alpha^*$ and $u = u^*$ furnishes the minimum time complexity of the algorithm.

Now let us discuss the complexity of the supercode decoding algorithm (see section 3.3). Given a code rate R, we choose the values of v, α, l and that furnish the minimum to the complexity exponent, found below in Theorem 3.7. This determines the algorithm parameters $y = v \cdot n$ and $e_1 = \alpha \cdot n$. Let

$$L_n(k, e_1) = (n \log n) \cdot \left[\binom{n}{d_{GV}} \middle/ \binom{k}{e_1} \cdot \binom{n - k}{d_{GV} - e_1}\right], \tag{3.20}$$

$$e_2 = \frac{(\delta_{GV} - \alpha) \cdot y}{1 - R - (l - 1) \cdot v}. \tag{3.21}$$

The supercode decoding implies that we need to find a partition of the set $\{0, 1, \ldots, n-1\}$ into a k-subset with at most e_1 errors and an $(n-k)$-subset. This $(n-k)$-subset must have the property that once it is partitioned into s consecutive segments, at least l of them have at most e_2 errors. This is done in two stages. In the first stage we choose a partition of $\{0, 1, \ldots, n-1\}$ into subsets of size k and $n-k$ randomly and independently many times. The probability that one such choice does not generate the required distribution of errors equals

$$1 - \binom{k}{e_1} \cdot \binom{n-k}{d_{GV} - e_1} \bigg/ \binom{n}{d_{GV}}.$$

By repeating this choice independently $L_n(k, e_1)$ times (see (3.20)) we ensure that the probability of failing to construct the mentioned above system decays as $e^{-n \log n}$.

Letting $k = Rn$ and $e_1 = \alpha \cdot n$, we obtain the estimate of (3.20) as follows

$$L_n(k, e_1) \leq q^{n \varepsilon_1(R, \alpha) \cdot (1 + o(1))}, \tag{3.22}$$

where

$$\varepsilon_1(R, \alpha) = (\log_q 2) \cdot \left[H_2(\delta_{GV}) - R \cdot H_2\left(\frac{\alpha}{R}\right) - (1 - R) \cdot H_2\left(\frac{\delta_{GV} - \alpha}{1 - u}\right) \right]. \tag{3.23}$$

The second stage is performed for each partition of $\{0, 1, \ldots, n-1\}$ generated in the first stage. Taking the second subset, we split it into $s = (n-k)/y = O(n-k)$ consecutive segments of length y (except maybe the last, shorter, one). Every choice of this segment supplies us with a desired partition of $\{0, 1, \ldots, n-1\}$ into three parts. The total number of partitions equals $O(n-k) \cdot L_n(k, e_1)$.

Each partition enables us to isolate the coordinates of $V(y)$. We intend to perform the decoding of $V(y)$ by solving (3.11), i.e., to form a list K_i, $1 \leq i \leq s$ of vectors $\{\mathbf{u} = (\mathbf{u}_1 | \mathbf{u}_2)\}$ that satisfy it. The goal of this decoding is that $wt(\mathbf{e}_1) \leq e_1$, provided that the 'true' error vector appears in K_i for at least l different indexes i. To achieve this, we compute the minimal value of e_2 such that $(s - l + 1)$ y-segments of the error vector cannot all be heavier than e_2. Since $s - l + 1 = \dfrac{n - k - (l-1) \cdot y}{y}$ we obtain for e_2 the inequality

$$e_2 > \frac{\delta_{GV} - e_1}{s - l + 1} = \frac{(\delta_{GV} - \alpha) \cdot y}{1 - R - (l-1) \cdot v}.$$

Note that the greater the value of e_2, the greater is the size of the resulting list of vectors that should be tested in further steps. Therefore, e_2 should be as small as possible. This justifies our choice of e_2 in (3.21).

Looking at the description of the supercode decoding algorithm, we see that Steps 1–3.1 have algebraic complexity. The most computationally involved parts are Steps 3.2–5. Let us estimate their asymptotic complexity.

We begin with Step 3.1. Suppose that the values of e_1 and e_2 are fixed. Let us describe the process of solving (3.11) with respect to the unknown vector $(\mathbf{u}_1 | \mathbf{u}_2)$ with

$wt(\mathbf{u}_1|\mathbf{u}_2) \leq (e_1|e_2)$. To find a list of solutions of (3.11) we apply a version of split syndrome decoding.

Lemma 3.3   The list of solutions of (3.11) with $wt(\mathbf{u}_1|\mathbf{u}_2) \leq (e_1|e_2)$, where $e_1 \leq \alpha \cdot n$ and e_2 satisfies (3.21), can be compiled using the split syndrome algorithm. Both time and space complexity of this stage are bounded as $q^{(1/2)\cdot n\varepsilon_2(R,v,\alpha,l)\cdot(1+o(1))}$, where

$$\varepsilon_2(R,\,v,\,\alpha,\,l) = R \cdot H_2\left(\frac{\alpha}{R}\right) + v \cdot H_2\left(\frac{\delta_{GV} - \alpha}{1 - R - (l-1)\cdot v}\right). \tag{3.24}$$

The proof can be found in [7].

It is quite essential for us not to write out the lists K_i explicitly but to store them as pairs of lists (X_l, X_r) since the size $|K_i|$ can be much greater than the size of its 'building blocks.' This happens because each entry in X_l can be coupled with many entries in X_r to form error vectors $\mathbf{u}$ in the list K_i. We remark without further discussion that writing out the lists K_i explicitly would yield an algorithm of complexity asymptotically equal to that of punctured split syndrome decoding (Theorem 3.6).

Thus we need to find intersections of the lists K_i each of which is specified by two 'halves,' i.e., lists X_l and X_r. Our goal is to find error vectors that appear in at least l out of the s lists K_i, where l is a constant that depends only on the rate of the code V. Therefore, we can afford to examine all the possible $\binom{s}{l}$ groups of l lists.

The complexity of constructing the intersection of a given group of l lists K_i is the sum of the number of operations needed to compute the intersection and the size of the resulting list. The number of operations is estimated in the following lemma.

Lemma 3.4   The intersection of given l lists K_i can be computed in time of order $q^{(1/2)\cdot n\varepsilon_2(R,v,\alpha,l)+vn}$. The size of the memory used by the computation is at most $q^{(1/2)\cdot n\varepsilon_2(R,v,\alpha,l)}$.

Proof   See in [7].
Let us estimate the size of $K(\gamma)$. For this we use the following lemma.

Lemma 3.5   Let V be an (n, k) code of rate $R = k/n$. Suppose $S \subset E_q^n$ is a set of size $q^{\sigma n}$, $1 - R < \sigma < 1$ and let $U_\sigma = |V \bigcap S|$ be the number of codewords of V in this set. Then

$$U_\sigma \leq \frac{|S|}{q^{n-k}} \cdot q^{n\cdot o(1)} = q^{n\cdot(\sigma-(1-R))\cdot(1+o(1))}. \tag{3.25}$$

for all codes except for an n^{-n} fraction of them.

Proof   See in [7].
Note that the decrease rate of the fraction of 'bad' codes, i.e., codes that do not satisfy the statement of this lemma, is quite important for us since we are going to choose the

order of $q^{O(n)}$ codes $V(y)$ and need the estimate (3.25) to hold for all of them at a time. A more accurate estimate shows that this decay rate can be brought down to $q^{-O(n^2)}$. The last lemma allows us to estimate the size of the list of those vectors $\mathbf{u}_1$ that appear in at least l of the codes K_i, see (3.12).

Corollary 3.1   For almost all codes V and almost all choices of l supercodes $V(y)$, the size of the list $K(\gamma)$ is at most $q^{n \cdot (\varepsilon_2(R, v, \alpha, l) - lv) \cdot (1 + o(1))}$, where the function ε_2 is defined in (3.24).

Proof   See in [7].
We are now able to estimate the complexity of Step 4 of the supercode decoding algorithm.

Lemma 3.6   The time complexity of implementing Step 4 of the supercode decoding algorithm has the exponential order $\max\left\{\frac{1}{2}\varepsilon_2(R, v, \alpha, l) + v, \ \varepsilon_2(R, v, \alpha, l) - lv\right\}$. The space complexity is bounded above by $q^{(1/2)n \cdot \varepsilon_2(R, v, \alpha, l)}$.

Proof   As said before Lemma 3.4, the complexity of Step 3.2 is a sum of two terms. The first term is estimated in Lemma 3.4. The second term is the size of the resulting list, estimated in the previous corollary. The complexity of this step is estimated from above by the sum of these two exponential functions. Therefore, the exponent of the time complexity of this step is at most the maximum of the two exponents. *Q.E.D.*

Combining Lemmas 3.3, 3.6, and formula (3.23), we can prove the following result.

Theorem 3.7   The supercode decoding algorithm for almost all long linear (n, k) codes of rate $R = k/n$ performs minimum-distance decoding. The time complexity of its sequential implementation for almost all codes is at most $q^{n\zeta^{(q)}(R) \cdot (1 + o(1))}$, where

$$\zeta^{(q)}(R) = \min_{v, \alpha, l}\left\{\varepsilon_1(R, \alpha) + \max\left(\frac{1}{2}\varepsilon_2(R, \alpha, v, l) + v, \ \varepsilon_2(R, \alpha, v, l) - lv\right)\right\}. \qquad (3.26)$$

and the functions ε_1 and ε_2 are defined in (3.23) and (3.24), respectively. The optimisation parameters are restricted to

$$\begin{aligned} \max\left(0, \ \delta_{GV} + R - 1\right) &< \alpha < \min\left(\delta_{GV}, \ R\right) \\ \delta_{GV} - \alpha &< 1 - R - (l - 1) \cdot v \end{aligned} \qquad (3.27)$$

The space complexity of the algorithm is estimated from above as $q^{(1/2) \cdot n \cdot \varepsilon_2(R, v, \alpha, l) \cdot (1 + o(1))}$. A parallel implementation of the algorithm requires a Boolean circuit of size $q^{n\zeta^{(q)}(R)(1 + o(1))}$ and depth $O(nl)$.

Proof   The complexity of Steps 2–3.1 is algebraic and contributes only to $o(1)$ terms in the exponent. Let us estimate the complexity of Step 5. As said above, for each vector $\mathbf{m} \in K$ we compute a list of at most $q^{O(n^{1/2})}$ code vectors that agree with it in the

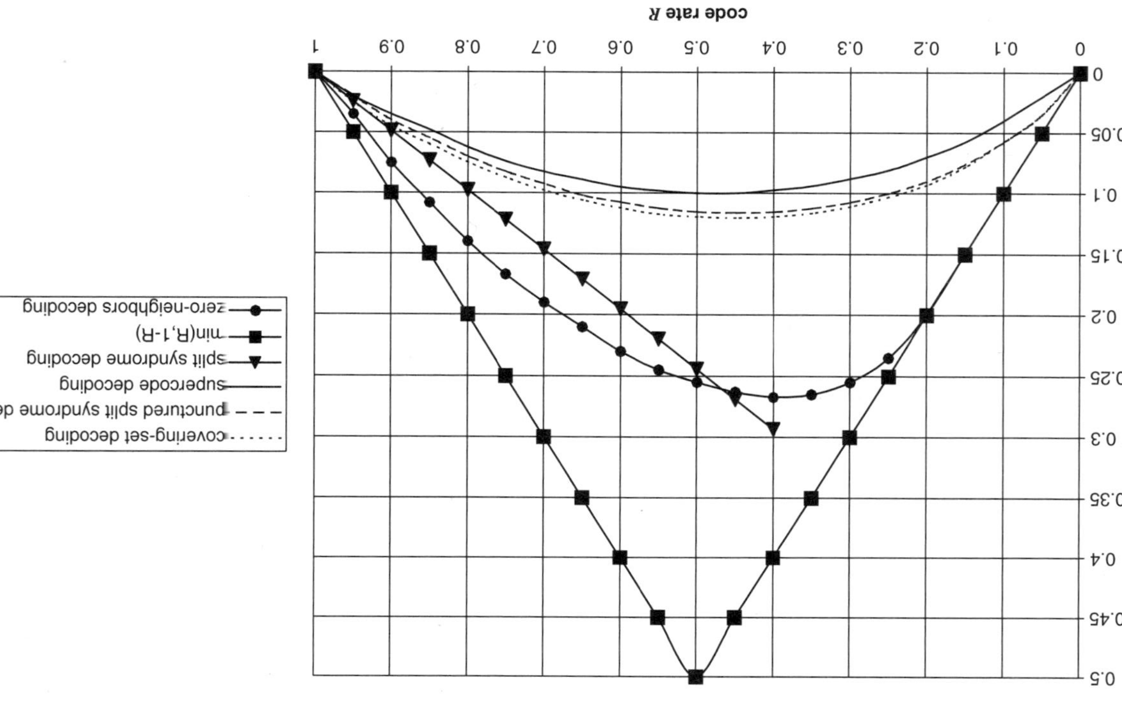

Figure 3.1 Complexity of the hard-decision decoding algorithms for binary codes

k-part. If one of these vectors, say $\mathbf{c}'$, is closer to the received vector $\mathbf{b}$ than the current decision $\hat{\mathbf{a}}$, we update it by assigning $\hat{\mathbf{a}} = \mathbf{c}'$. Thus Step 5 has time complexity of the same exponential order as Step 4. Therefore, by Lemma 3.6 we see that the most time-consuming steps of the algorithm are Steps 4 and 5. The entire sequence of steps is repeated $L_n(k, e_1)$ times. Therefore, the complexity exponent of the algorithm equals $\log_q L_n(k, e_1)$, given by (3.23), plus the exponent found in Lemma 3.6. The parameters α, v, l should be chosen to minimise this value. The first of inequalities (3.27) is obvious; the second is implied by the definition of e_2. $Q.E.D.$

The asymptotic complexity of the supercode decoding algorithm is exponentially smaller that the best known result [6] for any $q \geq 2$ and any code rate R, $0 < R < 1$. The complexity exponents of the algorithms mentioned in this section for binary codes are shown in Figure 3.1 (the complexity of the zero-neighbors decoding is represented in Theorem 3.3, covering-set decoding in Theorem 3.4, split syndrome decoding in Theorem 3.5, punctured split syndrome decoding in Theorem 3.6, and the complexity of the supercode decoding is represented in Theorem 3.7).

Thus the asymptotic complexity of the supercode decoding algorithm is less than the complexity of all other hard-decision decoding methods known.

REFERENCES

1. Gordon, D. M. (1982). Minimal permutation sets for decoding the binary Golay code, *IEEE Trans. Inform. Theory*, **IT-28**, 541–3.
2. Wolfmann, J. (1983). A permutation decoding of the (24, 12, 8) Golay code, *IEEE Trans. Inform. Theory*, **IT-29**, 748–51.
3. Krouk, E. A. and Fedorenko, S. V. (1995) Decoding by generalized information sets, *Probl. Inform. Transm.*, **31**, (2), 54–61 (in Russian) and 134–9 (English translation).
4. Barg, A. (1998). Complexity issues in coding theory, in *Handbook of Coding Theory*, vol. 1, (ed. V. Pless and Huffman, W. C.) Elsevier Science, Amsterdam, The Netherlands: pp. 649–754.
5. Dumer, I. (1989). Two decoding algorithms for linear codes, *Probl. Inform. Transm.*, **25**, (1), 24–32 (in Russian) and 17–23 (English translation).
6. Dumer, I. (1999). On minimum distance decoding of linear codes, in *Proc. 5th Joint Soviet–Swedish Int. Workshop Information Theory*, pp. 50–52, Moscow, Russia.
7. Barg, A., Krouk, E. and van Tilborg H. C. A. (1999). On the Complexity of Minimum Distance Decoding of Long Linear Codes, *IEEE Trans. Inform. Theory*, **IT-45**, 1392–1405.
8. Gallager, R. (1963). *Low-Density Parity-Check Codes*. MIT Press, Cambridge, MA.
9. Aho, A. V., Hopcroft, J. E., and Ullman, J. D. (1974). *The Design and Analysis of Computer Algorithms*. Addison-Wesley, London, U.K..
10. Savage, J. E. (1969 and 1971). The complexity of decoders, I, II, *IEEE Trans. Inform. Theory*, **IT-15**, 689–95, **IT-17**, 77–85.
11. Evseev, G. S. (1983). Complexity of decoding for linear codes, *Probl. Inform. Transm.*, **19**, (1), 3–8 (in Russian) and 1–6 (English translation).
12. Barg, A. and Dumer, I. (1986). Concatenated decoding algorithm with incomplete inspection of code vectors, *Probl. Inform. Transm.*, **22**, (1), 3–8 (in Russian) and 1–7 (English translation).
13. Coffey, J. T. and Goodman, R. M. F. (1990). The complexity of information set decoding, *IEEE Trans. Inform. Theory*, **35**, 1031–7.
14. Coffey, J. T., Goodman, R. M. F., and Farrell, P. (1991). New approaches to reduced complexity decoding, *Discr. Appl. Math.*, **33**, 43–60.

15. Blinovskii, V. M. (1987). Lower asymptotic bound on the number of linear code words in a sphere of given radius in F_q^n, *Probl. Inform. Transm.*, **23**, (2) 50–3 (in Russian) and 130–2 (English translation).

16. Hwang, T.-Y. (1979). Decoding linear block codes for minimizing word error rate, *IEEE Trans. Inform. Theory*, **IT-25**, 733–7.

17. Levitin, L. and Hartmann, C. R. P. (1985). A new approach to the general minimum distance decoding problem: The zero-neighbors algorithm, *IEEE Trans. Inform. Theory*, **IT-31**, 378–84.

18. Dumer, I. (1996). Suboptimal decoding of linear codes: Partition technique, *IEEE Trans. Inform. Theory*, **42**, 1971–86.

19. Dumer, I. (1996). Covering lists in maximum likelihood decoding, in *Proc. 34th Annu. Allerton Conf. Communications, Control, and Computing*, 683–92.

20. Ashikhmin, A. and Barg, A. (1998). Minimal vectors in linear codes, *IEEE Trans. Inform. Theory*, **44**, 2010–17.

21. Krouk, E. A. (1989). Decoding complexity bound for linear block codes, *Problems of Info. Trans.*, **25**, (3), 103–7 (in Russian) and 251–4 (English translation).

22. Cormen, T. H., Leiserson, C. E., and Rivest, R. L. (1990). *Introduction to Algorithms*. MIT Press, Cambridge, MA, USA.

23. Knuth, D. E. (1973). *The Art of Computer Programming*, vol. 3. Reading, MA: Addison-Wesley, Reading, MA, USA.

4

Codes with Algebraic Decoding

In this chapter we consider the most explored and often used block codes, BCH codes and their important particular case Reed-Solomon (RS) codes. We start with the simplest case: Hamming codes.

4.1 HAMMING CODES

Hamming codes are capable of correcting single errors. These codes were invented by R.W. Hamming and his paper [1] is considered to be the first work on coding theory. The code construction can be easily explained with the help of Bose criterion (Lemma 2.1). In accordance with this criterion a linear code is capable of correcting a single error if, and only if, a parity-check matrix $\mathbf{H}$ of this code does not contain collinear columns. In a binary case that means matrix $\mathbf{H}$ consists of distinct nonzero columns. The total number of different nonzero columns of dimension r is $2^r - 1$. Hence, any $r \times (2^r - 1)$ matrix $\mathbf{H}_r$ combined from these columns defines the linear *Hamming code* with length $n = 2^r - 1$, number of information symbols $k = 2^r - r - 1$ and minimum distance $d = 3$. Any (not necessarily linear) code of length $n = 2^r - 1$ capable of correcting a single error has in accordance with Hamming bound (see Section 2.5) a cardinality of no more than $\dfrac{2^n}{n+1} = 2^{n-r}$ code words. Therefore, the binary Hamming codes are perfect. Recall that codes achieving Hamming bound are called *perfect* and they are optimal. It is known that there are no perfect codes capable of correcting t-fold errors ($t > 1$) except of binary (23, 12, 7) Golay code, correcting triple errors, and ternary (11, 6, 5) Golay code, correcting double errors [2].

It is convenient to arrange the columns of matrix $\mathbf{H}$ in such a way that a column $\mathbf{h}_j$ be the binary representation of its own index j. Then to decode the received word $\mathbf{x}$ (to correct single errors) it is necessary to calculate the syndrome $\mathbf{s} = \mathbf{x} \cdot \mathbf{H}^T = (s_0, \ldots, s_{r-1})$ and the number $S = s_0 + 2s_1 + \ldots + 2^{r-1} s_{r-1}$ is the index of the corrupted symbol. If $S = 0$, i.e. $\mathbf{s}$ is the all-zero vector then it is assumed that no error occurs. If $n + 1$ is not the power of 2 then it is possible to consider the shortened Hamming code, i.e., a code defined by the parity-check matrix obtained from matrix $\mathbf{H}_r$ by deletion of some i ($i = 2^r - 1 - n$) from $2^r - 1$ columns, where $r = \lceil \log_2(n + 1) \rceil$. The shortened Hamming codes cannot be improved in the class of linear codes, i.e. $k(n, d) = n - \lceil \log_2(n + 1) \rceil$. Also it is known that the shortened Hamming codes of length $n = 2^r - 1 - i$, $i = 1, 2, 3$ are optimal in the class of all codes [14]. On the other hand for $i = \lambda \cdot 2^r$ ($0 < \lambda < 0.5$) and for large n there exist

Error Correcting Coding and Security for Data Networks   G. Kabatiansky, E. Krouk and S. Semenov
© 2005 John Wiley & Sons, Ltd   ISBN: 0-470-86754-X

nonlinear codes with $d = 3$, which have about $(1 - \lambda)^{-1}$ times more code words than the shortened Hamming codes of the same length. These nonlinear codes asymptotically achieve Hamming bound [15].

Now consider nonbinary codes capable of correcting single errors over the alphabet B the cardinality of which $q = |\,\text{B}\,|$ is the power of prime number, i.e., $q = p^m$ and $\text{B} = F_q$. Then, as was pointed above, matrix $\mathbf{H}$ should not contain the collinear columns. As an example of this kind of matrix with maximal possible number $n_r = \dfrac{q^r - 1}{q - 1}$ of r-dimensional columns consider the matrix

$$
\mathbf{H}_{r,q} = \begin{bmatrix} 1 & \dots & 1 & \vdots & 0 & \dots & 0 & \vdots & & \vdots & 0 \\ & & & \vdots & 1 & \dots & 1 & \vdots & & \vdots & \dots \\ & & & \vdots & & & & \vdots & \dots & \vdots & 0 \\ & & & \vdots & & & & \vdots & & \vdots & \\ \mathbf{F}_q^{r-1} & & & \vdots & \mathbf{F}_q^{r-2} & & & \vdots & & \vdots & 1 \end{bmatrix},
\tag{4.1}
$$

where $\mathbf{F}_q^l$ is $l \times q^l$ matrix the columns of which are all q^l l-dimensional vectors over the field F_q. Arrange the columns of matrix $\mathbf{F}_q^l$ in such a way that the column with index j be the q-ary representation of number j. A code defined by the matrix $\mathbf{H}_{r,q}$ is called Hamming code. The decoding algorithm of the q-ary Hamming code is as follows:

1. Calculate the syndrome of the received vector $\mathbf{x}$: $\mathbf{s} = (s_0, \ \dots, \ s_{r-1}) = \mathbf{x} \cdot \mathbf{H}_{r,\,q}^T$.

2. Find nonzero element s_i with minimal index i, then the error value $e = s_i$ and the error position j coincides with the position of column $\mathbf{s}' = \left(\dfrac{s_0}{s_i}, \ \dots, \dfrac{s_{r-1}}{s_i} \right)$ in matrix $\mathbf{H}_{r,\,q}$ (division in the field F_q). If $\mathbf{s} = \mathbf{0}$ it is assumed that no error occurred.

As in the binary case the q-ary Hamming codes are optimal because they reach the Hamming bound $|\,V\,| \leq \dfrac{q^n}{1 + (q-1) \cdot n}$, and the shortened Hamming codes are optimal in the class of linear codes. However, if $q = p^m$ and $m > 1$ then shortened Hamming codes are not optimal already in the class of group codes. In the class of all codes capable of correcting single errors the Hamming codes of length $n = \dfrac{q^r - 1}{q - 1} \cdot (1 - \lambda)$ have about $(1 - \lambda)$ times less code words than optimal codes $(0 < \lambda < 1 - 1/q)$, which, as is proved in [15], reach the Hamming bound asymptotically.

More frequently Hamming codes are used in cyclic representation. Recall that the minimum distance of a binary cyclic code is no less than 3 if, and only if, the code length n is equal to the smallest positive integer (called *period* of $g(x)$) for which generator polynomial $g(x)$ divides $x^n - 1$. Therefore, if $g(x)$ is the primitive polynomial of degree m we obtain the cyclic code $V_{g(x)}$ of length $n = 2^m - 1$ with the number of parity symbols m and with minimum distance $d \geq 3$. The parity-check matrix of this code contains all possible nonzero m-vectors as columns, i.e. it differs from the parity-check matrix $\mathbf{H}_m$ only by a permutation of columns. Thus, a binary Hamming code is equivalent to the cyclic code with the primitive generator polynomial.

For q-ary codes the answer on the corresponding question depends on whether numbers $(q - 1)$ and m are relatively prime or not. If they are, then the q-ary Hamming code of length

$n = \dfrac{q^m - 1}{q - 1}$ is equivalent to the cyclic code with the generator polynomial $g(x) = f_\beta(x)$, where $\beta \in F_{q^m}$ and the period of β is $\dfrac{q^m - 1}{q - 1}$ (i.e. $\beta = \alpha^{q-1}$, where α is the primitive element of F_{q^m}), $f_\beta(x)$ is the minimal polynomial over F_q of β (i.e. the monic polynomial of the lowest possible degree from $F_q[x]$ for which β is the root). In the other case the q-ary Hamming code is not equivalent to any cyclic code.

Consider now the construction of near optimal single-error-correcting codes. Very useful for construction of codes with better parameters than shortened Hamming codes is the structure called a *subcode over subset*.

Let alphabet B ($|\,B\,| = q$) be the subset of the alphabet B ($|\,B\,| = Q > q$) and V be a code of length n with minimum distance d and cardinal number $M = |\,V\,|$ over the alphabet B. Then q-ary code $V_B = \{\mathbf{v} = (v_0, \ldots, v_{n-1}) : v_i \in B, \mathbf{v} \in V\}$ consisting of B-ary words of code V is called a subcode of code V over subset B. Obviously, the length of the code V_B is equal to n and the distance is not less than d because V_B is the subcode, i.e. $V_B \subset V$. It is generally quite difficult to estimate the cardinal number of the particular code V_B. Instead, consider the family of q-ary codes $V_{B,\mathbf{b}} = \{V + \mathbf{b}\}_B = \{\mathbf{x} = (x_0, \ldots, x_{n-1}) : x_i \in B, \mathbf{x} - \mathbf{b} \in V\}$. Since code $\{V + \mathbf{b}\}$ has exactly the same parameters n, M, d as code V, the distance of subcode of this code $d(V_{B,\mathbf{b}}) \geq d$ and the mean cardinal number of code $\overline{M}$ in this family is equal to

$$\overline{M} = \frac{1}{Q^n} \cdot \sum_{\mathbf{b} \in E_Q^n} |\, V_{B,\mathbf{b}} \,| = \sum_{\mathbf{v} \in V} \sum_{\mathbf{a} \in E_Q^n} 1 = \left(\frac{q}{Q}\right)^n \cdot |\, V \,|, \tag{4.2}$$

Hence, there exists a code $V_{B,\mathbf{b}}$ in this family, such that $|\, V_{B,\mathbf{b}} \,| \geq \overline{M}$.

Example 4.1 [15, 16]. Construct a 4-ary single-error-correcting code of length 6. Let $q = 4$, $Q = 5$ and code V be the 5-ary Hamming code of length 6 with 5^4 code words. It follows from (4.2) that $\overline{M} \geq \dfrac{4^6}{5^2}$ and hence, there exists a code $\{V + \mathbf{b}\}$ such that its 4-ary subcode has at least $\left\lceil \dfrac{4^6}{5^2} \right\rceil = 164$ code words. Notice that the shortened 4-ary Hamming code of the same length has only $4^3 = 64$ code words.

With the help of subcode over subset construction it is possible to construct from the Hamming codes, over a large alphabet, the q-ary single-error-correcting codes with packing density asymptotically no less than $\dfrac{q - 1}{\hat{q} - 1}$, where $\hat{q}$ is the minimum integer greater than q that can be represented as the power of the prime number.

The other useful structure for the construction of error-correcting codes is the coset. Let $\mathbf{H}$ be the parity-check $r \times n$ matrix of some q-ary (n, $n-r$) code C with minimum distance $d = 2t + 1$ (now we again assume that q-ary alphabet is the field F_q). Then all linear combinations of t or fewer columns of matrix $\mathbf{H}$ are distinct vectors. Denote this set of r-dimensional vectors as $U_{\mathbf{H}}$ ($\mathbf{0} \in U_{\mathbf{H}}$ as a result of linear combination with all-zero coefficients), where $|\, U_{\mathbf{H}} \,| = \sum_{i=0}^{t} \binom{n}{i} \cdot (q - 1)^i$. Let V be a q-ary code of length r capable of correcting the set of errors $U_{\mathbf{H}}$, i.e., it follows from $\mathbf{u} + \mathbf{v} = \mathbf{u}' + \mathbf{v}'$, $\mathbf{u}, \mathbf{u}' \in U_{\mathbf{H}}$, $\mathbf{v}, \mathbf{v}' \in V$ that $\mathbf{u} = \mathbf{u}'$, $\mathbf{v} = \mathbf{v}'$. Then the following statement holds true:

Theorem 4.1   Code $C_V = \left\{ \mathbf{x} = (x_0, \ldots, x_{n-1}) \in E_q^n : \mathbf{H} \cdot \mathbf{x}^T \in V \right\}$ is capable of correcting t errors and has $|\,C_V\,| = |\,C\,| \cdot |\,V\,|$ code words.

Proof   Let us prove this statement by reductio ad absurdum. Let us consider vector $\mathbf{x}$ as vector-column. Assume that there exists the error vectors $\mathbf{e}$ and $\mathbf{e}'$: $\|\,\mathbf{e}\,\| \le t$, $\|\,\mathbf{e}'\,\| \le t$ and vectors $\mathbf{x} \ne \mathbf{x}' \in C_V$ such that $\mathbf{x} + \mathbf{e} = \mathbf{x}' + \mathbf{e}'$. Multiplying both parts of this equation by matrix $\mathbf{H}$ obtain $\mathbf{H} \cdot \mathbf{x} + \mathbf{H} \cdot \mathbf{e} = \mathbf{H} \cdot \mathbf{x}' + \mathbf{H} \cdot \mathbf{e}'$, where $\mathbf{H} \cdot \mathbf{x}, \mathbf{H} \cdot \mathbf{x}' \in V$, $\mathbf{H} \cdot \mathbf{e}, \mathbf{H} \cdot \mathbf{e}' \in U_{\mathbf{H}}$. Then $\mathbf{H} \cdot \mathbf{e} = \mathbf{H} \cdot \mathbf{e}'$, because the code V is capable of correcting errors from the set $U_{\mathbf{H}}$. Recall that all linear combinations of t or fewer columns of matrix $\mathbf{H}$ are distinct; hence $\mathbf{e} = \mathbf{e}'$, which means that $\mathbf{x} = \mathbf{x}'$ and this statement contradicts the assumption. The code C_V is the union over all $\mathbf{v}$ of solutions of the equation $\mathbf{H} \cdot \mathbf{x} = \mathbf{v}$, $\mathbf{v} \in V$, where r is the rank of matrix $\mathbf{H}$. Since for any $\mathbf{v}$ the number of solutions is the same and equals to $q^{n-r} = |\,C\,|$, we have that the overall number of solutions for all $\mathbf{v}$ equals to $|\,C\,| \cdot |\,V\,|$.   *Q.E.D.*

In the case of single-error-correcting codes it is possible to choose as $U_{\mathbf{H}}$ any homogeneous (i.e. $\mathbf{u} \in U_{\mathbf{H}} \Rightarrow \lambda \cdot \mathbf{u} \in U_{\mathbf{H}}$, $\lambda \in F_q$) subset of E_q^n. The corresponding construction is called the method of *homogeneous packing* [14], since it is possible to construct from the code V capable of correcting the homogeneous set of errors $U_{\mathbf{H}}$ the code $V_{\mathbf{H}}$ of length $\dfrac{|\,U_{\mathbf{H}}\,|}{q-1}$ capable of correcting single errors, and it easy to verify that the packing density in this case does not decrease.

The important example of the homogeneous set is the set $U_{l,N}$ of those vectors from the space $E_q^{N \cdot l}$ for which the nonzero positions are grouped in one of N phased packets of length l, i.e. $U_{l,N} = \bigcup\limits_{j=0}^{N-1} \{ \mathbf{x} = (x_0, \ldots, x_{N \cdot l-1}) : x_i = 0, \ i \notin \{ j \cdot l, \ldots, j \cdot l + l - 1 \} \}$. The code capable of correcting the set of errors $U_{l,N}$ is called the code capable of correcting single phased packets of length l. Since the elements of the field F_{q^l} can be regarded as l-dimensional vectors over field F_q, this correspondence specifies the isomorphism of the spaces $E_{q^l}^N$ and E_q^{Nl} where q-ary code of length $N \cdot l$ capable of correcting single phased packets of length l corresponds the q^l-ary code of length N capable of correcting single errors. The union of this structure with subcodes over subsets and with the method of homogeneous packing gives us the class of asymptotically optimal single-error-correcting codes [15].

Example 4.2   Construct the binary single-error-correcting code of length 18. Consider the 4-ary code of length 6 of Example 4.1 as a binary code of length 12 capable of correcting single phased error packets of length 2 and consisting of $M \ge 164$ code words. From the latter code with the help of cosets construction obtain a binary code of length $n = 6 \cdot 3$, consisting of $M \cdot 2^6$ code words. This is the single-error-correcting code of length 18 with $2^8 \cdot 41$ code words what is 1.28 times greater than the number of code words in the shortened Hamming code [15,16].

As usual for coding theory in the case when q is not a prime power there is a lack of information. During the long period of time among the single-error-correcting codes the best parameters have the codes of length n defined by the following system of equations:

$$\begin{cases} \sum_{i=0}^{n} x_i = \alpha \bmod (2q-1) \\ \sum_{i=0}^{n} ix_i = \beta \bmod p \end{cases}, \tag{4.3}$$

where p is a prime number greater than n. These codes are an analog of Varshamov-Tenengoltz codes for correcting single asymmetric errors [18]. The best of the codes defined by the system (4.3) has the number of code words M_{opt} no less than the mean over α and β, i.e.

$$M_{opt} \geq \overline{M}_{\alpha, \beta} = \frac{q^n}{p \cdot (2q-1)}. \tag{4.4}$$

Since $\lim_{n \to \infty} \frac{p(n)}{n} = 1$, where $p(n)$ is the minimum prime number greater than n, then the packing density of these codes (i.e. the ratio of the cardinal number to the value of Hamming bound) with large n is about $\frac{q-1}{2q-1}$, i.e. less than $\frac{1}{2}$.

As was mentioned above the structure of subcode over subset allows the construction of codes of a large size, i.e., with packing density asymptotically no less than $\frac{q-1}{\hat{q}-1}$ [15], where $\hat{q}$ is the minimal prime power such that $\hat{q} \geq q$. It is shown in [16] that if there is a perfect (reaching the Hamming bound) q-ary code single-error-correcting code of length $q+1$, then with $n \to \infty$ there are codes asymptotically reaching the Hamming bound, i.e. the packing density of the optimal codes also tends to unit with increasing the code length in a general q-ary case.

4.2 REED-SOLOMON CODES

The Reed-Solomon (RS) codes invented in 1960 [3] are still one of the most applicable class of codes and despite (or maybe due to) their simplicity are the basis of deep generalisations. Let us start the discussion of RS codes from the following simple statement: The rank of a parity-check matrix $\mathbf{H}$ of the $(n, n\text{-}r)$ code with minimum distance d is no less than $d-1$ since any its $d-1$ columns are linearly independent. On the other hand, the rank of a matrix is no more than the number of rows, and hence for any $(n, n\text{-}r)$ code the following *Singleton bound* holds true

$$d \leq r + 1. \tag{4.5}$$

Note that the bound (4.5) holds true for any (not necessary linear) codes, namely

$$M_q(n, d) \leq q^{n-d-1}. \tag{4.6}$$

The code reaching the bound (4.6) is an optimal code and it is called a *maximum distance separable (MDS) code*. Thus, in the parity-check $r \times n$ matrix of MDS code any r columns are linearly independent. The example of such kind of matrix is the matrix $\mathbf{H} = \left[h_{ij}\right] = \left[\alpha_j^i\right]$, $i = 0, \ldots, r - 1, j = 0, \ldots, n - 1$, where $\alpha_0, \ldots, \alpha_{n-1}$ are distinct elements of the field F_q $(n \leq q)$. In other words, the ith row of matrix $\mathbf{H}$ consists of the values of the term z^i calculated in the points $\alpha_0, \ldots, \alpha_{n-1}$ of the field F_q:

$$\mathbf{H}_r = \begin{matrix} 1 \\ z \\ \vdots \\ z^{r-1} \end{matrix} \left| \begin{bmatrix} 1 & 1 & \cdots & 1 \\ \alpha_0 & \alpha_1 & \cdots & \alpha_{n-1} \\ \cdots & \cdots & \cdots & \cdots \\ \alpha_0^{r-1} & \alpha_1^{r-1} & \cdots & \alpha_{n-1}^{r-1} \end{bmatrix} \right. . \tag{4.7}$$

Let us show that any r columns in matrix (4.7) are linearly independent. Should this not be the case then the columns $\mathbf{h}_{j_1}, \ldots, \mathbf{h}_{j_r}$ are linearly dependent. This means $\det\left(\mathbf{h}_{j_1}, \ldots, \mathbf{h}_{j_r}\right) = 0$. Hence, the rows of this minor determinant $\mathbf{H}_{j_1,\ldots,j_r}$ also are linearly dependent.

The linear dependence of rows of this minor determinant with coefficients $\lambda_0, \ldots, \lambda_{r-1}$ means that the polynomial $\lambda_0 + \lambda_1 \cdot z + \ldots + \lambda_{r-1} \cdot z^{r-1}$ of degree less than r has r different roots $\alpha_{j_1}, \ldots, \alpha_{j_r}$, and that is impossible.

It is easy to demonstrate from the statement about matrix (4.7) that in matrices $\mathbf{H}_{a,r}$ of more general form with the rows defined by the values of term z^j $(j = a, a + 1, \ldots, a + r - 1)$ on elements $\alpha_j \in F_q \backslash 0$

$$\mathbf{H}_{a,r} = \begin{bmatrix} \alpha_0^a & \cdots & \cdots & \alpha_{n-1}^a \\ \cdots & \cdots & \cdots & \cdots \\ \cdots & \cdots & \cdots & \cdots \\ \alpha_0^{a+r-1} & \cdots & \cdots & \alpha_{n-1}^{a+r-1} \end{bmatrix} \tag{4.8}$$

any r columns are linearly independent. The codes defined by the matrices (4.8) are called *Reed-Solomon (RS) codes*. More often RS codes are defined as cyclic codes. Let elements $\alpha_0, \ldots, \alpha_{n-1}$ form a subgroup. This subgroup should be a cyclic subgroup because the group under multiplication of the whole finite field is the cyclic group. Hence, there exists element $\beta \in F_q$ such that $\alpha_i = \beta^i, i = 0, \ldots, n - 1$. Then the matrix (4.8) can be converted to the form

$$\mathbf{H}_{a,r,\beta} = \begin{bmatrix} 1 & \beta^a & \cdots & \beta_{n-1}^{a(n-1)} \\ 1 & \beta^{a+1} & \cdots & \beta^{(a+1)(n-1)} \\ \cdots & \cdots & \cdots & \cdots \\ 1 & \beta^{a+r-1} & \cdots & \beta_{n-1}^{(a+r-1)(n-1)} \end{bmatrix} \tag{4.9}$$

Since the cyclic shift of any row of matrix $\mathbf{H}_{a,r,\beta}$ (this shift can be obtained by the multiplication of the row by $\beta^{a(n-1)}$) belongs to the linear space of rows of this matrix, the code defined by the matrix $\mathbf{H}_{a,r,\beta}$ is the cyclic code (as well as the corresponding dual code). The condition that vector-column $\mathbf{v}$ is the codeword of the code defined by the matrix $\mathbf{H}_{a,r,\beta}$

$$\mathbf{H}_{a,r,\beta} \cdot \mathbf{v} = \mathbf{0}$$

is equivalent to the following condition

$$v(\beta^j) = 0, \quad j = a, \ldots, a + r - 1, \tag{4.10}$$

where $v(x)$ is the polynomial corresponding to vector $\mathbf{v}$. Therefore, the generator polynomial of this code is

$$g(x) = \text{L.C.M.}\{f_{\beta^j}(x)\}, \quad j = a, \ldots, a + r - 1, \tag{4.11}$$

where $f_\gamma(x)$ is the minimal polynomial of the element γ. In the considered case $f_\gamma(x) = x - \gamma$ and therefore

$$g(x) = \prod_{j=a}^{a+r-1} (x - \beta^j). \tag{4.12}$$

The most interesting case is when β is the primitive element of field F_q. Then $n = q - 1$ and such kinds of code are most often called RS codes [2]. The code defined by the matrix (4.7) has the length q when $\{\alpha_0, \ldots, \alpha_{n-1}\} = F_q$ and is called *1-extended RS code*. It is possible to extend this kind of code by one symbol more if we add symbol ∞ to the set $\{\alpha_0, \ldots, \alpha_{n-1}\} = F_q$ with formal definition $\infty^j = 0$ for $j = 0, 1, \ldots, r - 2$ and $\infty^{r-1} = 1$. The question about the possibility of constructing nontrivial MDS codes (i.e. $2 < d < n$) of length $n > q + 1$ is the unsolved problem throwing back to the projective geometry. A well-accepted hypothesis maintains that these codes do not exist except in the cases when $q = 2^m$, $n = q + 2$ and either code with distance $d = 4$ or its dual code with distance $d = q$.

Example 4.3 Specify the code of length $q + 2$ with the help of systematic encoding that is matched up to the set of information symbols $a_0, \ldots, a_{q-2}$ three parity-check symbols b_0, b_1, b_2 computed in accordance with formula

$$b_0 = \sum_{i=0}^{q-2} a_i; \quad b_1 = \sum_{i=0}^{q-2} a_i \cdot \alpha^{i+1}; \quad b_2 = \sum_{i=0}^{q-2} a_i \cdot \alpha^{2(i+1)}.$$

where α is the primitive element of the field F_q. It is easy to verify that for $q = 2^m$ the distance of this code is 4, i.e. this is the nontrivial MDS code of length $q + 2$.

Let V be an $(n, n\text{-}r)$ RS code defined by the parity-check matrix (4.8) over the field F_Q, $Q = q^m$. Since the field F_q may be defined as the subfield of the field F_Q, consider a subcode V_q of the code V over subfield F_q. This code is no more than q-ary *BCH code* [2]. It follows from the general relation of subcode over subset that $| V_q | \geq q^{n-rm}$. In this particular case the given estimation can be improved by considering the field F_Q as an m-dimensional vector space over F_q and the elements of the matrix $\mathbf{H}_{a,r}$ as m-dimensional columns (vectors) over the field F_q. Then the rank (q-ary) of the new matrix $\mathbf{H}_{a,r}^{(q)}$ is equal to the number of parity-check symbols of code V_q. It is known [2] that there exists a normal basis in the field F_Q, i.e. the element γ such that the elements $\gamma, \gamma^q, \ldots, \gamma^{q^{m-1}}$ are the basis of the vector space F_Q over the field F_q. In this basis the 'block' of m rows of the matrix $\mathbf{H}_{a,r}^{(q)}$ corresponding to the row $z^{j \cdot q}$ of the matrix $\mathbf{H}_{a,r}$ can be obtained by the cyclic shift (downward) of the rows of the 'block' corresponding to the row z^j. From this follow the estimations of number of parity-check symbols $r^{(q)}$ of the q-ary code V_q of length $n \leq q^m$ and with minimum distance no less than d, which we will write for the two most interesting cases:

$$r^{(q)} \leq m \cdot \left(d - 1 - \left\lfloor \frac{d-1}{q} \right\rfloor \right) \quad \text{for } a = 1. \tag{4.13}$$

$$r^{(q)} \leq 1 + m \cdot \left(d - 2 - \left\lfloor \frac{d-2}{q} \right\rfloor \right) \quad \text{for } a = 0. \tag{4.14}$$

Substituting $q = 2$ and $d = 2t + 1$ in (4.13) demonstrates that the number of parity-check symbols of BCH code of length n capable of correcting t errors is no more than $t \cdot \lfloor \log_2 n \rfloor$. It follows from this statement, and from Hamming bound, that BCH codes are asymptotically optimal 'on redundancy', i.e. $\lim\limits_{n \to \infty} \dfrac{r_{BCH}}{r(n,\, 2t + 1)} = 1$.

Substituting $q = 3$ and $d = 5$ in (4.14) gives a result showing that the number of parity-check symbols of the corresponding 3-ary BCH codes of length n capable of correcting 2 errors is no more than $1 + 2 \cdot \lfloor \log_3 n \rfloor$. In particular, Hamming bound shows that for $n = 3^n$ these codes are optimal in the class of linear codes. One more class of codes is also known- 4-ary codes capable of correcting 2 errors [19, 20], which almost satisfies the Hamming bound for the number of parity-check symbols. There are no known classes of codes with parameters q and t except the above mentioned that satisfy the following equation

$$\lim\limits_{\substack{n \to \infty \\ q,\, t = const}} \frac{r}{t \cdot \log_q n} = 1. \tag{4.15}$$

The BCH codes do not satisfy (4.15) with $q > 2$ (except the case $q = 3$, $t = 2$) and for the case $t = 2$ there exist codes asymptotically better than BCH codes, i.e. for these codes the left part of (4.15) has value 7/6, which is better than 1.5 for BCH codes [21]. For an extended discussion of the BCH codes see the next section.

4.3   BCH CODES

For the estimation of the minimum distance the following description of the cyclic codes is very useful. Let $g(x)$ be the generator polynomial of the cyclic (n, k) code V over the field F_q. As before we assume that the code is nontrivial, i.e., $2 < d < n$ and, therefore, n is the period of the polynomial $g(x)$, i.e., $g(x) \mid x^n - 1$ and $g(x)$ does not divide $x^{n'} - 1$ for $n' < n$. Let the code length n and the characteristic p of field F_q be relatively prime. Then the polynomial $x^n - 1$ has no repeated roots since $LCD\left(x^n - 1, \dfrac{d}{dx}(x^n - 1) \right) = 1$. Hence the polynomial $g(x)$ also has no repeated roots because it is the divisor of the polynomial $x^n - 1$. Let m be the minimal positive integer such that n divides $q^m - 1$. Then F_{q^m} is the minimal field containing all roots of $g(x)$. The following theorem called *BCH bound* holds true.

Theorem 4.2   Let α be an element of the field F_{q^m} and let α^a, α^{a+1}, $\ldots$, α^{a+s-1} be the roots of the polynomial $g(x)$. Then the minimum distance of the cyclic code V with the generator polynomial $g(x)$ is no less than $s + 1$.

Proof Any codeword $\mathbf{v} \in V$ can be considered as a polynomial $v(x)$ and the elements α^a, α^{a+1}, ..., α^{a+s-1} are the roots of this polynomial because $g(x)$ is the divisor of $v(x)$, i.e. $v(\alpha^j) = 0$ for $j = a$, ..., $a + s - 1$. In accordance with (4.10) that means the codeword $\mathbf{v}$ belongs to the cyclic RS code with the minimum distance $s + 1$. *Q.E.D.*

Usually this theorem is used when α is the primitive element of the field F_{q^m}. The corresponding statement says that the minimum distance of a cyclic code is more than the length of the maximal 'set of consecutive roots', i.e. the maximum number of consecutive powers j_0, ..., j_{s-1}, where α^{j_0}, ..., $\alpha^{j_{s-1}}$ are the roots of the polynomial $g(x)$. The parameter $s + 1$ usually is called the *designed distance* of a BCH code.

Example 4.4 Consider binary cyclic code V with the generator polynomial $g(x) = x^7 + x^6 + x^5 + x^2 + x + 1$. Find the parameters of this code. To obtain the code length we have to find x^n of minimal power such that the remainder of division x^n by $g(x)$ is equal 1. The remainder of x^7 divided by $g(x)$ is

$$x^7 \bmod g(x) = x^6 + x^5 + x^2 + x + 1$$

Now multiply by x on modulo $g(x)$ both left and right part of this equation

$$x^8 \bmod g(x) = x^7 + x^6 + x^3 + x^2 + x \bmod g(x) = x^6 + x^5 + x^2 + x + 1 + x^6 + x^3 + x^2 + x = x^5 + x^3 + 1$$

In the same manner we will calculate remainders of division terms x^9, x^{10}, ... by $g(x)$ until the remainder become equal to 1.

$$\left. \begin{aligned}
x^9 &= x^6 + x^4 + x \\
x^{10} &= x^6 + x + 1 \\
x^{11} &= x^6 + x^5 + 1 \\
x^{12} &= x^5 + x^2 + 1 \\
x^{13} &= x^6 + x^3 + x \\
x^{14} &= x^6 + x^5 + x^4 + x + 1 \\
x^{15} &= 1
\end{aligned} \right\} \bmod g(x)$$

Thus, the code length is $n = 15$. The number of information symbols k equals to the difference of code length and the degree of the generator polynomial

$$k = n - \deg g(x) = 15 - 7 = 8.$$

Since $g(x)$ is the divisor of $x^n - 1$, all roots of $g(x)$ belong to the field $GF(2^4)$. Let γ be the primitive element of this field such that $\gamma^4 + \gamma + 1 = 0$. It is easy to verify that the roots of $g(x)$ are $\gamma^0 = 1$, γ^5, γ^6, γ^9, γ^{10}, γ^{12}. The longest set of consequent powers consists of two elements γ^5, γ^6 or γ^9, γ^{10}. It follows from the lower BCH bound that the minimum distance of given code is $d \geq 3$. Use the theorem 4.2 with $\alpha = \gamma^5$. Then α^0, $\alpha^1 = \gamma^5$, $\alpha^2 = \gamma^{10}$ are the roots of the polynomial $g(x)$. Hence, the minimum distance of the code is no less than 4 and the exhaustive search shows that it is the real minimum distance of the code.

Now let us introduce the classical definition of the BCH code [2]. Let n be the divisor of $q^m - 1$ and n is not divisor of $q^i - 1$ for $i < m$. Then the q-ary *BCH code* with the designed distance d is the cyclic code of length n with the generator polynomial

$$g(x) = \text{LCM}\{\varphi_{\alpha^a}(x), \varphi_{\alpha^{a+1}}(x), \ldots, \varphi_{\alpha^{a+d-2}}(x)\}, \tag{4.16}$$

where α is the primitive nth root of unit in the field F_{q^m}, $\varphi_\gamma(x)$ is the minimal polynomial of the element γ. If $n = q^m - 1$ and $a = 1$, then the code is called a *primitive* BCH *code*.

Using (4.16), (4.12) and (4.11) it is easy to verify that this definition of BCH code coincides with that given in the section 4.2 definition of BCH code as a q-ary subcode of a q^m-ary RS code. The theorem 4.2 gives only the lower bound of the distance of BCH code and the real minimum distance can be greater than the design distance. The meaning of the design distance is in the fact that there exist effective algorithms capable of correcting the corresponding to the design distance number of errors.

4.4 DECODING OF BCH CODES

Let V be a cyclic (n, k) BCH code over the field F_q with the design distance $d = 2t + 1$ and $g(x)$ be the generator polynomial of this code. Let $\gamma, \gamma^2, \ldots, \gamma^{2t}$ be the roots of the generator polynomial $g(x)$. Then the parity-check matrix of the code can be written in the form

$$\mathbf{H} = \begin{bmatrix} 1 & \gamma & \gamma^2 & \cdots & \gamma^{n-1} \\ 1 & \gamma^2 & (\gamma^2)^2 & \cdots & (\gamma^{n-1})^2 \\ \cdots & \cdots & \cdots & \cdots & \cdots \\ 1 & \gamma^{2t} & (\gamma^2)^{2t} & \cdots & (\gamma^{n-1})^{2t} \end{bmatrix}. \tag{4.17}$$

Let $b(x)$ be the received word equal to the sum of the codeword $a(x)$ and the error pattern $e(x)$

$$b(x) = a(x) + e(x), \tag{4.18}$$

and the error polynomial

$$e(x) = e_0 + e_1 x + \ldots + e_{n-1} x^{n-1} \tag{4.19}$$

contains exactly $\nu \le t$ nonzero coefficients $e_{i_1}, \ldots, e_{i_\nu}$. Multiplying $e(x)$ by $\mathbf{H}$ obtain the syndrome vector with the components

$$s_j = b(\gamma^j), \quad j = 1, \ldots 2t \tag{4.20}$$

Since $a(\gamma^j) = 0$ for any codeword, the equation (4.20) can be written as follows:

$$s_j = e_{i_1} \cdot (\gamma^j)^{i_1} + e_{i_2} \cdot (\gamma^j)^{i_2} + \ldots + e_{i_\nu} \cdot (\gamma^j)^{i_\nu}. \tag{4.21}$$

The error polynomial (4.19) is fully defined by the set of pairs $\{e_{i_1}, i_1\}$, $\{e_{i_2}, i_2\}$, $\ldots, \{e_{i_\nu}, i_\nu\}$. Denote e_{i_m} by Y_m and γ^{i_m} by X_m, then

$$s_j = X_1^j Y_1 + X_2^j Y_2 + \cdots + X_\nu^j Y_\nu, \quad j = 1, \ldots, 2t. \tag{4.22}$$

Y_m is called the *error-value* and X_m is called the *error-location number*. The syndrome components s_j can be calculated directly from the received vector, therefore (4.22) can be regarded as the system of $2t$ nonlinear equations relative to 2ν unknowns $X_1, X_2, \ldots, X_\nu$, $Y_1, Y_2, \ldots, Y_\nu$. If $t \geq \nu$, system (4.22) has a unique solution that should be found to decode the received word.

However, it is not so easy to solve the system of nonlinear equations directly. To avoid difficulties with the direct solving of this system consider the polynomial $\sigma(x)$

$$\sigma(x) = 1 + \sigma_1 x + \cdots + \sigma_\nu x^\nu = (1 - xX_1) \cdot (1 - xX_2) \cdot \cdots \cdot (1 - xX_\nu). \tag{4.23}$$

The polynomial $\sigma(x)$ called the *error-location polynomial* is the polynomial of minimal degree such that the roots of this polynomial are values X_1^{-1}, X_2^{-1}, $\ldots$, X_ν^{-1}, i.e. the reciprocals of the error-locations.

If the coefficients $\sigma_1, \ldots, \sigma_\nu$ of the polynomial $\sigma(x)$ are known then we should find the roots of this polynomial to calculate the error-locations. Hence the problem of calculation of the error-locations can be solved in two steps: first, find the coefficients of the polynomial $\sigma(x)$ and second, find the roots of $\sigma(x)$.

To find the coefficients $\sigma_1, \ldots, \sigma_\nu$ it is necessary to show the relation of these coefficients (unknown values) with the known syndrome components s_j, $j = 1, \ldots, 2t$. Substituting values X_1^{-1}, X_2^{-1}, $\ldots$, X_ν^{-1} in $\sigma(x)$ obtain the system of equations

$$1 + \sigma_1 \cdot (X_j^{-1}) + \sigma_2 \cdot (X_j^{-1})^2 + \ldots + \sigma_\nu \cdot (X_j^{-1})^\nu = 0, \quad j = 1, \ldots, \nu. \tag{4.24}$$

Multiplying the left and right parts of the equations (4.24) by $X_1^{\nu+1} Y_1$, $X_2^{\nu+1} Y_2$, $\ldots$, $X_\nu^{\nu+1} Y_\nu$ correspondingly we obtain

$$\begin{cases} Y_1 X_1^{\nu+1} + \sigma_1 Y_1 X_1^\nu + \sigma_2 Y_1 X_1^{\nu-1} + \ldots + \sigma_\nu Y_1 X_1 = 0 \\ \cdots\cdots\cdots\cdots\cdots\cdots\cdots\cdots\cdots\cdots\cdots\cdots\cdots\cdots\cdots \\ Y_\nu X_\nu^{\nu+1} + \sigma_1 Y_\nu X_\nu^\nu + \sigma_2 Y_\nu X_\nu^{\nu-1} + \ldots + \sigma_\nu Y_\nu X_\nu = 0 \end{cases} \tag{4.25}$$

Adding equations (4.25) gives

$$(Y_1 X_1^{\nu+1} + \ldots + Y_\nu X_\nu^{\nu+1}) + \sigma_1(Y_1 X_1^\nu + \ldots + Y_\nu X_\nu^\nu) + \ldots + \sigma_\nu(Y_1 X_1 + \ldots + Y_\nu X_\nu) = 0. \tag{4.26}$$

Taking in account (4.22) we can write (4.26) as follows

$$s_{\nu+1} + \sigma_1 s_\nu + \ldots + \sigma_\nu s_1 = 0. \tag{4.27}$$

Now multiplying the left and right parts of the equations (4.24) by $X_j^{\nu+2} Y_j$, $X_j^{\nu+3} Y_j$, $\ldots$ and with the same manipulations as above, obtain the system of equations that defines the relation between the syndrome components and the coefficients of the error-location polynomial

$$\begin{cases} s_1\sigma_\nu + s_2\sigma_{\nu-1} + \ldots + s_\nu\sigma_1 = -s_{\nu+1} \\ s_2\sigma_\nu + s_3\sigma_{\nu-1} + \ldots + s_{\nu+1}\sigma_1 = -s_{\nu+2} \\ \ldots\ldots\ldots\ldots\ldots\ldots\ldots\ldots\ldots\ldots\ldots\ldots\ldots \\ s_\nu\sigma_\nu + s_{\nu+1}\sigma_{\nu-1} + \ldots + s_{2\nu-1}\sigma_1 = -s_{2\nu} \end{cases} . \tag{4.28}$$

The system (4.28) unlike (4.22) is the system of linear equations and there are well-known methods of solving it.

However to obtain the system (4.28) it is necessary to know the value of ν. The following theorem allows us to obtain ν.

Theorem 4.3 The determinant of the matrix $\mathbf{M}_\mu$

$$\mathbf{M}_\mu = \begin{bmatrix} s_1 & s_2 & \ldots & s_\mu \\ s_2 & s_3 & \ldots & s_{\mu+1} \\ \ldots & \ldots & \ldots & \ldots \\ s_\mu & s_{\mu+1} & \ldots & s_{2\mu-1} \end{bmatrix}$$

is nonzero $|\mathbf{M}_\mu| \neq 0$ if μ is equal to the number of errors $\mu = \nu$. If $\mu > \nu$ then $|\mathbf{M}_\mu| = 0$.

Proof Let $X_j = 0$ for $j > \nu$ and let

$$\mathbf{W}_\mu = \begin{bmatrix} 1 & 1 & \ldots & 1 \\ X_1 & X_2 & \ldots & X_\mu \\ \ldots & \ldots & \ldots & \ldots \\ X_1^{\mu-1} & X_2^{\mu-1} & \ldots & X_\mu^{\mu-1} \end{bmatrix}$$

be the Vandermonde matrix. Denote by $\mathbf{D}_\mu$ the diagonal matrix

$$\mathbf{D}_\mu = \begin{bmatrix} Y_1X_1 & 0 & \ldots & 0 \\ 0 & Y_2X_2 & \ldots & 0 \\ \ldots & \ldots & \ldots & \ldots \\ 0 & 0 & \ldots & Y_\mu X_\mu \end{bmatrix}$$

It is easy to verify that in accordance with (4.22)

$$\mathbf{M}_\mu = \mathbf{W}_\mu \cdot \mathbf{D}_\mu \cdot \mathbf{W}_\mu^{\mathrm{T}},$$

where $\mathbf{W}_\mu^{\mathrm{T}}$ is the transposed Vandermonde matrix. Then the determinant of matrix $\mathbf{M}_\mu$ is

$$|\mathbf{M}_\mu| = |\mathbf{W}_\mu| \cdot |\mathbf{D}_\mu| \cdot |\mathbf{W}_\mu^{\mathrm{T}}| . \tag{4.29}$$

If $\mu = \nu$ then all the elements $X_1, \ldots, X_\mu, \quad Y_1, \ldots, Y_\mu$ differ from zero and $X_1, \ldots, X_\mu$ are the distinct elements. Then all determinants in the right part of (4.29) differ from zero and $| \mathbf{M}_\mu | \neq 0$. If $\mu > \nu$ then $| \mathbf{W}_\mu | = 0$ and $| \mathbf{M}_\mu | = 0$.   *Q.E.D.*

With the help of Theorem 4.3 it is possible to find value of ν in the following way. Consider matrices $\mathbf{M}_\mu$ for $\mu = t, t - 1, t - 2$, etc. until find μ_0 such that $| \mathbf{M}_{\mu_0} | \neq 0$. Then $\mu_0 = \nu$ is the true number of errors.

For given ν it is possible to solve the system (4.28) (e.g. with the help of the Gauss method) and obtain the coefficients of the error-location polynomial $\sigma(x)$. Now it is enough to find the roots of $\sigma(x)$ and we can obtain the error-location numbers just by inverting these roots. To find the roots of the polynomial $\sigma(x)$ in the finite field we can just substitute in $\sigma(x)$ all field elements in turn (this procedure for searching the roots of a polynomial is called *Chien's search*).

To calculate error values $Y_1, \ldots, Y_\nu$ it is enough to solve the system of first ν equations (4.22) after substituting the known values of $X_1, \ldots, X_\nu$.

Thus, the algorithm of decoding a BCH code, which is called usually the *Peterson-Gorenstein-Zierler algorithm* or the direct method of decoding a BCH code, consists of four steps:

- syndrome calculation;

- finding the coefficients of the error-location polynomial;

- finding the roots of the error-location polynomial;

- calculation of error-values.

The formal algorithm is as follows:

1. Calculate the syndrome components $s_j = b(\gamma^j)$, $\quad j = 1, \ldots, 2t$, where $b(x)$ is the received polynomial.

2. Consider determinants $| \mathbf{M}_t |$, $| \mathbf{M}_{t-1} |$, $\ldots$ until find ν such that $| \mathbf{M}_\nu | \neq 0$.

3. Calculate the coefficients of the error-location polynomial $\sigma_1, \ldots, \sigma_\nu$

$$(\sigma_\nu, \ldots, \sigma_1) = (-s_{\nu+1}, \ldots, -s_{2\nu}) \cdot \left(\mathbf{M}_\nu^{-1}\right)^{\mathrm{T}}.$$

4. Find the roots of the error-location polynomial $\sigma(x)$ with the help of Chien's search.

5. Calculate error-values

$$
\begin{bmatrix} Y_1 \\ Y_2 \\ \vdots \\ Y_\nu \end{bmatrix}
=
\begin{bmatrix}
X_1 & X_2 & \ldots & X_\nu \\
X_1^2 & X_2^2 & \ldots & X_\nu^2 \\
\ldots & \ldots & \ldots & \ldots \\
X_1^\nu & X_2^\nu & \ldots & X_\nu^\nu
\end{bmatrix}
\cdot
\begin{bmatrix} s_1 \\ s_2 \\ \vdots \\ s_\nu \end{bmatrix}
$$

The most complex computation in this algorithm has step 3 associated with solving the system (4.28). However, (4.28) is not the arbitrary system, as the coefficients of the equalities in this system are well structured. All known methods of simplification of the direct decoding are based on this structure. However, these simplified methods do not give any clarification in understanding the process of decoding a BCH code.

Because of this, we will give known, more efficient, simplified algorithms for decoding a BCH code without the whys and wherefores.

One of the best methods of implementation of step 3 is the iterative Berlekamp-Massey algorithm. The idea of this algorithm is that the polynomial $\sigma(x)$ is calculated with the help of the sequential approximations $\sigma^{(0)}(x)$, $\sigma^{(1)}(x)$, etc. until $\sigma^{(\nu)}(x) = \sigma(x)$ and in doing so $\sigma^{(j)}(x)$ is chosen as the refinement of $\sigma^{(j-1)}(x)$.

It follows from (4.28) that the syndrome components can be expressed in the form of the recursive equation

$$s_j = -\sum_{i=1}^{\nu} \sigma_i \cdot s_{j-i}, \quad j = \nu + 1, \ldots, 2\nu. \tag{4.30}$$

We will say that the polynomial $\sigma^{(j)}(x)$ generates $s_1, \ldots, s_j$ if it holds true, the equation

$$s_j = -\sum_{i=1}^{L} \sigma_i^{(j)} \cdot s_{j-i},$$

where L is the degree of the polynomial $\sigma^{(j)}(x)$ and $\sigma_i^{(j)}$ are the coefficients of this polynomial.

Then the solution of the system (4.28) is equivalent to the finding of the polynomial $\sigma(x)$ of minimal degree that generates the syndrome components $s_1, s_2, \ldots, s_{2t}$. The iterative search of $\sigma(x)$ starts with $\sigma^{(0)}(x)$ and is as follows: for given $\sigma^{(j-1)}(x)$ that generates $s_1, s_2, \ldots, s_{j-1}$ verify the capability of $\sigma^{(j-1)}(x)$ to generate s_j; if it is so then $\sigma^{(j)}(x)$ is assumed to be equal $\sigma^{(j-1)}(x)$, otherwise $\sigma^{(j)}(x)$ is chosen as

$$\sigma^{(j)}(x) = \sigma^{(j-1)}(x) + \Delta_j(x),$$

where $\Delta_j(x)$ is the correction polynomial for the jth step. The process is repeated until the polynomial that generates all the syndrome components is found.

Let us present the formal Berlekamp-Massey algorithm of search of the polynomial $\sigma(x)$. It contains some auxiliary polynomials $B(x)$ and $T(x)$, which do not have a meaningful interpretation. L is the degree of the current polynomial $\sigma(x)$ at the jth step.

1. Set the initial parameters $\sigma^{(0)}(x) = 1$, $j = 0$, $L = 0$, $B(x) = 1$.

2. $j = j + 1$.

3. Calculate the jth discrepancy $\Delta_j = s_j + \sum_{i=1}^{L} \sigma_i^{(j-1)} \cdot s_{j-i}$.

4. Verify if $\sigma^{(j-1)}(x)$ generates s_j or not: compare Δ_j with zero. If $\Delta_j = 0$, then $\sigma^{(j)}(x) = \sigma^{(j-1)}(x)$ and go to step 9.

5. Calculate $T(x) = \sigma^{(j-1)}(x) - \Delta_j \cdot x \cdot B(x)$.

6. Check is it necessary to increase the degree of the current polynomial. Compare $2L$ with $j - 1$. If $2L > j - 1$ then go to step 8.

7. Calculate new $B(x)$, $\sigma(x)$ and L. $B(x) = \Delta_j^{-1} \cdot \sigma^{(j-1)}(x)$; $\sigma^{(j)}(x) = T(x)$; $L = L - j$. Go to step 10.

8. Calculate $\sigma^{(j)}(x) = T(x)$.

9. Calculate $B(x) = x \cdot B(x)$.

10. Check the condition $j = 2t$. If $j < 2t$ go to step 2.

11. Check the condition $\deg \sigma(x) = L$. If $\deg \sigma(x) = L$ then go to step 13.

12. Stop the Berlekamp-Massey algorithm, $\sigma(x) = \sigma^{(j)}(x)$. Go to the next decoding stage.

13. Stop the Berlekamp-Massey algorithm; the uncorrectable error pattern is detected.

Example 4.5 Consider Berlekamp-Massey algorithm for decoding 16-ary (15, 9) Reed-Solomon code capable of correcting triple errors. Let the transmitted polynomial be $a(x) = 0$, the received polynomial be $b(x) = a(x) + e(x) = e(x) = \alpha x^7 + \alpha^5 x^5 + \alpha^{11} x^2$. $s_1 = \alpha \cdot \alpha^7 + \alpha^5 \cdot \alpha^5 + \alpha^{11} \cdot \alpha^2 = \alpha^{12}$, $s_2 = 1$, $s_3 = \alpha^{14}$, $s_4 = \alpha^{13}$, $s_5 = 1$, $s_6 = \alpha^{11}$. The steps of the Berlekamp-Massey algorithm are listed in the Table 4.1.

Table 4.1 The steps of the Berlekamp-Massey algorithm

j	Δ_j	$T(x)$	$B(x)$	$\sigma^{(j)}(x)$	L
0			1	1	0
1	α^{12}	$1 + \alpha^{12}x$	α^3	$1 + \alpha^{12}x$	1
2	α^7	$1 + \alpha^3 x$	$\alpha^3 x$	$1 + \alpha^3 x$	1
3	1	$1 + \alpha^3 x + \alpha^3 x^2$	$1 + \alpha^3 x$	$1 + \alpha^3 x + \alpha^3 x^2$	2
4	1	$1 + \alpha^{14} x$	$x + \alpha^3 x^2$	$1 + \alpha^{14} x$	2
5	α^{11}	$1 + \alpha^{14}x + \alpha^{11}x^2 + \alpha^{14}x^3$	$\alpha^4 + \alpha^3 x$	$1 + \alpha^{14}x + \alpha^{11}x^2 + \alpha^{14}x^3$	3
6	0	$1 + \alpha^{14}x + \alpha^{11}x^2 + \alpha^{14}x^3$	$\alpha^4 + \alpha^3 x$	$1 + \alpha^{14}x + \alpha^{11}x^2 + \alpha^{14}x^3$	3

Then $\sigma(x) = 1 + \alpha^{14}x + \alpha^{11}x^2 + \alpha^{14}x^3 = (1 + \alpha^7 x) \cdot (1 + \alpha^5 x) \cdot (1 + \alpha^2 x)$, the roots are $\alpha^{-7}, \alpha^{-5}, \alpha^{-2}$; and the error-location numbers are $\alpha^7, \alpha^5, \alpha^2$.

4.5   THE SUDAN ALGORITHM AND ITS EXTENSIONS

The Sudan algorithm for decoding some low rate Reed-Solomon codes beyond half of their minimum distance d was invented in 1997 [4]. Later Guruswami and Sudan [5] managed to significantly improve this algorithm to make it capable of decoding almost all

Reed-Solomon (RS) codes beyond $d/2$ limit. Let us note that if more than $\left\lfloor \dfrac{d-1}{2} \right\rfloor$ errors have occurred in a received word, then the decoding may not be unique, so the decoder may output a *list* of codewords within a certain distance from the received word. The final decision can be carried out using some additional information, e.g. 'soft' information from the demodulator. Some other authors Kötter, Roth-Ruckenstein, and Nielsen succeeded in decreasing the complexity of the Guruswani-Sudan (GS) algorithm.

Following the P.Elias [22] definition, a list decoding algorithm of decoding radius T should produce for any received vector $\mathbf{y}$ the list $L_T(\mathbf{y}) = \{\mathbf{c} \in C : d(\mathbf{y},\mathbf{c}) \le T\}$ of all vectors $\mathbf{c}$ from a code C, which are at distance at most T apart from vector $\mathbf{y}$. Bounded distance decoding, i.e., correcting up to $\left\lfloor \dfrac{d-1}{2} \right\rfloor$ errors, is a particular case of list decoding when $T = \left\lfloor \dfrac{d-1}{2} \right\rfloor$ and any list contains no more than one code vector.

Let us consider the problem of list decoding of Reed-Solomon codes, one of the most widely used and well-studied classes of error-correcting codes. There are many ways to define the Reed-Solomon code, one of them was considered in section 4.2. Here it is convenient to use the 'dual' definition:

Consider finite field $GF(q)$ and some set $X = \{x_1, \ldots, x_n\}$ of its distinct elements. The (n, k) RS code consist of all vectors $\mathbf{f} = (f(x_1), \ldots, f(x_n))$, where $f(x) = f_0 + f_1 x + \ldots + f_{k-1}x^{k-1}$ is a polynomial over $GF(q)$ of degree less than k. Since the number of roots of $f(x)$ does not exceed $\deg f(x)$ we have that Hamming weight $wt(\mathbf{f}) \ge n - \deg f(x) \ge n - k + 1$. Hence, the minimum distance of RS code $d \ge n - k + 1$, and by the Singleton bound (4.5) we have $d = n - k + 1$.

According to the general definition a list decoding algorithm for (n, k) RS code can be reformulated in the following way. For any given received vector $y = (y_1, \ldots, y_n)$ find all polynomials $p(x)$ of degree $\deg p(x) < k$ such that $p(x_s) = y_s$ for at least $n - T$ values x_s. We call such $p(x)$ T-consistent. Denote $\hat{k} = k - 1$, which is more convenient to use in below given formulae.

The original breakthrough Sudan's algorithm [4] exploits two very simple mathematical facts:

- a homogenous system of linear equations has a nontrivial, i.e., a nonzero solution if the number of equations is less than the number of variables;

- if a polynomial has more roots than its degree, then this is an identically zero polynomial.

Geometrically list decoding of RS code means finding all curves $y - p(x) = 0$, which pass through at least $n - T$ points (x_s, y_s) of the "plane" F_q^2. Consider instead some general algebraic curve

$$Q(x,y) = \sum q_{ij}x^i y^j,$$

which passes through all n points (x_s, y_s), i.e.,

$$Q(x_s, y_s) = 0, \quad 1 \le s \le n. \tag{4.31}$$

Definition 4.1 Define weighted degree of $Q(x,y)$ as

$$\deg_{\{1,\hat{k}\}} Q(x,y) = \max_{\{i,j\}:q_{ij}\neq 0} i + \hat{k}j.$$

This definition becomes clear due to the following

Lemma 4.1 If $n - T > \deg_{\{1,\hat{k}\}} Q(x,y)$ and $p(x)$ is T-consistent, then $y - p(x)$ divides $Q(x,y)$.

Proof Consider a univariate polynomial $g(x) = Q(x,p(x))$. The degree of $g(x)$ does not exceed weighted degree $\deg_{\{1,\hat{k}\}} Q(x,y)$ of $Q(x,y)$. On the other hand, $g(x_s) = 0$ for every s such that $y_s = p(x_s)$. Hence, $g(x)$ has at least $n - T$ roots and under Lemma's condition it leads to $g(x) \equiv 0$. The latter means that $(y - p(x))|Q(x,y)$ or, saying in other words, that $p(x)$ is a root of $Q(x,y)$ considered as a univariate polynomial $\hat{Q}(y)$ on y with coefficients from the ring $F_q[x]$.   *Q.E.D.*

The next lemma shows that desired $Q(x,y)$ of relatively low weight degree exists.

Lemma 4.2 For any l such that $l^2 \geq 2\hat{k}n$ there exists $Q(x,y)$ of weighted degree $\deg_{\{1,\hat{k}\}} Q(x,y) \leq l$ satisfying (4.31).

Proof Consider (4.31) as a system of linear equations for unknown 'variables' q_{ij}. Namely,

$$\sum_{i,j} q_{ij} \cdot x_s^i \cdot y_s^j = 0, \quad 1 \leq s \leq n. \tag{4.32}$$

Since $i + \hat{k}j \leq l$ there are

$$l + 1 \text{ 'variables' } q_{i,0}, \quad 0 \leq i \leq l,$$
$$l + 1 - \hat{k} \text{ 'variables' } q_{i,1}, \quad 0 \leq i \leq l - \hat{k},$$
$$\dots\dots\dots\dots\dots\dots\dots\dots\dots\dots\dots\dots\dots\dots\dots\dots$$
$$l + 1 - \hat{k}\left\lfloor\frac{l}{\hat{k}}\right\rfloor \text{ 'variable' } q_{i,\lfloor l/\hat{k}\rfloor}, 0 \leq i \leq l - \hat{k}\left\lfloor\frac{l}{\hat{k}}\right\rfloor.$$

Hence, there are totally M variables, where $M = \sum_{\nu=0}^{\lfloor l/\hat{k}\rfloor} (l + 1 - \nu\hat{k}) = \left(\left\lfloor\frac{l}{\hat{k}}\right\rfloor + 1\right) \cdot (l + 1) - \hat{k}\sum_{\nu=0}^{\lfloor l/\hat{k}\rfloor} \nu = \left(\left\lfloor\frac{l}{\hat{k}}\right\rfloor + 1\right) \cdot \left(l + 1 - \frac{1}{2}\hat{k} \cdot \left\lfloor\frac{l}{\hat{k}}\right\rfloor\right) > \frac{l^2}{2\hat{k}}$. Therefore, if $\frac{l^2}{2\hat{k}} > n$ then the number of unknown variables is greater than the number of equations and hence, there exist nontrivial, i.e., nonzero solution of (4.31).   *Q.E.D.*

Combining these two lemmas obtain the *original Sudan algorithm*. Namely, for any T and l such that $n - T > l$ and $l^2 \geq 2\hat{k}n$

1. Find a bivariate polynomial $Q(x,y)$ of $\deg_{\{1,\hat{k}\}} Q(x,y) \leq l$ satisfying (4.31). It can be done by solving the linear system (4.32), for instance with the help of Gauss elimination procedure.

2. Find all 'roots' $p(x)$ of $\hat{Q}(y)$, i.e., all divisors $(y - p(x)) \mid Q(x, y)$.

3. Output only T-consistent roots $p(x)$, i.e., those that satisfy $d(y, p(x)) \leq T$.

The best choice $l = \left\lceil \sqrt{2\hat{k}n} \right\rceil$ gives the list decoding algorithm with decoding radius $T = n - 1 - \left\lceil \sqrt{2\hat{k}n} \right\rceil$.

The main difference between the GS and the original Sudan algorithm is that in the GS algorithm it does not suffice that $Q(x_i, y_i) = 0$. It is required that every point (x_i, y_i) is a singularity of Q. Informally, a singularity is a point where the curve given by $Q(x, y) = 0$ intersects itself. Then in the first phase of the GS algorithm the additional constraints will force us to increase the allowed degree of Q. However, we gain in the second phase. In this phase we look for the roots of Q and now we know that p passes through many singularities of Q, rather than just points on Q. In such a case we need only half as many singularities as regular points, and this is where the advantage comes from [5].

The singularities of a bivariate polynomial over a finite field can be defined as follows.

Definition 4.2 A polynomial $Q(x, y)$ has a singularity of order r at point (α, β) if the 'shifted' polynomial $Q(x + \alpha, y + \beta)$ has no monomials of ordinary total degree ((1, 1) weighted degree) less than r.

To calculate the order of $Q(x, y)$ at point (α, β), we need to be able to express $Q(x + \alpha, y + \beta)$ as a polynomial in x and y. The following propositions, due to H. Hasse [6], tell us one way to do this [7].

Proposition 4.1 If $Q(x) = \sum_i q_i x^i \in F[x]$, then for any $\alpha \in F$, we have

$$Q(x + \alpha) = \sum_j Q_j(\alpha) x^j, \tag{4.33}$$

where

$$Q_j(x) = \sum_i \binom{i}{j} q_i x^{i-j}. \tag{4.34}$$

The function $Q_j(x)$ in the left part of equation (4.34) is called the jth *Hasse derivative* of $Q(x)$.

Note that equation (4.33) is Taylor's formula (without reminder) when field F has characteristic 0, since in that case,

$$Q_j(x) = \frac{1}{j!} \cdot \frac{d^j}{dx^j} Q(x).$$

Proposition 4.2 Let $Q(x, y) = \sum_{i,j} q_{i,j} x^i y^j \in F[x, y]$. For any (α, β), we have

$$Q(x + \alpha, y + \beta) = \sum_{u, v_1} Q_{u,v}(\alpha, \beta) x^u y^v, \tag{4.35}$$

where

$$Q_{u,v}(x, y) = \sum_{i,j} \binom{i}{u}\binom{j}{v} q_{i,j} x^{i-u} y^{j-v}. \tag{4.36}$$

The function $Q_{u,v}(x, y)$ in the left part of equation (4.36) is called the (u, v)th *Hasse (mixed partial) derivative* of $Q(x, y)$.

Proof Using the binomial theorem, we express $Q(x + \alpha, y + \beta)$ as a polynomial in x and y:

$$Q(x + \alpha, y + \beta) = \sum_{i,j} q_{i,j}(x + \alpha)^i (y + \beta)^j = \sum_{i,j} q_{i,j} \left(\sum_u \binom{i}{u} x^u \alpha^{i-u} \right) \left(\sum_v \binom{j}{v} y^v \beta^{j-v} \right)$$

$$= \sum_{u,v} x^u y^v \left(\sum_{i,j} \binom{i}{u}\binom{j}{v} q_{i,j} \alpha^{i-u} \beta^{j-v} \right) = \sum_{u,v} Q_{u,v}(\alpha, \beta) x^u y^v \quad Q.E.D.$$

Corollary A polynomial $Q(x, y)$ has a root of multiplicity (order) r at a point (α, β) if all its Hasse derivatives at (α, β) of total order less than r are equal to zero, i.e. $Q_{u,v}(\alpha, \beta) = 0$ for all u and v such that $0 \le u + v \le r$.

Now we can give a formal description of the *Guruswami-Sudan algorithm*.

The inputs of the algorithm are the code length n, code dimension k, the interpolation points $\{(x_s, y_s)\}, s = 1..n$, and the required root multiplicity r.

1. Compute a non-zero polynomial $Q(x, y)$ of minimal possible $(1, k - 1)$ weighted degree l such that

$$Q_{j_1,j_2}(x_i, y_i) = \sum_{j_1' \ge j_1} \sum_{j_2' \ge j_2} \binom{j_1'}{j_1}\binom{j_2'}{j_2} q_{j_1',j_2'} x_i^{j_1'-j_1} y_i^{j_2'-j_2} = 0, \quad j_1 + j_2 < r, \quad i = 1..n. \tag{4.37}$$

2. Find all polynomials $p(x) \in GF(q)[x]$ of degree $\deg(p(x)) < k$ such that p is a root of Q, i.e., $Q(x, p(x)) = 0$ or $y - p(x)$ is a factor of $Q(x, y)$. For each of these polynomials check if $p(x_i) = y_i$ for at least t values of $i = 1..n$, and if so, include p in output list.

Obviously, the algorithm can be executed in time polynomial in n since the underlying problems are solving the system of linear equations (4.37) and factorisation (or, which is even simpler, finding roots) of a polynomial. The last problem is not a simple one since we need to factorise bivariate polynomials.

The next two lemmas are generalisations of Lemma 4.1.

Lemma 4.3 If (x_i, y_i) is an input point of the algorithm, and $p(x)$ is a polynomial such that $p(x_i) = y_i$, then $(x - x_i)^r$ divides $g(x) = Q(x, p(x))$.

Proof Let $p'(x) = p(x + x_i) - y_i$. Since $p'(0) = 0$, $p'(x) = xp''(x)$ for some polynomial $p''(x)$. Let $Q^{(i)}(x, y) = Q(x + x_i, y + y_i)$ and

$$g'(x) = Q^{(i)}(x, p'(x)).$$ (4.38)

Compute $g(x) = Q(x, p(x)) = Q^{(i)}(x - x_i, p(x) - y_i)) = Q^{(i)}(x - x_i, p'(x - x_i)) = g'(x - x_i)$. Since $Q^{(i)}(x, y)$ must not have any coefficients of total degree less than r, substitution of $p'(x) = xp''(x)$ into (4.34) leads to a polynomial $g'(x)$ divisible by x^r. Thus, $(x - x_i)r$ divides $g(x)$. Q.E.D.

Lemma 4.4 If $p(x)$ is a polynomial of degree less than k such that $y_i = p(x_i)$ for at least t values of x_i and $rt > l = \deg_{\{1,\hat{k}\}} Q(x, y)$, $(\hat{k} = k - 1)$ then $y - p(x)$ divides $Q(x, y)$.

Proof Consider the polynomial $g(x) = Q(x, p(x))$. Obviously, $\deg g(x) \le l = \deg_{\{1,\hat{k}\}} Q(x, y)$. In accordance with Lemma 4.3, for every i such that $y_i = p(x_i)$, $(x - x_i)^r$ divides $g(x)$. Therefore, $\pi(x) = \prod_{i:p(x_i)=y_i} (x - x_i)^r$ divides $g(x)$. It follows from $\deg \pi(x) \ge rt > l = \deg g(x)$ that $g(x) = 0$, which means that $y = p(x)$ is a root of $Q(x, y)$. Q.E.D.

The only problem remaining is the selection of such parameters (actually, the only parameter is r) of the algorithm that the polynomial $Q(x, y)$ does exist. In fact, the only requirement is that the number of equations in (4.34), which equals to $n\binom{r + 1}{2}$, must be greater than the number of variables (i.e., coefficients in the polynomial), which is at least $\frac{l^2}{2\hat{k}}$ (see Lemma 4.2). It leads to a simple quadratic inequality. In [5] authors show that

$$r = 1 + \left\lfloor \frac{(k - 1)n + \sqrt{((k - 1)n)^2 + 4(t^2 - (k - 1)n)}}{2(t^2 - (k - 1)n)} \right\rfloor$$

guarantees the existence of polynomial $Q(x, y)$ of small enough degree $(rt - 1)$.

Theorem 4.4 The Guruswami-Sudan algorithm returns all $p(x)$ such that $p(x_i) = y_i$ for at least $n - T > \sqrt{(k - 1)n}$ values of x_i.

Proof If $t > \sqrt{(k - 1)n}$, one can always select r to be large enough to ensure the existence of $Q(x, y)$ of small degree. By Lemma 4.4 all polynomials of degree less than k such that $y_i = p(x_i)$ for at least t values of of x_i divide such a polynomial and should be thus discovered at the second step of the algorithm. Q.E.D.

Since r affects the running time of the algorithm one can set r to be less than is required to handle as many errors as possible. This will significantly reduce the computational complexity at the cost of some performance degradation.

Now let us consider some developments of the GS algorithm.

One of the serious improvements of the GS algorithm deals with invention of the efficient algorithm for finding roots of a bivariate polynomial. It is possible to show [8], that the roots of a bivariate polynomial can be expressed as roots of a number of univariate polynomials. The algorithm exploiting this fact is presented below. This algorithm uses as a subroutine a procedure for finding roots of a univariate polynomial. The algorithm takes as input a bivariate polynomial $Q(x, y)$ and positive integer k, and returns as output ϕ the set of all y-roots of $Q(x, y)$ of degree $\leq k$. The pseudocode of the Roth-Ruckenstein algorithm of finding roots of a bivariate polynomial can be written as follows

```
RECONSTRUCT(Q(x, y), k, i, φ)

1   Find the largest r such that xʳ divides Q(x, y)
2   M(x, y): = Q(x, y)/xʳ;
3   Find all roots mⱼ of a univariate polynomial M(0, y);
4   for Each root mⱼ
5     do φ[i] := mⱼ;
6       if i = k-1
7         then return φ;
8       else M̂(x, y): = M(x, xy + mⱼ);
9   RECONSTRUCT( M̂(x, y), k, i+1, φ).
```

Another natural extension of the GS algorithm is in the case of weighted curve fitting. In the GS algorithm one does not need to make all points to be singularities of the same order. Each point (x_i, y_i) may be assigned with an integer weight w_i and (4.37) should be modified as follows:

$$Q_{j_1, j_2}(x_i, y_i) = \sum_{j'_1 \geq j_1} \sum_{j'_2 \geq j_2} \binom{j'_1}{j_1} \binom{j'_2}{j_2} q_{j'_1, j'_2} x^{j'_1 - j_1} y^{j'_2 - j_2} = 0, \quad j_1 + j_2 < r \cdot w_i, \quad i = 1..n$$

$$(4.39)$$

for some integer r. Again, one can select such parameters of the algorithm (weights w_i and r) that the polynomial $Q(x, y)$ does exist and the second step of the algorithm returns all polynomials $p(x)$: deg $p(x) < k$. such that

$$\sum_{i: p(x_i)=y_i} w_i \geq t \tag{4.40}$$

In [5] it is shown that $t > \sqrt{k \sum_{i=1}^{n} w_i^2}$ is sufficient to solve this problem. Note, that x_i values

need NOT to be distinct ones and there may be as many interpolation points n as one needs provided the number of constraints is less than the number of coefficients in the polynomial $Q(x, y)$.

This algorithm can be used for the *soft decision decoding* of Reed-Solomon and some other codes. In [9] the authors suggested an algorithm which can be used for soft-decision decoding of Reed-Solomon codes. As input the algorithm accepts a matrix $\Pi = [\pi_{ij}]$: $\pi_{ij} = \Pr\{y_i = \alpha_j\}$, i.e. the cells of the matrix define a posteriori probability distribution of each symbol in the received word. The algorithm presented below computes integer weights m_{ij}, $i,j = 1 .. n$ (*multiplicity matrix*) for the polynomial reconstruction problem as stated above. The main idea of it is to set greater weights m_{ij} for most probable pairs $(x_i; y_j)$.

The algorithm must be provided with code parameters n, k, $n \times n$ matrix Π and a total number of interpolation points to be generated. The algorithm has not been proved to be optimal (in fact, the authors in [10] prove that the problem of construction of an optimal multiplicity matrix M is *NP*-hard), however, with carefully selected parameters it provides significant performance gain [11]. The pseudocode of the algorithm of soft decision decoding of RS code can be presented as follows:

```
SoftRSDecode(n, k, Π, s)
```

```
1  Π* := Π;
2  M: = 0;
3  while  s > 0
4  do
5      Find a position (i, j) of the greatest entry π*ᵢⱼ in Π*;
6      π*ᵢⱼ := πᵢⱼ / (mᵢⱼ + 1);
7      mᵢⱼ: = mᵢⱼ + 1;
8      s: = s - 1;
9      Solve the system (4.39) with weights defined by M and find a
         polynomial Q(x, y);
10     Find its roots pᵢ(x) : deg pᵢ(x) < k and corresponding codewords Yᵢ
         = {y₁, ..., yₙ}.
11     Select as output the most probable codeword Y = arg maxᵢ ∏ₖ₌₁ⁿ Πₖyᵢₖ.
```

One more improvement of the GS algorithm is the Nielsen interpolation algorithm [12,13]. The algorithm presented in [12] exploits the structure of the system of linear equations (4.37). Here we describe an improved version of the algorithm found in [13].

The main idea of the algorithm is to split the set of all possible solutions of the interpolation problem into a number of disjoint classes and iteratively construct an interpolation polynomial of minimal degree for each class. Finally one can select the smallest one as a solution of the interpolation problem.

Let us introduce *lexicographic* monomial ordering as

$$x^\alpha y^\beta \leq_{lex} x^a y^b \Leftrightarrow \alpha < a \vee (\alpha = a \wedge \beta \leq b).$$

Since the GS algorithm requires construction of an interpolation polynomial having minimal possible weighted degree, one has to introduce the *weighted-degree monomial ordering*:

$$f \leq_{w \deg} g \Leftrightarrow w \deg_{(1,k)}(f) < w \deg_{(1,k)}(g) \vee (w \deg_{(1,k)}(f) = w \deg_{(1,k)}(g) \wedge f \leq_{w \deg} g).$$

Let $\mathrm{LT}f(x, y)$ denote the leading term of a polynomial $f(x, y)$ with respect to $\leq_{w \deg}$ monomial ordering. Then

$$f(x, y) \leq_{w \deg} g(x, y) \Leftrightarrow \mathrm{LT}f(x, y) \leq_{w \deg} \mathrm{LT}g(x, y).$$

In order to obtain a non-zero solution of the system (4.37) it is sufficient to consider polynomials of form $Q(x, y) = \sum_{i=0}^{N} q_i m_i$, where $N = n\binom{r+1}{2}$ is the total number of equations in the system and m_i are all distinct monomials $x^{j1} y^{j2}$ ordered by their weighted degree:

$$m_0 \leq_{w\deg} m_1 \leq_{w\deg} \cdots \leq_{w\deg} m_i \leq_{w\deg} \cdots$$

It can be easily shown that

$$w\deg_{(0,1)} m_i \leq \rho_r - 1, \quad i \leq N,$$

where

$$\binom{\rho_r}{2} \leq \frac{n\binom{r+1}{2}}{k} < \binom{\rho_r + 1}{2},$$

i.e. one can consider only polynomials $Q(x, y)$ with degree in y less than ρ_r. Let us split the set of polynomials $Q(x, y)$: $w\deg_{(0,1)} m_i \leq \rho_r - 1$ into a number of disjoint classes $G_j = \left\{ Q(x, y) \mid w\deg_{(0,1)}(\mathrm{LT}Q) = j \right\}$, $k = 0 .. \rho_r - 1$. Let us sequentially process constraints (4.37) and at each step for each class G_j construct the minimal with respect to $w\deg_{(1;k)}$ polynomial $Q_j(x, y) \in G_j$ satisfying all processed constraints. Allowed operations are:

1. Add to $Q_j(x, y) \in G_j$ another polynomial $\gamma \cdot Q_{j_0}(x, y)$, $Q_{j_0}(x, y) \leq_{w\deg} Q_j(x, y)$, $Q_{j_0}(x, y) \in G_{j_0}$, $j \neq j_0$, $\gamma \in GF(q)$. Clearly, this operation does not increase the order of $Q_j(x, y)$ but keeps it in the same class G_j.

2. Multiply $Q_{j_0}(x, y) \in G_{j_0}$ by $(x - \delta)$, $\delta \in GF(q)$. This operation introduces the minimal possible increase in degree of the polynomial but keeps it in the same class.

The pseudocode of the Nielsen interpolation algorithm is as follows:

Iterative Interpolation $(n, \{(x_i, y_i), i = 1 .. n\}, r, \rho_r)$

```
1    for i: = 0 to ρ_r - 1
2    do Q_i (x, y): = y^i;
3    for i: = 1 to n
4    do for β := 0 to r - 1
5       do for α := 0 to r - β - 1
6          do Compute Δ_j := coeff(Q_j(x + x_i, y + y_i), x^α y^β),   j = 0..ρ_r - 1
7             Find j_0 = arg min Q_j(x, y)
                        j: Δ_j≠0
8             for j ≠ j_0
9             do Q_j(x, y) := Q_j(x, y) - (Δ_j/Δ_{j_0}) Q_{j_0}(x, y);
10               Q_{j_0}(x, y) := Q_{j_0}(x, y)(x - x_i);
11   return min Q_j(x, y)
              i
```

The proof of the algorithm can be found in [11].

The complexity of the algorithm can be estimated as follows. At each step (a) evaluation of Hasse derivatives is performed and (b) polynomials $Q_j(x, y)$ are multiplied by scalar values and summed. From (4.36) one can see that for a bivariate polynomial having s terms $O(s)$ operations are required to compute its Hasse derivative at any point. Number of terms in polynomials $Q_s(x, y)$ grows from 1 at algorithm startup to $O\left(n\binom{r+1}{2}\right)$. Thus, the overall complexity of evaluation of Hasse derivatives is $O\left(n\left(\frac{r(r+1)}{2}\right)^2 \rho_r\right) = O\left(n^2\left(\frac{n}{k}\right)^{1/2} r^5\right)$.

Similarly, complexity of manipulations with polynomials at each step is $O(\rho_r s)$ and the overall complexity is $O\left(n^2\left(\frac{n}{k}\right)^{1/2} r^5\right)$,

REFERENCES

1. Hamming, R. W. (1950). Error Detecting and Error Correcting Codes, *Bell Syst. Tech. J.*, **29**, 147–60.
2. MacWilliams, F. J. and Sloan, J. J., (1977). *The Theory of Error-Correcting Codes*, North-Holland, Amsterdam, The Netherlands.
3. Reed, I. S. and Solomon, G. (1960). Polynomial codes over certain finite fields, *J. SIAM*, **8** 300–4.
4. Sudan, M. (1997). Decoding of Reed–Solomon Codes beyond the Error-Correction Bound, *J. Complexity*, **13**, 180–93.
5. Guruswami, V. and Sudan, M. (1999). Improved Decoding of Reed–Solomon Codes and Algebraic Geometry Codes, *IEEE Trans. Inform. Theory*, **45**, (6), 1757–67.
6. Hasse, H. (1936). Theorie der höheren Differentiate in einem algebroishen Funcktionenk örper mit vollkommenem Konstantkörper nei belie beger Charakteristic, *J. Reine. Ang. Math.* **75**, 50–4.
7. McEliece, R. J. (2003). The Guruswami-Sudan Decoding Algorithm for Reed-Solomon Codes, in *IPN Progress Report*, May.
8. Roth, R. and Ruckenstein, G. (2000). Efficient decoding of Reed-Solomon codes beyond half the minimum distance. *IEEE Transactions on Information Theory*, 46(1):246–57.
9. Kötter, R. (1996). Fast Generalized Minimum-Distance Decoding of Algebraic Geometry and Reed-Solomon Codes, *IEEE Trans. Inform. Theory*, **42**, (3), 721–36.
10. Kötter, R. and Vardy, A. (2000). Algebraic soft-decision decoding of Reed-Solomon codes. in *Proceedings of 38th Annual Allerton Conference on Communication, Control and Computing*.
11. Gross, W. J., Kschischang, F. R., Koetter, R. and Gulak, P. (2002). Simulation results for algebraic soft-decision decoding of Reed-Solomon codes. In *Proceedings of the 21st Biennial Symposium on Communications*, pp. 356–60.
12. Nielsen, R. R. and Hoholdt, T. (1998). Decoding Reed-Solomon codes beyond half the minimum distance. In *Proceedings of the International Conference on Coding Theory and Cryptography, Mexico*.
13. Nielsen, R. R. (2001). *List decoding of linear block codes*. PhD thesis, Technical University of Denmark.
14. Best, M. R. and Brouwer, A. E. (1977). Triply shortened binary Hamming code is optimal. *Discr. Math.*, **17**, 235–45.
15. Kabatyanskii, G. A. and Panchenko, V. I. (1988). Unit-sphere packings and coverings of the Hamming space, *Problems of Information Transmissions*, **24**, (4), 3–16.
16. Hamalainen, H. (1988). Two new binary codes with minimum distance three, *IEEE on Information Transmission*, **34**, 885.

17. Panchenko, V. I. (1988). Packings and coverings over an arbitrary alphabet, *Problems of Information Transmissions*, **24**, (4), 93–6.

18. Varshamov, R. R. and Tenengolts, G. M. (1965). Codes which correct single asymmetric errors, *Automation and Remote Control* (in Russian), **26**, (2), 286–90.

19. Gevorkian, D. N. Avetisian, A. M. and Tigranyan, V. A. (1975). On the construction of codes correcting two errors in Hamming metric over Galois fields, *Vychislitelnia Technika, Kuibishev*, (3), 19–21. (in Russian).

20. Dumer, I. I. and Zinoviev, V. A. (1978). New maximal code over the Galois field GF(4), *Problems of Information Transmission*, **14**, (3), 24–34.

21. Dumer, I. I. (1995). Nonbinary double-error-correcting codes designed by means of algebraic varieties, *IEEE Trans. on of Information Theory*, **41**, (6), 1550–60.

22. Elias, P. (1957). List decoding for noisy channels, *Tech. Report 335*, Research Lab. of Electronics, MIT, USA.

5

Decoding of LDPC Codes

5.1 LOW-DENSITY PARITY-CHECK CODES

5.1.1 Basic Terms and Definitions

Low-density parity-check (LDPC) codes were first suggested by R.Gallager [1,2], and were investigated further in [3,4,5,6]. Traditionally, LDPC-code is defined by its parity-check matrix $\mathbf{H}$, which has the sparse property, i.e., its rows and columns have a low number of non-zero elements comparing to matrix size. More precisely, we define (n, γ, ρ)-code as a linear code of length n, with parity-check matrix containing the columns of weight γ and the rows of weight ρ.

The parity-check matrix $\mathbf{H}$ contains

$$r = n\gamma/\rho \tag{5.1}$$

rows, and therefore, the code rate is lower-limited as

$$R \geq 1 - \gamma/\rho \tag{5.2}$$

[1]. The example of $(16, 3, 4)$-code is shown in Figure 5.1.

Besides the traditional definition of a code as a zero-space of its parity-check matrix, LDPC-codes are often defined by means of the *incidence graph* of $\mathbf{H}$ matrix (the so-called *Tanner graph* [7]). Such an incidence graph is a bipartite graph with two sets of nodes: n symbol nodes, which correspond to columns, and r check nodes, which correspond to the rows of the parity-check matrix. The edges of the graph correspond to non-zero positions in $\mathbf{H}$. An example of such a graph is presented in Figure 5.2.

The quality of LDPC construction is defined by different characteristics: minimum Hamming distance d_0, the minimum length of the cycle in Tanner graph (*girth*) g_0, weights distributions of rows and columns in parity-check matrix.

Error Correcting Coding and Security for Data Networks   G. Kabatiansky, E. Krouk and S. Semenov
© 2005 John Wiley & Sons, Ltd   ISBN: 0-470-86754-X

$$
\left[\begin{array}{cccc|cccc|cccc|cccc}
1&&&&1&&&&1&&&&1&&&\\
&1&&&&1&&&&1&&&&1&&\\
&&1&&&&1&&&&1&&&&1&\\
&&&1&&&&1&&&&1&&&&1\\
\hline
1&&&&&&&1&&&1&&&1&&\\
&1&&&1&&&&&&&1&&&1&\\
&&1&&&1&&&1&&&&&&&1\\
&&&1&&&1&&&1&&&1&&&\\
\hline
1&&&&&&1&&&&&1&&&&1\\
&1&&&&&&1&1&&&&&&1&\\
&&1&&1&&&&&1&&&&1&&\\
&&&1&&1&&&&&1&&1&&&
\end{array}\right]
$$

Figure 5.1 Example of LDPC matrix

LDPC-codes with the equal number of ones in rows and columns are called *regular* [1,3], while the codes with unequal number of ones are called *irregular* [8]. Weights distributions can be defined by means of *generating functions* $\lambda(x)$ and $\rho(x)$ [8]:

$$
\begin{aligned}
\lambda(x) &= \sum_{i=2}^{d_v} \lambda_i x^{i-1} \\
\rho(x) &= \sum_{i=2}^{d_c} \rho_i x^{i-1}
\end{aligned}
\tag{5.3}
$$

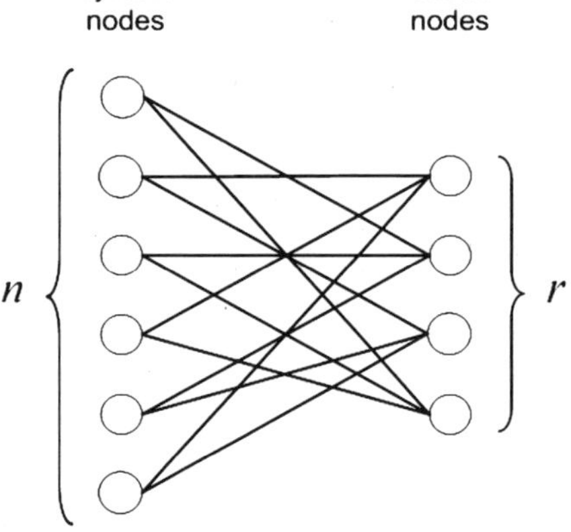

Figure 5.2 Tanner graph of LDPC code

where λ_i is the ratio of columns in $\mathbf{H}$ with weight i, ρ_i is the ratio of rows in $\mathbf{H}$ with weight i, and d_v, d_c are the maximum weights of columns and rows, correspondingly. For example in Figure 5.1, $\lambda(x) = x^2, \rho(x) = x^3$.

Let us define [8]

$$\sum_{i \geq 2} \lambda_i/i = \int_0^1 \lambda(x)dx$$

Then (5.1) and (5.2) can be written as

$$r = n\frac{\int_0^1 \rho(x)dx}{\int_0^1 \lambda(x)dx} \tag{5.4}$$

$$R \geq 1 - \frac{\int_0^1 \rho(x)dx}{\int_0^1 \lambda(x)dx} \tag{5.5}$$

Usually, constructing the good irregular codes uses probability methods, analysis of such codes is made asymptotically, while the regular constructions are based on the objects with known properties, and can be analysed using these properties. Suitably chosen, weight distributions of irregular codes can give gain especially with low SNRs (in AWGN channel), when the quality of the code is defined by its average characteristics. By increasing SNR, the distance properties of the code become determinative in the error probability, and here the gain can be obtained using regular constructions, which allow analysing their minimum distance, and constructing codes with better spectral properties.

5.1.2 General Results

LDPC-codes are linear codes with a parity-check matrix having a low number of non-zero elements. However, such a definition gives no specific methods for constructing the parity-check matrix. Moreover, the probability methods are often used when constructing LDPC-codes. Nevertheless, having no particular construction, there are results analysing LDPC codes as a separate class. These results are based on considering the ensembles of codes and estimating their characteristics as average on ensemble [5]. The LDPC-codes ensemble is defined primarily by weight distributions (γ, ρ) of parity-check matrix rows and columns (i.e., parameters defining the number of non-zero elements in parity-check matrix), the codelength n is considered as the parameter estimated in asymptotic for given weights.

In [1] R. Gallager showed that for ensemble of (n, γ, ρ)-codes the parameter δ exists such that with increasing of n almost all the codes in the ensemble have the minimum distance $n\delta$, where δ does not depend on n, and therefore, the minimum distance of most codes from the ensemble grows linearly with n.

Pinsker and Zyablov [6] showed that among LDPC-codes there are codes with decoding algorithms that can correct up to δn errors with complexity of $n \log n$.

In the works of Richardson, Urbanke and others [5,8,9] the analysis of LDPC decoding is shown, and the problem of weights distributions $\lambda(x)$ and $\rho(x)$ optimising for the ensemble of codes is considered. It is shown that for LDPC codes with belief propagation decoder (see Section 5.5) and several communication channels (BSC, BEC and AWGN are considered) the parameter σ exists, called threshold, and usually defined as some 'channel parameter' (i.e., noise dispersion for AWGN), for which the following statement holds. If the data is transmitted over a channel with parameter σ, then the error probability tends to zero as the number of decoding iterations increases, otherwise, the error probability is lower-bounded by positive constant. Optimisation of rows and columns weight allows the obtaining of codes with asymptotically better (higher) thresholds, which benefit with low SNRs. The procedure of calculating the thresholds for given $(\lambda(x), \rho(x))$ is called *density evolution* and can be found in [5,8,9].

However, this analysis is asymptotic, and gives no method for constructing codes for given weight distributions, on particular code lengths. Construction that allows the code to be obtained with given weights distributions $(\lambda(x), \rho(x))$, with maximisation of girth g_0 (to provide correctness of density evolution procedure), is suggested in [10] and is considered in Section 5.2.4.

In the following section we consider a review of some LDPC constructions showing good results in AWGN channels.

5.2 LDPC CONSTRUCTIONS

In this section the review and description of some LDPC constructions are presented. These LDPC constructions are known at the present time and give lower error probability in AWGN channel.

The following characteristics can be selected to compare different constructions:

1. minimum distance d_0;

2. girth g_0;

3. flexible selection of parameters — code length n and rate R;

4. error probability in AWGN channel.

The basis of our consideration is regular constructions. The properties of regular constructions are presented in Table. 5.1. Additional property 3 assures that the girth g_0 is greater than 4.

Table 5.1   Properties of regular LDPC codes

	LDPC Properties
1.	each row of parity-check matrix $\mathbf{H}$ contains exactly ρ ones
2.	each column of parity-check matrix $\mathbf{H}$ contains exactly γ ones
3.	the number of nonzero positions common to any two columns, no more than 1

We shall consider the following LDPC constructions: Euclidean-geometry codes (EG-LDPC), codes based on Reed-Solomon codes (RS-LDPC), generalised block codes based on Vandermonde matrix (W-LDPC), PEG construction. All these constructions except the latter are regular.

5.2.1 LDPC Codes Based on Finite Geometries

Error-correcting codes based on finite geometries — projective geometry *PG* or Euclidean geometry *EG*, were described in [11,12,13]. However, as low-density codes these codes were considered not so long ago, in [14,15], and comparing to other known regular LDPC constructions these codes based on finite geometries show good performance in an AWGN channel.

The drawback of these constructions is their inflexibility in parameter selection, which is the consequence of finite geometries properties. The decoding of LDPC codes based on finite geometries can be done by the majority-logic decoder [11,16,17,18] with hard or soft decisions. Besides, any common LDPC decoding procedure is suitable for decoding these codes (see also section 5.5).

Here we describe the constructions of Euclidean-geometry low-density codes and give their known basic characteristics and results.

Euclidean-geometry codes are defined as incidence system of geometry $EG(m,q)$, $q = p^s$ (a brief description of Euclidean geometries is given in Appendix 5.A). Since the number of ones in parity-check matrix of Euclidean-geometry code is small compared with matrix size, this code can be considered as LDPC-code.

The LDPC-code based on Euclidean geometry with parity-check matrix $\mathbf{H}_{EG}$, is constructed in the following way. The rows of parity-check matrix correspond to lines in Euclidean geometry, while the columns correspond to points in $EG(m,p^s)$. The elements of $\mathbf{H}_{EG}$ are defined from incidence vectors of Euclidean geometry lines (Figure 5.3):

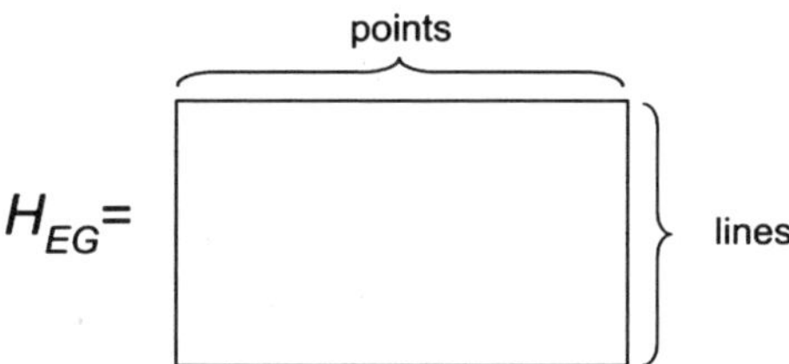

Figure 5.3 Parity-check matrix of *EG*-code

$$\mathbf{H}_{EG}(i,j) = \begin{cases} 1, & \text{if point } j \text{ lies on the line } i, \\ 0 & \text{otherwise} \end{cases} \tag{5.6}$$

The parity-check matrix $\mathbf{H}_{EG}$ has

$$n = q^m \tag{5.7}$$

columns and

$$r = q^{m-1}(q^m - 1)/(q - 1) \tag{5.8}$$

rows. Each column contains

$$\gamma = (q^m - 1)/(q - 1) \tag{5.9}$$

ones, each row contains

$$\rho = p^s \tag{5.10}$$

ones (all parameters follow from the properties of Euclidean geometry, see (5.A1)–(5.A5)).

Usually, the Euclidean-geometry codes with $p = 2$, not including zero point, are considered [14,15]. Such codes are sometimes called the EG-codes of type 0, and are cyclic codes [11,19] with the parameters

$$n = 2^{ms} - 1$$
$$r = (2^{(m-1)s} - 1)(2^{ms} - 1)/(2^s - 1)$$

The number of information symbols of such codes is estimated in [20].

Besides defining the parity-check matrix as in Figure 5.3, the EG-code with parity-check matrix transposed to (5.6) can be considered. Then the rows correspond to geometry points, and the columns to geometry lines. In both cases the geometry properties and (5.7)–(5.10) provide the required characteristics of regular LDPC-code, formulated in Table 5.1:

1. each row contains ρ ones;

2. each column contains γ ones;

3. any two columns have no more than one non-zero position in common (since there is only one line that can be passed through two points);

4. any two rows have no more than one non-zero position in common (since two lines intersect at no more than one point).

Additional property 4 means that the code with the parity-check matrix $\mathbf{H}_{EG}^T$ is also a regular LDPC-code satisfying all required parameters.

Since the columns of parity-check matrix (5.6) have no more than one non-zero position in common, any γ columns of parity-check matrix are linearly independent, and hence, can not form the zero syndrome. The minimum distance of the code with parity-check matrix (5.6) is estimated as

$$d_0 \geq \gamma + 1. \tag{5.11}$$

The girth of EG-LDPC codes is

$$g_0 = 6.$$

Some additional results on Euclidean-geometry codes are presented in Section 5.3.

5.2.2 Construction Based on Reed-Solomon Codes

One more scheme based on incidence system, was suggested in [21]. While the combinatorial objects — finite geometries, were used for preceding construction, here the words of Reed-Solomon code are used. The code parameters are selected in such a way as to ensure the properties of Table 5.1.

Let us consider the Reed-Solomon codes (RS-codes) over $GF(q)$. RS-codes are the MDS codes, and therefore, they have minimum distance

$$d_{RS} = n - k + 1,$$

where $n = q - 1$ is code length, k is the number of information symbols. Shortening RS-code on $n - \rho$ information symbols gives $(\rho, 2, \rho - 1)$-code C_ρ of length ρ with two information symbols. All non-zero code words of this code has weight ρ or $\rho - 1$. Let us choose codeword a of weight ρ and form the sub-space of the code

$$C_\rho^{(1)} = \{\beta a : \beta \in GF(q)\}$$

The set $C_\rho^{(1)}$ consists of q vectors, and each non-zero vector from $C_\rho^{(1)}$ has weight ρ, and each two vectors differ in all positions.

Let us construct the cosets $C_\rho^{(1)}, \ldots, C_\rho^{(q)}$ of space C_ρ, based on subspace $C_\rho^{(1)}$. Any two vectors in any coset differ in all positions, two vectors from different cosets differ in $\rho - 1$ or ρ positions.

The constructed cosets give the base for an incidence system, from which the parity-check matrix $\mathbf{H}_{RS-LDPC}$ of LDPC code is constructed. The parity-check matrix consists of γ horizontal stripes-submatrices, $\gamma \in \{1, \ldots, q\}$, and has a form

$$\mathbf{H}_{RS-LDPC} = \begin{bmatrix} \mathbf{H}^{(1)} \\ \mathbf{H}^{(2)} \\ \cdots \\ \mathbf{H}^{(r)} \end{bmatrix}, \tag{5.12}$$

where $\mathbf{H}^{(t)}$ is defined by coset $C_\rho^{(t)}$ as follows. Let $a_j^{(t)}(s)$ be the j-th symbol of s-th vector from the set $C_\rho^{(t)}, j \in \{0, \ldots, q-1\}, s \in \{1, \ldots, q\}$. Let $\mathbf{c}(\alpha)$ be the incidence vector of field element $\alpha \in GF(q)$, i.e.,

$$c_j(\alpha) = \begin{cases} 1, & \text{if } j = \alpha \\ 0, & \text{otherwise} \end{cases} \tag{5.13}$$

Then $\mathbf{H}^{(t)}$ can be defined as

$$\mathbf{H}^{(t)} = \begin{bmatrix} \mathbf{c}(\alpha_0^{(t)}(1)) & \mathbf{c}(\alpha_1^{(t)}(1)) & \cdots & \mathbf{c}(\alpha_{q-1}^{(t)}(1)) \\ \mathbf{c}(\alpha_0^{(t)}(2)) & \mathbf{c}(\alpha_1^{(t)}(2)) & \cdots & \mathbf{c}(\alpha_{q-1}^{(t)}(2)) \\ \cdots & \cdots & \cdots & \cdots \\ \mathbf{c}(\alpha_0^{(t)}(q)) & \mathbf{c}(\alpha_1^{(t)}(q)) & \cdots & \mathbf{c}(\alpha_{q-1}^{(t)}(q)) \end{bmatrix}, \tag{5.14}$$

or, in other words, it follows from (5.13) and (5.14) that elements of $\mathbf{H}^{(t)}$ are

$$\mathbf{H}^{(t)}(i, j) = \begin{cases} 1, & \text{if } \alpha_{\lfloor j/q \rfloor}^{(t)}(i) \equiv j \bmod q \\ 0, & \text{otherwise} \end{cases} \tag{5.15}$$

It follows from the properties of shortened RS-code C_ρ, cosets $C_\rho^{(t)}$ and construction method (5.12)–(5.15) that LDPC-code defined by the parity-check matrix $\mathbf{H}_{RS-LDPC}$ has the properties formulated in the Table 5.1.

In [21] the estimate of minimum distance is given

$$d_0 \geq \begin{cases} \gamma + 1, & \text{if } \gamma \text{ is odd} \\ \gamma + 2, & \text{if } \gamma \text{ is even} \end{cases}, \tag{5.16}$$

which is the same estimate as for finite-geometries constructions relative to the weight γ of columns in parity-check matrix. This estimation is also based on the number of orthogonal parity-checks [11,19]. In practice the true minimum distance of RS-LDPC code can be much higher than the estimate (5.16).

The parameters of RS-LDPC codes are rather flexible. However, there are no known expressions for the rates of RS-LDPC codes.

5.2.3 Gilbert Codes and Their Generalisations

The Gilbert codes are low-density parity-check codes (LDPC-codes). They were suggested by Gilbert [22] in order to correct error bursts. The burst-correcting capability of these codes was considered in [23,24,25], where the estimation of maximal correctable burst length was obtained. In Section 5.4 the Gilbert codes capable of correcting error bursts are considered, and the exact expressions of burst-correcting capability of these codes are presented. In this section we consider the Gilbert codes as an example of regular LDPC structure.

A Gilbert code is defined by its parity-check matrix $\mathbf{H}_l$,

$$\mathbf{H}_l = \begin{bmatrix} \mathbf{I}_m & \mathbf{I}_m & \mathbf{I}_m & \cdots & \mathbf{I}_m \\ \mathbf{I}_m & \mathbf{C} & \mathbf{C}^2 & \cdots & \mathbf{C}^{l-1} \end{bmatrix}, \tag{5.17}$$

where $\mathbf{I}_m$ is $(m \times m)$-identity matrix, $\mathbf{C}$ is the $(m \times m)$-matrix of cyclic permutation:

$$\mathbf{C} = \begin{bmatrix} 0 & 0 & 0 & \cdots & 0 & 1 \\ 1 & 0 & 0 & \cdots & 0 & 0 \\ 0 & 1 & 0 & \cdots & 0 & 0 \\ \cdots & \cdots & \cdots & \cdots & \cdots & \cdots \\ 0 & 0 & 0 & \cdots & 1 & 0 \end{bmatrix} \tag{5.18}$$

where $l \leq m$ [24]. Clearly, a Gilbert code is $(2, l)$ regular LDPC-code with the $\gamma = 2$ ones in column and $\rho = l$ ones in row. The code length is $n = ml$, the number of redundant symbols is estimated as $r = 2m - 1$ [24].

The minimum distance d_0 of Gilbert code is connected with girth g_0:

$$d_0 = g_0/2 \tag{5.19}$$

The spectral and distance properties of Gilbert codes are estimated using the following statements.

Theorem 5.1 Let $\mathbf{H}_l$ be the matrix of (5.17), $Z_\ell = \{0, 1, \ldots, \ell - 1\}$ be the set of residues modulo ℓ. Then, if the sets of integers $\{a_i\}$, $\{b_i\}$ exist such that equation

$$\sum_{i=0}^{\omega-1} (-1)^i (a_i - b_i) = 0 \bmod m$$

holds, where

$$a_i \in Z_\ell, \qquad b_i \in Z_\ell,$$
$$a_0 \neq b_0, \qquad a_{\omega-1} \neq b_{\omega-1},$$
$$a_i \neq a_{i-1}, \qquad b_i \neq b_{i-1},$$

then the code with parity-check matrix $\mathbf{H}_l$ contains the codeword of weight 2ω.

Theorem 5.2 The minimum distance and girth of Gilbert code with $\ell \geq 3$ are

$$d_0 = 4,$$
$$g_0 = 8.$$

Following from Theorem 5.2, the Gilbert code has a very low minimum distance, and hence, cannot be used to correct independent errors. However, the generalisations of Gilbert codes can be defined as follows. Consider the parity-check matrix

$$\mathbf{H}_{s,l} = \begin{bmatrix} \mathbf{I}_m & \mathbf{I}_m & \mathbf{I}_m & \cdots & \mathbf{I}_m \\ \mathbf{C}^0 & \mathbf{C}^1 & \mathbf{C}^2 & \cdots & \mathbf{C}^{l-1} \\ \mathbf{C}^{i_0^{(3)}} & \mathbf{C}^{i_1^{(3)}} & \mathbf{C}^{i_2^{(3)}} & \cdots & \mathbf{C}^{i_{l-1}^{(3)}} \\ \cdots & \cdots & \cdots & \cdots & \cdots \\ \mathbf{C}^{i_0^{(s)}} & \mathbf{C}^{i_1^{(s)}} & \mathbf{C}^{i_2^{(s)}} & \cdots & \mathbf{C}^{i_{l-1}^{(s)}} \end{bmatrix}, \tag{5.20}$$

where $\mathbf{H}_{s,l}$ is $s \times l$-matrix, $i_j^{(k)} \in \{0, \ldots, m\}$ is the degree of the cyclic permutation matrix $\mathbf{C}$ in j-th block of k-th stripe. Since one of parameters of LDPC-code is the girth, the numbers $i_j^{(k)}$ of any stripe k should not repeat. Then the set $\{i_j^{(k)} : j = 0, \ldots, \ell - 1\}$ is defined by the permutation of different residues modulo m.

Construction (5.20) can be the basis for defining regular LDPC codes. Notice that not only the cyclic permutation matrix $\mathbf{C}$ can be used as a block, but any generator of the cyclic group of order no less than l as well.

As an example of such construction consider $\mathbf{H}_{s,l}$, where the degrees of cyclic permutation matrix can be selected corresponding to Vandermonde matrix [11,19,23].

In this case we get (γ, ρ) LDPC-code with the parity-check matrix

$$\mathbf{H}_V = \begin{bmatrix} \mathbf{I}_m & \mathbf{I}_m & \cdots & \mathbf{I}_m \\ \mathbf{I}_m & \mathbf{C} & \cdots & \mathbf{C}^{\rho-1} \\ \cdots & \cdots & \cdots & \cdots \\ \mathbf{I}_m & \mathbf{C}^{\gamma-1} & \cdots & \mathbf{C}^{(\gamma-1)(\rho-1)} \end{bmatrix}. \tag{5.21}$$

With m prime there are no columns having more than one non-zero position in common in this kind of matrix, and the minimum distance of such code can be estimated as

$$\gamma + 1 \leq d_0 \leq 2m. \tag{5.22}$$

The upper bound of minimum distance does not depend on parameters γ and ρ. If we wish to construct the code with this minimum distance, it means that in the given (γ, ρ)-ensemble we provide minimum distance increasing linearly with code length, increasing only m. This task is perspective.

5.2.4 PEG Construction

In [10] the empiric procedure is suggested to construct the Tanner graph maximising the girth g_0.

PEG construction is based on pre-calculated weight distributions of symbol and check nodes in a Tanner graph, for example, using a 'density evolution' procedure (see section 5.1.2). However, the algorithm can use any other distribution as well, including regular.

The algorithm of constructing a graph with given $\lambda(x)$ and $\rho(x)$ is based on iterative edge-by-edge steps, maximising local girth for given nodes. The result may be in regular or irregular form, depending on weights distribution. In [10] the lower bounds on minimum distance and girth for PEG codes are obtained.

In Figure 5.4 the procedure of constructing the Tanner graph is presented [10]. Let the Tanner graph consist of n symbol nodes v_i, $1 \leq i \leq n$, and r check nodes c_j, $1 \leq j \leq r$. Let

for $j = 0$ to $n - 1$ **do**
begin
    for $k = 0$ to $d_{s_j} - 1$ **do**
    begin
      if $k = 0$
          $E_{s_j}^0 \leftarrow \text{edge}(c_i, s_j)$, where $E_{s_j}^0$ is the first edge incident to s_j and c_i is one check node such that it has the lowest check degree under the current graph setting $E_{s_0} \cup E_{s_1} \cup \ldots \cup E_{s_{j-1}}$.
      else
          expanding a tree from symbol s_j up to depth ℓ under the current graph setting such that $\overline{N_{s_j}^\ell} \neq \varnothing$ but $\overline{N_{s_j}^{\ell+1}} = \varnothing$, or the cardinality of $\overline{N_{s_j}^\ell}$ stops increasing but is less than m, then $E_{s_j}^k \leftarrow \text{edge}(c_i, s_j)$, where $E_{s_j}^k$ is the k-th edge incident to s_j and c_i is one check node picked from the set $\overline{N_{s_j}^\ell}$ having the lowest check-node degree.
    end
end

Figure 5.4 PEG Construction

d_{v_i}, d_{c_j} be the degree of symbol node v_i and check node c_j, respectively, where node degree means the number of edges incident to it (this value is defined by weights distributions $\lambda(x)$ and $\rho(x)$), E_{v_i} is the set of edges incident to symbol node v_i, $N_{v_i}^{\ell}$ is the set of check nodes that can be reached from symbol node v_i by ℓ edges or less, $\overline{N_{v_i}^{\ell}}$ is the complementary set of $N_{v_i}^{\ell}$, in other words, $N_{v_i}^{\ell} \cup \overline{N_{v_i}^{\ell}} = V_c$, where V_c is the set of all check nodes in the graph.

In [10] the estimations of girth g_0 and minimum distance d_0 of PEG codes are presented. Let d_v and d_c be the maximal weights of symbol and check nodes in Tanner graph, respectively. Then the girth is lower bounded as

$$g_0 \geq 2(\lfloor t \rfloor + 2),$$

where

$$t = \frac{\log\left(rd_c - \dfrac{rd_c}{d_v} - r + 1\right)}{\log((d_v - 1)(d_c - 1))} - 1$$

Let us consider Tanner graphs with regular symbol nodes, having the constant degree d_v, and let the graph have the girth g_0. Then the minimum distance of the code defined by such a graph, is estimated as

$$d_0 \geq \begin{cases} 1 + \dfrac{d_v\left((d_v - 1)^{\lfloor(g_0-2)/4\rfloor} - 1\right)}{d_v - 2} & \text{if } g_0/2 \text{ is odd} \\[2em] 1 + \dfrac{d_v\left((d_v - 1)^{\lfloor(g_0-2)/4\rfloor} - 1\right)}{d_v - 2} + (d_v - 1)^{\lfloor(g_0-2)/4\rfloor} & \text{if } g_0/2 \text{ is even} \end{cases}$$

In PEG construction the main attention is given to absence of short cycles in the Tanner graph. However, in practice it is not clear how the presence of short cycles can degrade the decoder's performance. In [26] it is considered not to be a problem of presence or absence of short cycles, but how these cycles are connected to each other. The idea of this work is that if there are many edges leading from the nodes that form short cycle, then the decoder can work well even in the presence of short cycles. The parameter ACE is suggested, and the procedure of constructing the code, maximising ACE is considered in [26]. As the result, the irregular codes with improved performance with high SNRs were obtained.

5.3   ESTIMATING THE MINIMUM DISTANCE OF EG-LDPC CODES

In this section the results of minimum distance analysis for Euclidean-geometry codes, considered in Section 5.2.2, and their shortenings are presented. In [14,27] some methods for shortening EG-LDPC codes are described. By means of simulation it is shown that the error probability using such shortenings can be decreased, but there are no analytical results of shortened codes properties.

We consider the Euclidean-geometry codes with parity-check matrix transposed to (5.6):

$$\mathbf{H}_{EG}(i,\, j) = \begin{cases} 1, & \text{if point } i \text{ lies on the line } j, \\ 0, & \text{otherwise} \end{cases} \qquad (5.23)$$

Let us consider the Euclidean-geometry space, i.e. the geometry $EG(3, q)$, $q = p^s$. Such codes have length

$$n = q^2(q^3 - 1)/(q - 1) \qquad (5.24)$$

and their parity-check matrix H consists of

$$r_H = q^3 \qquad (5.25)$$

rows. Note that $\mathbf{H}$ is not necessarily full-rank, so r_H can be used only as an upper bound on the number of check symbols.

Consider the line in such Euclidean geometry. The line contains q points. If we take the point not on the line, there is the only line containing this point and parallel to the initial line. Every such line also contains q points. Since there are q^m points in geometry, there are q^{m-1} lines in total, parallel to each other. We shall call such a set of lines a *parallel class*. For the case $m = 3$ the geometry contains $q^2(q^3 - 1)/(q - 1)$ lines that can be divided into $(q^3 - 1)/(q - 1)$ parallel classes, each class containing q^2 lines. For a Euclidean plane, i.e., the case $m = 2$, $q(q + 1)$ lines can be divided into $q + 1$ parallel classes with q lines in each class. Each point from the plane is presented in parallel class exactly once. The example of parallel classes $P_1, \ldots, P_5$ from the plane $EG(2, 2^2)$ is presented in Figure 5.5.

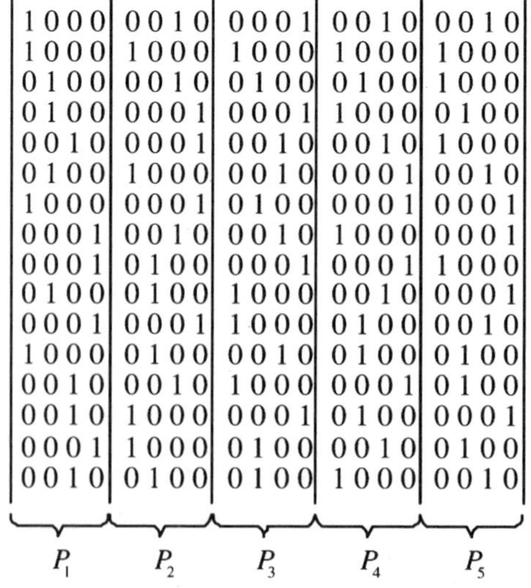

```
| 1 0 0 0 | 0 0 1 0 | 0 0 0 1 | 0 0 1 0 | 0 0 1 0 |
| 1 0 0 0 | 1 0 0 0 | 1 0 0 0 | 1 0 0 0 | 1 0 0 0 |
| 0 1 0 0 | 0 0 1 0 | 0 1 0 0 | 0 1 0 0 | 1 0 0 0 |
| 0 1 0 0 | 0 0 0 1 | 0 0 0 1 | 1 0 0 0 | 0 1 0 0 |
| 0 0 1 0 | 0 0 0 1 | 0 0 1 0 | 0 0 1 0 | 1 0 0 0 |
| 0 1 0 0 | 1 0 0 0 | 0 0 1 0 | 0 0 0 1 | 0 0 1 0 |
| 1 0 0 0 | 0 0 0 1 | 0 1 0 0 | 0 0 0 1 | 0 0 0 1 |
| 0 0 0 1 | 0 0 1 0 | 0 0 1 0 | 1 0 0 0 | 0 0 0 1 |
| 0 0 0 1 | 0 1 0 0 | 0 0 0 1 | 0 0 0 1 | 1 0 0 0 |
| 0 1 0 0 | 0 1 0 0 | 1 0 0 0 | 0 0 1 0 | 0 0 0 1 |
| 0 0 0 1 | 0 0 0 1 | 1 0 0 0 | 0 1 0 0 | 0 0 1 0 |
| 1 0 0 0 | 0 1 0 0 | 0 0 1 0 | 0 1 0 0 | 0 1 0 0 |
| 0 0 1 0 | 0 0 1 0 | 1 0 0 0 | 0 0 0 1 | 0 1 0 0 |
| 0 0 1 0 | 1 0 0 0 | 0 0 0 1 | 0 1 0 0 | 0 0 0 1 |
| 0 0 0 1 | 1 0 0 0 | 0 1 0 0 | 0 0 1 0 | 0 1 0 0 |
| 0 0 1 0 | 0 1 0 0 | 0 1 0 0 | 1 0 0 0 | 0 0 1 0 |
     P_1       P_2       P_3       P_4       P_5
```

Figure 5.5 Parallel classes of the plane $EG(2, 4)$

Shortening the parity-check matrix of EG-code on the columns correspondent to the parallel classes, we obtain the code with the number of ones in each row and each column remaining the same (regular code), since each geometry point can be contained in a parallel class only once. The minimum distance of such a shortened code is no less than that of the original code, and the number of ones in the row becomes less. This can improve the functionality of an iterative decoder.

The described shortening method allows optimising the code parameters to required code lengths by selecting the geometry parameters and the number of shortened classes.

For the Euclidean-geometry code with the parity-check matrix (5.23) there are the following statements concerning minimum distance of this code.

Let us consider the case $p = 2$ and $p \neq 2$ separately. Let $p \neq 2$. Next, let

$$W(x, y) = \sum_{i=0}^{n} A_i x^{n-i} y^i \tag{5.26}$$

be the weight function of the code [32], where A_i is the number of code words of weight i presented in the code, x denotes the number of zeros, y the number of ones.

Then the following statements hold true.

Theorem 5.3 If the parity-check matrix (5.23) of Euclidean-geometric code with $m = 2, q = p^s, p \neq 2$, is full-rank, then the coefficient A_i of the weight function (5.26) can be calculated as

$$A_i = \begin{cases} \binom{q+1}{i/q}, & i|2q \\ 0, & \text{otherwise} \end{cases} \tag{5.27}$$

Corollary 5.1 The minimum distance of the code of Theorem 5.3 is

$$d_0 = 2q \tag{5.28}$$

For every particular code the rank of the parity-check matrix can be calculated. Tests show that for considered geometries with $m = 2, p \neq 2$ the parity-check matrices are indeed full-rank. Estimation (5.28) significantly differs from (5.11), obtained using the number of orthogonal parity-checks.

Now consider the case $p = 2$. For the field of characteristics 2 the parity-check matrix (5.23) contains linearly dependent rows. For these codes the following statements can be formulated.

Theorem 5.4 If the code with parity-check matrix (5.23) with $m = 2, p = 2$, contains the word of weight $q + 1$, then the non-zero positions of this word correspond to $q + 1$ lines from different parallel classes.

Corollary 5.1 Shortening EG-code (5.23) with $m = 2, p = 2$ on the lines from any parallel class, gives the code with minimal distance

$$d_0 \geq q + 2 \tag{5.29}$$

Using results of this section, the EG-codes with known distance properties can be constructed, as well as shortening methods for EG-codes can be suggested. In Section 5.6 some of these methods, as well as the simulation results in AWGN channel, are considered.

5.4 BURST-ERROR-CORRECTING LDPC-CODES

In this section the class of binary codes capable of correcting (single) error bursts is considered. Gilbert proposed this class of codes in 1960 [22]. Since this time many works have been published estimating the correcting capability of these codes.

We consider the definition of Gilbert codes according to (5.17), as they were presented in [24,28,29]:

$$\mathbf{H}_l = \begin{bmatrix} \mathbf{I}_m & \mathbf{I}_m & \mathbf{I}_m & \cdots & \mathbf{I}_m \\ \mathbf{I}_m & \mathbf{C} & \mathbf{C}^2 & \cdots & \mathbf{C}^{l-1} \end{bmatrix}$$

In [24] the estimation of maximum correctable burst length for code with parity-check matrix (5.17) is presented, which is

$$b \le \min_{\gamma \in \{0, \ell-2\}} \max\{\gamma - 1, m - \gamma - 1\}. \tag{5.30}$$

The estimation (5.30) is not exact, giving the lower bound on the maximal length of correctable burst. The exactness of this estimation decreases with the growth of ℓ. In [28] the method of calculation the exact burst-correcting capability of such kind of codes is obtained (Theorem 5.5).

Theorem 5.5 Code with parity-check matrix $\mathbf{H}_l$, given by (5.17), can correct single error bursts of length b_ℓ, where b_ℓ is calculated by first complying condition:

1. $b_3 = m - 1$, m is odd

2. If $\ell > \lceil m/2 \rceil + 1$, then

$$\begin{cases} b_\ell = m - \lceil m/2 \rceil + 1, & m \text{ odd} \\ b_\ell = m/2 - 1, & m \text{ even} \end{cases}$$

3. If $\ell \le \lceil m/2 \rceil + 1$, then

$$b_\ell = m - \ell + 1, \quad m \,\vdots\, (\ell - 1)$$

$$b_\ell = m - \ell + 1, \quad \exists k > 0 : (m - \ell + 3 - k \cdot (\ell - 1)) \,\vdots\, (\ell - 2)$$

$$b_\ell = m - \ell + 2, \quad \exists k > 0 : (m - k \cdot (\ell - 3)) \,\vdots\, (\ell - 2)$$

$$b_\ell = m - \ell + 2, \quad \exists k > 0 : (m - k \cdot (\ell - 2)) \,\vdots\, (\ell - 1)$$

$$b_\ell = m - \ell + 2, \quad \exists k > 0 : (m - k \cdot (\ell - 1)) \,\vdots\, (\ell - 2)$$

4. If there are no complying conditions, then:

$$b_\ell = b_{\ell-1}$$

In [30] the generalised Gilbert codes were considered, i.e. codes with parity-check matrix (5.20):

$$\mathbf{H}_{s,l} = \begin{bmatrix} \mathbf{I}_m & \mathbf{I}_m & \mathbf{I}_m & \cdots & \mathbf{I}_m \\ \mathbf{C}^0 & \mathbf{C}^1 & \mathbf{C}^2 & \cdots & \mathbf{C}^{l-1} \\ \mathbf{C}^{i_0^{(3)}} & \mathbf{C}^{i_1^{(3)}} & \mathbf{C}^{i_2^{(3)}} & \cdots & \mathbf{C}^{i_{l-1}^{(3)}} \\ \cdots & \cdots & \cdots & \cdots & \cdots \\ \mathbf{C}^{i_0^{(s)}} & \mathbf{C}^{i_1^{(s)}} & \mathbf{C}^{i_2^{(s)}} & \cdots & \mathbf{C}^{i_{l-1}^{(s)}} \end{bmatrix}$$

It was shown that for $s = 3$ there are codes that can correct bursts of maximal possible length for this construction.

Denote as $P = \{i_1, i_2, \ldots, i_\ell\}$ the set of permutations of integers from 0 to $\ell - 1$. Define the matrix

$$\mathbf{H}_{3,l} = \begin{bmatrix} \mathbf{I}_m & \mathbf{I}_m & \mathbf{I}_m \\ \mathbf{I}_m & \mathbf{C} & \mathbf{C}^{l-1} \\ \mathbf{C}^{i_0} & \mathbf{C}^{i_1} & \mathbf{C}^{i_{l-1}} \end{bmatrix} \tag{5.31}$$

Codes with parity-check matrix (5.31) have

$$r = 3m - 2$$

check symbols, and length

$$n = \ell \cdot m$$

Such codes can correct single bursts of length larger than codes given by parity-check matrix (5.17). The correcting capability of these codes, clearly, depends on selecting the permutation P. In some cases codes defined by (5.31) can correct bursts of length $m - 1$, with $\ell \to m$. This is the maximal length correctable by these codes.

Denote as P' the set of permutations such that there are no elements i_k and i_{k+1} in permutation P' (i.e. neighbor elements) for which equation $i_{k+1} - i_k = 1$ is hold. In other words, none of the powers of cyclic permutation matrix $\mathbf{C}$ from the third stripe of (5.31) is more than preceding power exactly on one. Then the following theorem can be stated.

Theorem 5.6 If m is prime and $P = P'$ is permutation described above, then the code with parity-check matrix (5.31) can correct single error bursts of length $m - 1$ and less.

The Gilbert codes and their generalisations can be decoded with the help of a very effective algorithm presented in [24]. The general decoding procedures for LDPC codes are presented in the next section.

5.5 DECODING SCHEMES OF LDPC CODES

Decoding algorithms for low-density parity-check (LDPC) codes were first introduced by Gallager in 1963 [1] both for hard and soft decision cases (bit-flip and belief propagation algorithms, respectively). The soft-decision belief propagation iterative algorithm can operate with both probabilities and log-likelihood ratios and it gives good results in an AWGN channel as is shown in [1,3,31].

In this section we review the LDPC decoding schemes and suggest the approach of a multi-threshold decoding algorithm based on reliabilities that can reduce the decoding complexity and increase the decoding speed.

5.5.1 Decoding in Discrete Cannel (Bit-Flip Decoding)

The idea of decoding in discrete channel (hard-decision decoding) is that for some received symbol c_i any other symbol can be no more than in one parity-check of symbol c_i, because of parity-check matrix sparsity and absence of short cycles in Tanner graph. In other words, the set of parity-checks is orthogonal on symbol c_i [11,18,19]. Then the columns of parity-check matrix have less non-zero positions in common, and hence, unsatisfied parity-check (syndrome position) more probably consists of one erroneous symbol, than of sum of three or more. This leads to the following decoding procedure.

1. Calculate syndrome from the received word on zero-th iteration, or from the result of preceding iteration. If the syndrome is zero, or the maximum number of iterations is reached, the procedure is finished.

2. Calculate the number ℓ_i of unsatisfied parity-checks for each symbol c_i.

3. Flip the symbol or symbols with the largest ℓ_i.

4. Go to step 1.

The scheme of decoding the algorithm is presented in Figure 5.6. The algorithm is processed iteratively, until the codeword is obtained, or the maximum number of iterations is reached.

The complexity of described decoding procedure is very low, because for every bit the syndrome update is needed, the update complexity consists of few XOR complexities. This decoder also has simple implementation.

5.5.2 Decoding in Soft Channel (Belief Propagation Decoding)

For decoding in soft channels (soft-decision decoding) the task is to maximise the conditional probability $P(\mathbf{C}_m|\mathbf{Y})$, where $\mathbf{C}_m$ is codeword, and $\mathbf{Y}$ is the block of symbols observed on the channel output.

In the case of LDPC codes, the decoder gives symbol-by-symbol decisions, and in fact, to make the decision on a particular symbol, calculates the likelihood ratio

$$\mathrm{LR}(c_i) = \frac{P(c_i = 1|\mathbf{Y})}{P(c_i = 0|\mathbf{Y})} \qquad (5.32)$$

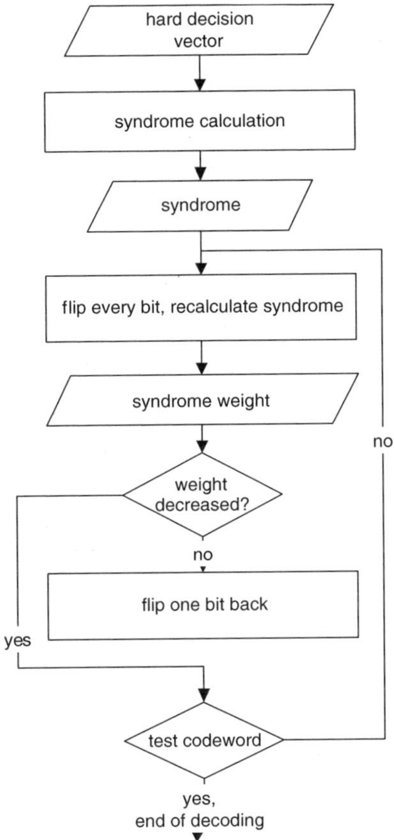

Figure 5.6 Hard decoding of LDPC-code

where c_i is code bit in position i. It is more convenient to use the log-likelihood ratio instead of likelihood ratio (5.32):

$$\mathrm{LLR}(c_i) = \log \frac{P(c_i = 1|\mathbf{Y})}{P(c_i = 0|\mathbf{Y})} \tag{5.33}$$

Values of LR or LLR are often called the symbol reliability.

Algorithm 'belief propagation,' the standard procedure for decoding LDPC-codes, suggested by Gallager [1,33,34], can be described as follows. The LLRs of symbols of the received 'soft' word is set to correspondent symbol nodes of Tanner graph. Then the decoder processes iterations, each consisting of two stages. During the first, 'vertical' stage, each i-th symbol node, $1 \leq i \leq n$, sends to each incident check node j, $1 \leq j \leq \gamma$, some value, called 'message,' that depends on all values received by i-th symbol node from all incident check nodes besides j-th.

The second, 'horizontal', stage operates similarly, the only difference is that the messages are calculated and sent from check nodes to symbol nodes. The one iteration of decoder is shown in Figure 5.7. Here the message flow between nodes v_1 and c_2 during one iteration is shown. The function f^c denotes message calculation by check node, f^v by symbol node.

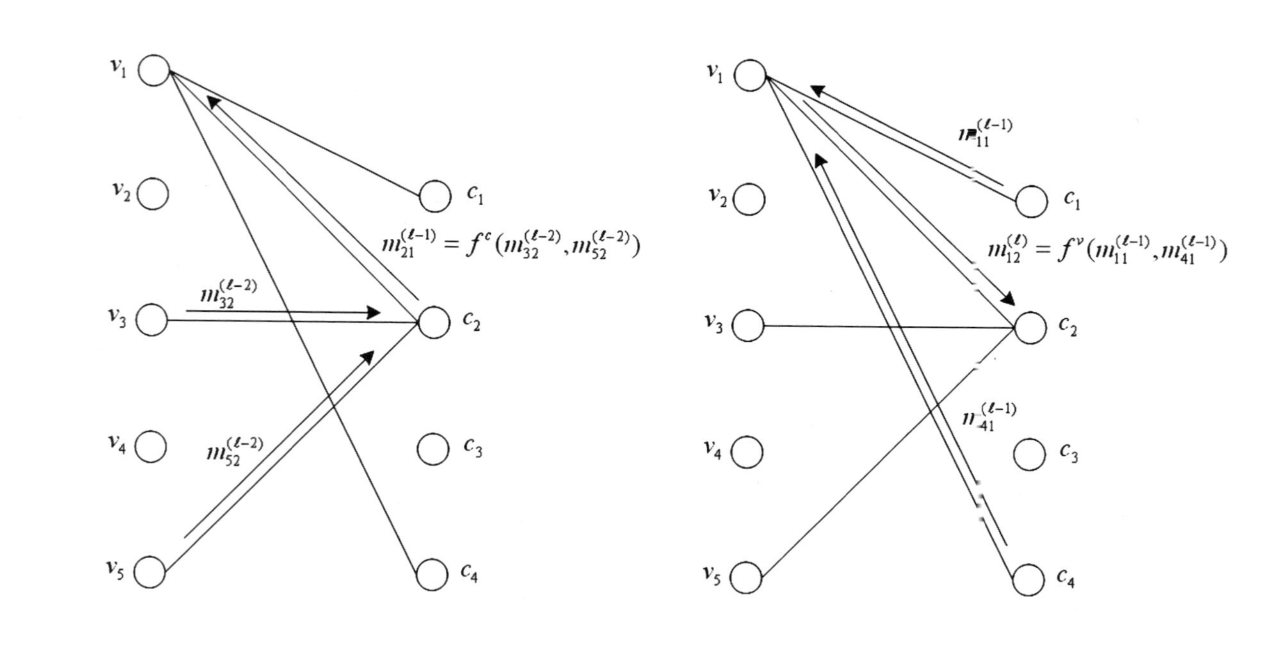

Figure 5.7 Iteration of LDPC-decoder

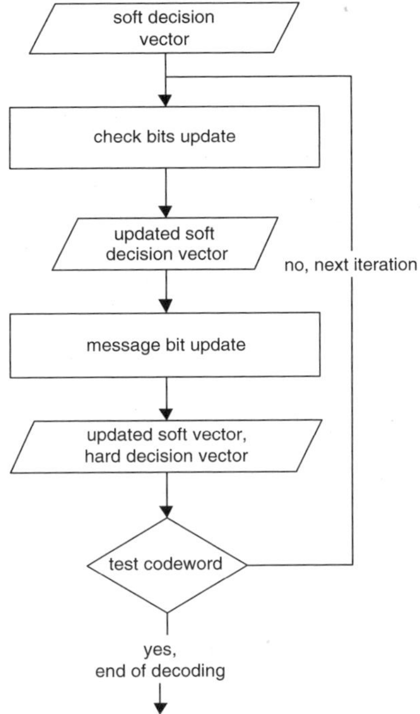

Figure 5.8   Soft decoding of LDPC-code

After each iteration the algorithm makes a hard decision on every symbol, corresponding to the sign of the current message in symbol node. If an obtained hard vector is a code word, or the maximum number of iterations is reached, the algorithm stops. The scheme of the algorithm is presented in Figure 5.8.

The maximal number of decoding iterations is selected depending on code length, required error probability and decoding complexity requirements. It is shown in [6] that $\log n$ iterations can be enough for decoding that gives complexity $n \log n$.

In practice, the maximal number of iterations performed by the decoder is a trade-off between the decoding speed and the error probability requirements. Note that different LDPC constructions may have different *convergence speed*. The convergence speed of on iterative decoder for a given construction is one of the parameters for selecting LDPC-code in a particular communication system. This problem is considered in more detail in section 5.6.

The complexity of belief propagation decoding is higher than the complexity of hard bit-flip decoding due to sophisticated probabilities update and float number operations. If the belief propagation decoder works in log-likelihood domain it requires LOG() calculation or making a lookup table to avoid exact LOG() calculation.

5.5.3 Multi-Threshold Decoder

The effective and powerful method of traditional LDPC decoding (belief propagation) can be simplified to improve the complexity effectiveness with low degradation in error-correcting performance. This section describes some fast methods of decoding LDPC codes.

The most known fast LDPC decoders are 'min-sum' algorithm [35] and UMP algorithm [31].

'Min-sum' algorithm is the simplification of 'sum-product' (belief propagation) LDPC decoding with the following rule: the Gallager function that is used for updating the probabilities in parity nodes is approximated by the minimum value of bits in the parity check line. It significantly accelerates the algorithm. The same acceleration is used in the UMP algorithm, which uses the minimum principle for likelihood recalculating as well, but in this case the minimum should be found among positive values, which is computationally faster.

Multi-threshold decoder (MT-decoder) has the same advantages as UMP in functioning speed, but improves the decoding quality. Here we describe the procedure of MT-decoding.

The multi-threshold decoder is an iterative decoder that uses a soft input vector to produce hard decisions and a reliabilities vector describing the absolute likelihood value of every bit.

At each iteration the decoder computes the reliability of every parity-check equation using the bit reliabilities; then for every bit and every parity-check, if the parity-check equation fails, the corresponding bit reliability is decreased, otherwise bit reliability is increased. When all equations are checked the hard decision is made using the following rule: if the updated bit reliability is greater than some threshold value, the bit and its reliability stay unchanged, otherwise the bit is flipped (inverted) and its reliability is changed as a function of updated reliability and threshold value. If the updated reliability is less then 0 but greater than threshold value then the new bit reliability is set to 0 (Figure 5.9).

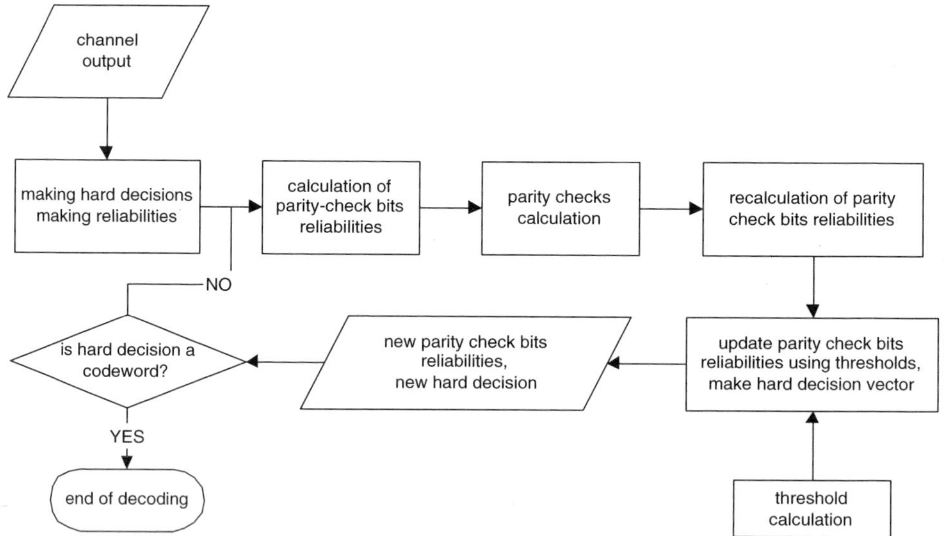

Figure 5.9 Multi-threshold decoding scheme

The process is repeated until the codeword is obtained or the decoder reaches the maximum number of iterations.

The idea of described threshold changing is that high threshold values at first iterations lead to absence of error propagation, and if the decoder corrects even a small number of errors, the decoding at following iterations becomes significantly easier. Experiments show that the decoding bit-error rate using a multi-threshold decoding scheme is close to MAP decoding.

The detailed description of the MT-decoder is as follows:

Let $N(m)$ be the set of codeword positions used in m-th parity check: $N(m) = \{n : \mathbf{H}_{m,n} = 1\}$, where $\mathbf{H}$ is the parity check matrix. Lets $M(n)$ be the set of parity checks that includes n-th codeword position: $M(n) = \{m : \mathbf{H}_{m,n} = 1\}$.

- *Initialisation.* For every element of the received vector Y_i the hard decision X_i and reliability R_i is computed. The reliability R_i is the absolute value of Y_i. For every $m \in M(n)$ $Y_{mn} = R_n$, $X_{mn} = X_i$

- *Step 1.* For each n and each $m \in M(n)$ calculate check sums:

$$S_{mn} = X_i \oplus \sum_{n' \in N(m) \backslash m} X_{mn}$$

and identify

$$Y_{mn\ \min} = \min_{n' \in N(m) \backslash m} \{Y_{mn'}\}$$

- *Step 2.* For each n and each $m \in N(m)$ calculate

$$Z_{mn} = R_n + \sum_{m' \in M(n) \backslash n} (-1)^{S_{m'n}} Y_{m'n\ \min}$$

- *Step 3.* For each n

$$Y_{mn} = \begin{cases} Z_{mn}, & Z_{mn} > 0 \\ -Z_{mn}, & Z_{mn} < \text{threshold} \\ 0, & \text{else} \end{cases}$$

$$X_{mn} = \begin{cases} X_{mn}, & Z_{mn} > \text{threshold} \\ 1 - X_{mn}, & Z_{mn} < \text{threshold} \end{cases}$$

$$Z_n = R_n + \sum_{m \in M(n) \backslash n} (-1)^{S_{mn}} Y_{mn\ \min}$$

$$X_i = \begin{cases} X_i, & Z_n > 0 \\ 1 - X_i, & Z_n < 0 \end{cases}$$

- Repeat steps 1–3 until $X \cdot H \neq 0$

The bit error rate curves show that the multi-threshold decoder gives a better result than known fast-decoding schemes, the multi-threshold decoder gives 0.5 dB gain compared to fast decoding having practically the same complexity, the thresholds used are:

- Minimum Z_{mn} at first iterations

- 0 at final iterations

These thresholds work with every LDPC code in every situation, but for a predefined code and situation in channel the thresholds can be selected more precisely. Experiments show

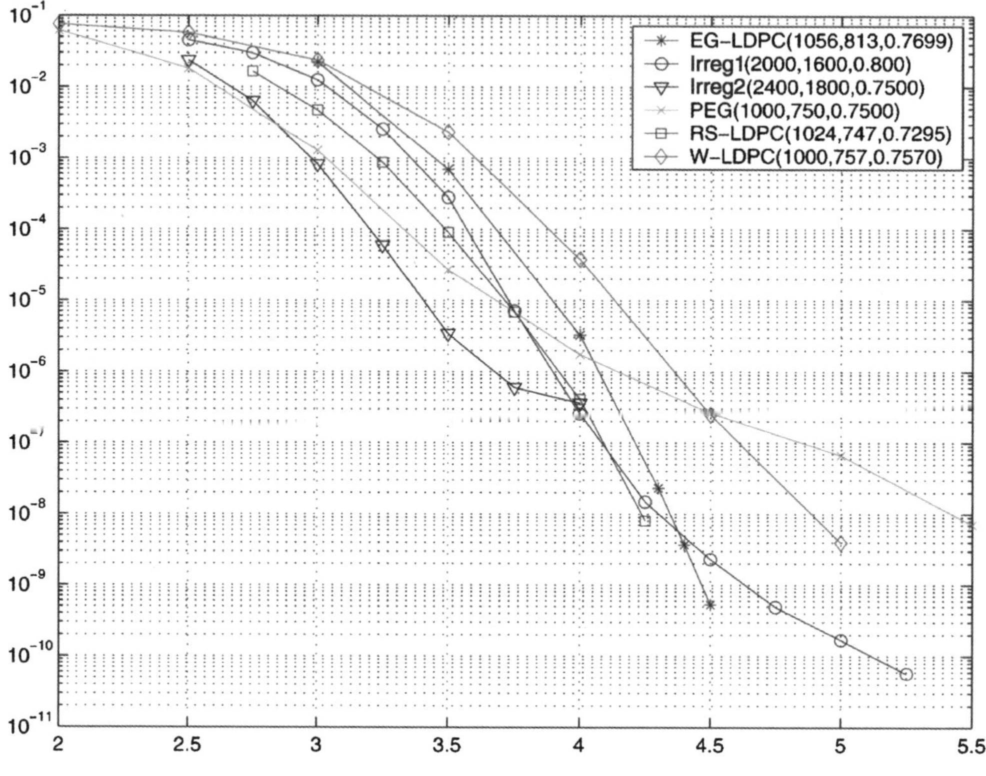

Figure 5.10   LDPC Codes performance

that threshold tuning and optimisation can give more BER gain, but finding the exact dependency of thresholds, code structure, and channel situation remains an open problem.

In Figure 5.10 the decoding quality of different LDPC constructions is presented, where EG-LDPC is an Euclidean-geometry code [15], Irreg1 and Irreg2 are irregular LDPC codes, PEG is a PEG construction [10], RS-LDPC is a construction based on Reed-Solomon codes [21], and W-LDPC is a LDPC-code based on Vandermonde matrix. LDPC constructions were described in section 5.2.

As can be seen from the plots, fast algorithms provide different error-correcting performances on different constructions under the same conditions. Best quality results in AWGN channel using fast decoding procedures with low BER (10^{-10} and below) are shown by EG-LDPC, with high BER by Irreg2 code and RS-LDPC.

5.5.4 Multi-Threshold Decoder Complexity

To estimate and compare the complexity of different decoding algorithms assume that LDPC code parity-check matrix has ρ ones at every row and γ ones at every column, the length of code is n and the rate of code is R. Then the minimum number of basic operations required by decoder at every iteration can be defined in terms of ρ, γ, n and R. Table 5.2 shows the number of operations of best possible decoder implementation.

Table 5.2 Minimum number of basic operations required for decoding one data block of length K (one iteration)

Decoder	Multiplications	Divisions	Exclusive ORs	Additions (subtractions)	Comparisons [32]
UMP decoder			$3\rho K + (\gamma + 1)n$	$3\gamma n$	$K(\rho + \log_2 \rho - 2) + 2\gamma n$
multi-threshold (first iteration)			$3\rho K + (\gamma + 1)n$	$3\gamma n$	$K(\rho + \log_2 \rho - 2) + (3\gamma + 1)n$
multi-threshold (other iterations)			$3\rho K + (\gamma + 1)n$	$3\gamma n$	$K(\rho + \log_2 \rho - 2) + 3\gamma n$
belief propagation	$n(11\gamma - 9)$	$n(\gamma + 1)$		$n(3\gamma + 1)$	

The complexity of multi-threshold decoding is very small and one should have in mind that the fast and multi-threshold decoders can operate only with integer numbers without fixed-point instructions at all. Also the implementation scheme can use both iteration pipeline scheme and parallel computing inside the iteration to achieve the maximum operation speed.

Table 5.3 shows the number of basic operations required for multi-threshold decoder on some LDPC constructions using the best possible decoders implementation.

Since LDPC decoding is iterative, decoding complexity in general case can be presented as the product of one iteration complexity by the total number of iterations:

$$C_{complete} = C_{one\ iteration} \times N_{iterations}$$

Table 5.3 MT-Decoding, number of basic operations

	N Iterations	XOR	ADD	IF	Total
PEG(2048,1018)	11	664576	982016	1130362	2776954
PEG(2048,1720,0.8398)	11	2312715	1193286	1785050	5291053
EG(2107,1764,0.8372)	11	2706321	689399	1397645	4793366
EG(2548,2205,0.8654)	12	4372441	952958	2118503	7443903
EG(2793,2450,0.8772)	12	5295568	1086480	2495130	8877179
EG(2048,1688,0.8242)	11	1985239	671996	1192585	3849821
RSLDPC(2048,1467)	11	1999712	1349152	1850939	5199803
RSLDPC(2048,1649)	11	1944096	668256	1177585	3789937
RSLDPC(2048,1681)	11	1955360	607840	1122321	3685521
RSLDPC(2048,1723)	11	1977184	548384	1070907	3596475
RSLDPC(2048,1807)	11	2020832	429472	968079	3418383
RSLDPC(2048,1919)	11	2094048	313248	876031	3283327

Thus, to analyse the decoding complexity, not only the complexity of one iteration should be taken into account, but also the number of iterations needed for providing the required decoding complexity. To estimate the decoding complexity, estimate the average number of operations that are required to transmit the packet of 1000 message bits. The total decoding complexity per one data packet using a multi-threshold decoder taking into consideration the required number of iterations is shown in the Table 5.4.

In the upper part of the table the number of basic operations (additions (ADDs), comparisons (IFs) and exclusive ORs (XORs)) needed by a decoder to complete one decoding iteration for each of codes presented in the table is shown.

In the central part of the table the total number of operations needed by a decoder to decode one word of each code, for different numbers of iterations, is shown. Taking into account the convergence speed, a different number of iterations is needed to decode the word of different codes, and therefore, a different total number of basic operations. The number of basic operations needed for decoding with a sufficient number of iterations is set in bold.

In the lower part of the table the number of codewords containing in one data packet of length 1000, and the total number of basic operations needed for one packet decoding, are calculated.

The table shows that, taking into consideration the number of iterations required, the multi-threshold decoder has the minimal complexity on W-LDPC code, then on RS-LDPC, and then on Irreg2, EG-LDPC, PEG and Irreg1 codes in the order of complexity increasing.

5.5.5 Calculating Thresholds for MT-Decoder

The central point of a multi-threshold decoder is using of the thresholds. Thresholds can be computed during first decoder iteration, and then used for the other iterations.

In this approach, the decoder needs no additional information about the channel (noise dispersion, etc.); however, during decoding time is needed to calculate thresholds, and in hardware implementations a large number of elementary blocks is used. Thus the work of the decoder with predefined thresholds, as well as procedure for threshold selecting, should be analysed.

The thresholds of a multi-threshold decoder are not constant and depend on the following parameters:

1. Channel SNR.

2. Code construction.

3. Codelength.

Thus fixed thresholds should be selected separately for each LDPC construction and codelength. We propose the following procedure of calculating fixed thresholds:

1. Simulate the functioning of multi-threshold decoder in AWGN channel for some SNR (or noise dispersion). Calculate the average value on all used thresholds.

2. Form the thresholds table for different SNRs.

Table 5.4 MT-Decoding complexity for different LDPC code construction

	Irreg2(2400,1800)		W-LDPC(1000,757)		Irreg1(2000,1600)		RS(1024,747)		EG(1056,813)		PEG(1000,750)	
	Ist iter	other	Ist iter	other	Ist iter	other	Ist iter	other	Ist iter	other	Ist iter	other
XORs	102600	102600	51420	51420	96400	96400	42000	42000	111840	111840	47376	47376
ADDs	57600	57600	15000	15000	24000	24000	15360	15360	98208	98208	19008	19008
IFs	90432.4	88032.4	32897.7	31897.7	58271.88	56271.88	29830	28806	127719	126663	35058	34002
Total approximate number of basic operations	250632.4	248232.4	99317.7	98317.7	178671.9	176671.9	87190	86166	337767	336711	101442	100386
Number of iterations 4	995330		394271		708688		345688		**1347900**		402600	
6	1491794		**590906**		1062031		518020		2021322		603372	
10	**2484724**		984177		1768719		**862684**		3368166		1004916	
14	3477654		1377448		2475406		1207348		4715010		**1406460**	
20	4967048		1967354		**3535438**		1724344		6735276		2008776	
Codeword length	1800		757		1600		747		813		750	
Packet size	1000		1000		1000		1000		1000		1000	
Number of words per packet	0.555555556		1.321003963		0.625		1.338688086		1.2300123		1.333333333	
Overall operations relative to packet	1380402.239		780589.4285		2209648.5		1154864.793		1657933.579		1875280	

Table 5.5   Fixed thresholds for MT decoder

SNR	EG (255,175)	Irreg1 (2000,1600)	Irreg2 (2400,1800)	PEG (504,252)	RS (2048,1649)	Vandermonde (1000,757)
2.00	−1.703209	−1.013369	−1.473104	−2.696182	−0.830832	−0.918977
2.25	−1.797118	−1.049672	−1.539172	−2.738346	−0.869202	−0.960916
2.50	−1.919731	−1.096213	−1.598608	−2.873956	−0.896573	−0.987151
2.75	−2.035528	−1.121660	−1.686542	−2.973981	−0.951840	−1.025731
3.00	−2.147337	−1.170183	−1.754481	−3.105872	−1.003469	−1.069663
3.25	−2.283049	−1.209998	−1.818938	−3.250985	−1.069735	−1.123508
3.50	−2.440478	−1.250196	−1.913103	−3.353540	−1.150105	−1.180198
3.75	−2.623968	−1.297393	−2.004750	−3.489512	−1.237858	−1.229442
4.00	−2.804447	−1.352606	−2.083417	−3.634091	−1.330557	−1.304540
4.25	−2.999468	−1.397764	−2.167832	−3.753314	−1.433518	−1.354703
4.50	−3.205640	1.441435		−3.866233	−1.546978	−1.419868

Fixed thresholds are shown in Table 5.5.

In Figure 5.11 the plots of decoding error probability for different constructions are shown, using a multi-threshold decoder with adaptive and fixed thresholds. It follows from the plots that the usage of fixed thresholds in an AWGN channel does not lead to an increase of error probability.

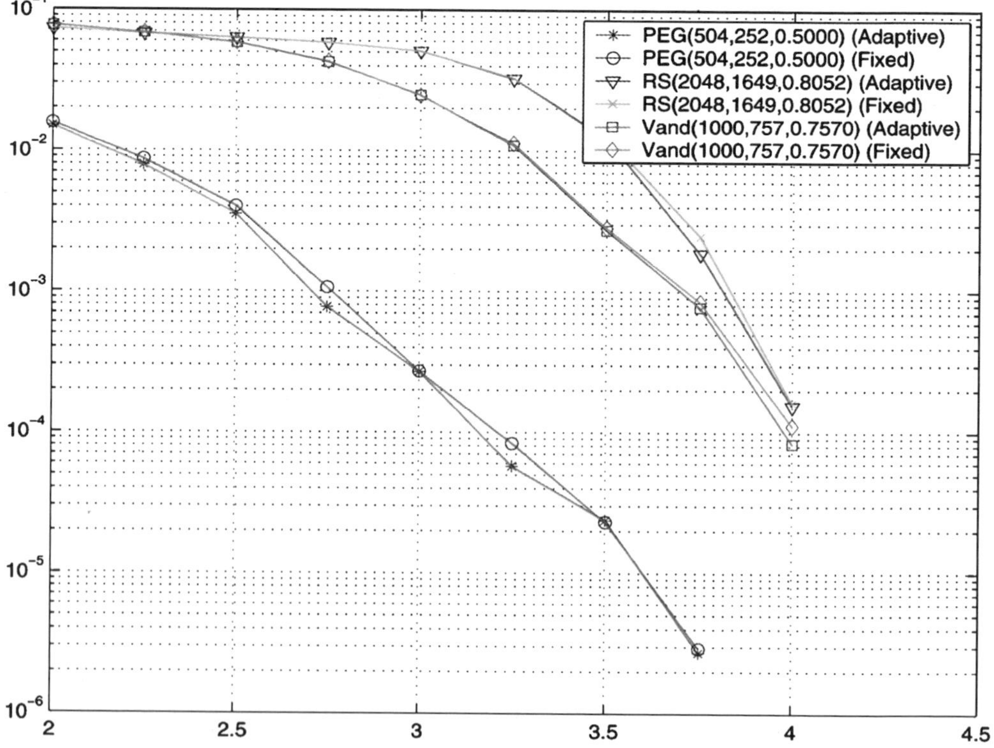

Figure 5.11   MT-decoder performance with fixed and adaptive thresholds

5.5.6   Convergence of LDPC Constructions Using MT-Decoder

Denote as the convergence speed the dependence of decoding quality on the number of iterations that should be done to provide this quality.

In Figure 5.12 the histogram of multi-threshold decoder convergence for different LDPC constructions is shown, assuming the maximum number of iterations is 20. The histogram

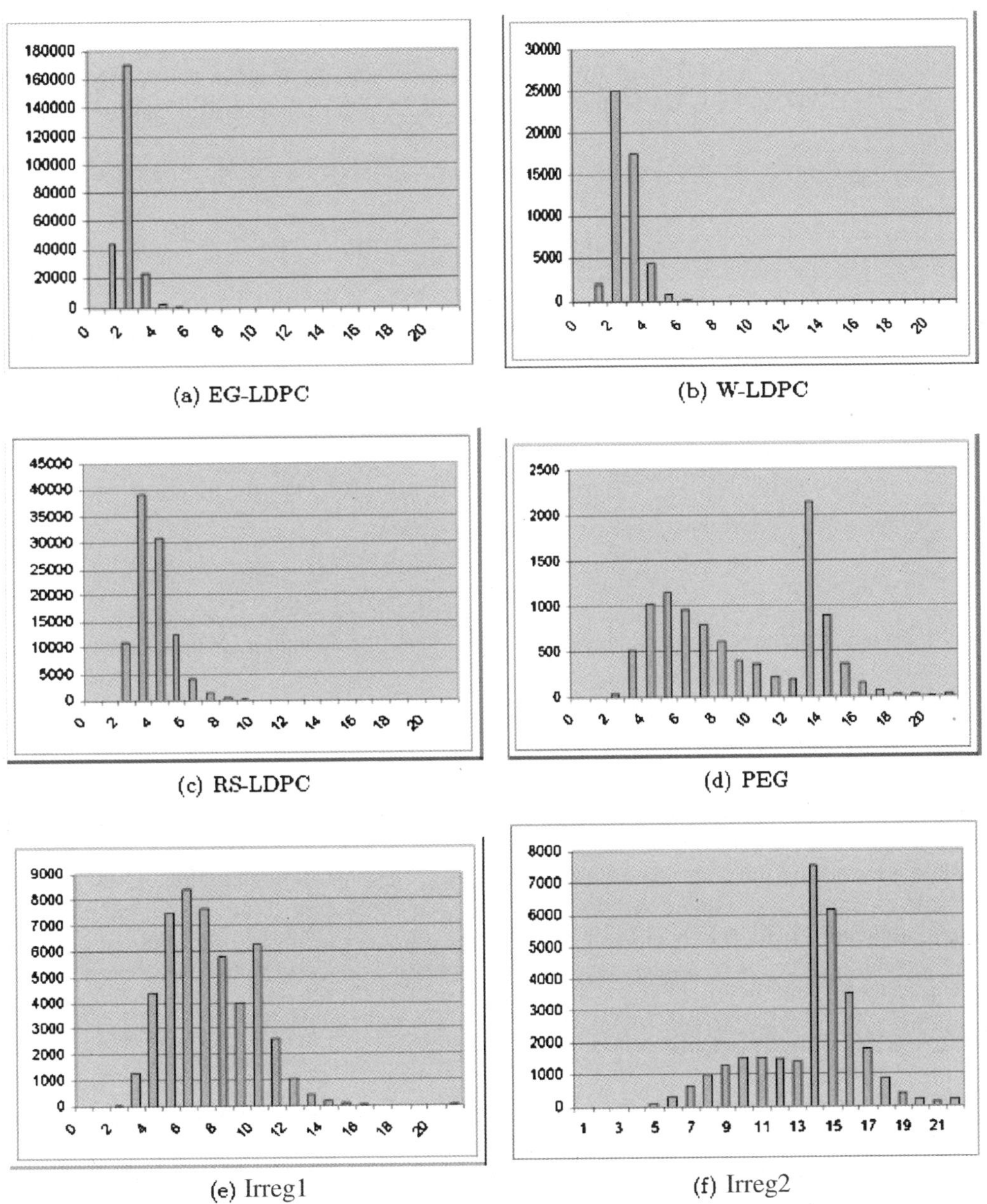

(a) EG-LDPC (b) W-LDPC

(c) RS-LDPC (d) PEG

(e) Irreg1 (f) Irreg2

Figure 5.12   LDPC Codes convergence

axes are the number of iterations and the number of decoding attempts, when decoder indeed finished its work in given number of iterations. All constructions were analysed with SNR providing the BER of 10^{-5}.

It follows from the histograms that the best convergence speed was obtained on EG-LDPC and W-LDPC codes.

5.6 SIMULATION RESULTS IN AWGN

In this section the results of simulating different LDPC constructions in an AWGN channel are presented. AWGN channel is described by SNR, which defines noise dispersion in the channel.

For comparison of different LDPC constructions, the SNR on information bit was used,

$$\text{SNR} = 10 \log_{10} \frac{E_b}{RN_0}, \text{dB} \tag{5.34}$$

where E_b is the energy of transmitted symbol per bit (which is 1 using BPSK modulation), N_0 is the noise power spectral density, R is the code rate.

All numeric results are obtained using soft belief propagation decoder, described in Section 5.5, with the maximum number of iteration equal to 30. The decoder used LLR of received data as input (5.33).

The signal-to-noise ratio (SNR) was calculated using (5.34), the bit error probability (BER) was calculated as

$$\text{BER} = \frac{N_e}{N},$$

where N_e is the number of erroneous information bits after decoding, N is the total number of transmitted information bits.

5.6.1 RS-LDPC Construction

The plots of simulation results for LDPC codes based on Reed-Solomon codes (RS-LDPC, see section 5.2.2), are shown in Figure 5.13.

Here we consider the codes based on RS over $GF(2^6)$, construction parameters are defined as $RS(6, \rho, \gamma)(n, k, R)$. Obtained codes have length 2048 and 1024, and estimations of minimal distance from 6 to 8. As can be seen from the curves, increasing minimal distance (i.e. γ parameter) leads to a decreasing error probability for codes with the same rate.

5.6.2 PEG Construction

Curves of simulation results for PEG construction, described in section 5.2.4, are shown in Figure 5.14.

Construction parameters are weights distributions $\rho(x)$ and $\lambda(x)$ of rows and columns of parity-check matrix, obtained from 'density evolution' procedure. Here we consider the code lengths of 2048 and 1056 bits, and rates from 0.8 and above.

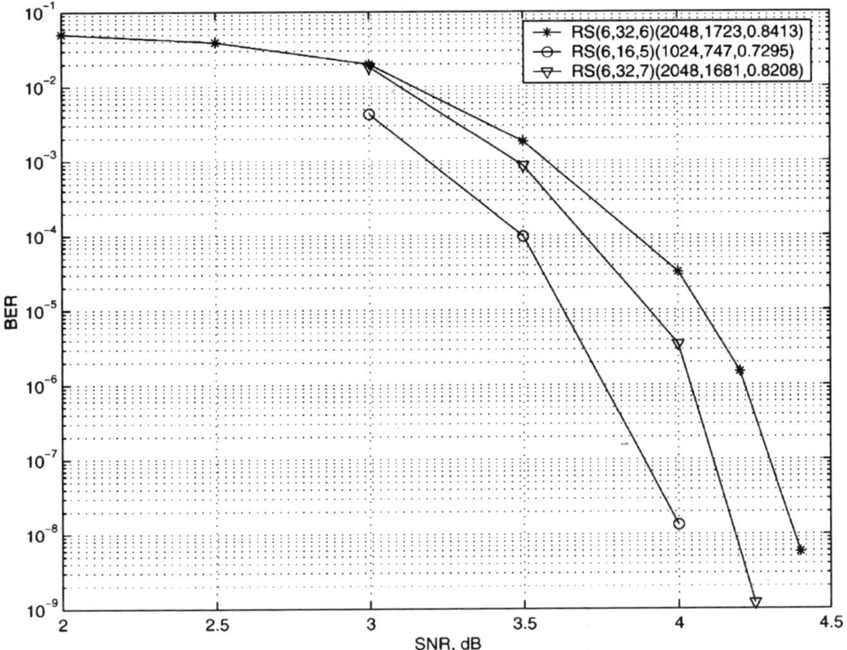

Figure 5.13   RS-LDPC Codes in AWGN channel

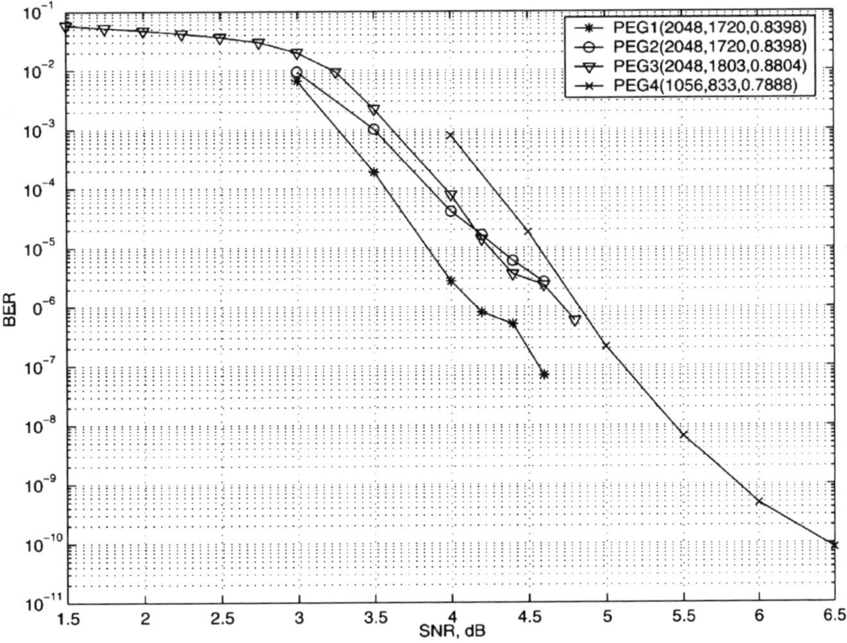

Figure 5.14   PEG Construction in AWGN channel

This construction is irregular, it gives gain with low SNRs, but with decreasing noise level, the 'error-floor' effect appears, i.e. a decrease of curve falling speed. This effect is common to many LDPC codes because of their poor minimum distance. However, it is especially so in the case for irregular constructions. Besides, PEG construction works worse with decreasing the code length, because the weights distribution optimisation bases on asymptotic analysis, and it can be incorrect for relatively small code length.

5.6.3 EG Construction

Here we present the results of simulations EG-LDPC code with the parity-check matrix (5.23). Correspondent curves are shown in Figure 5.15.

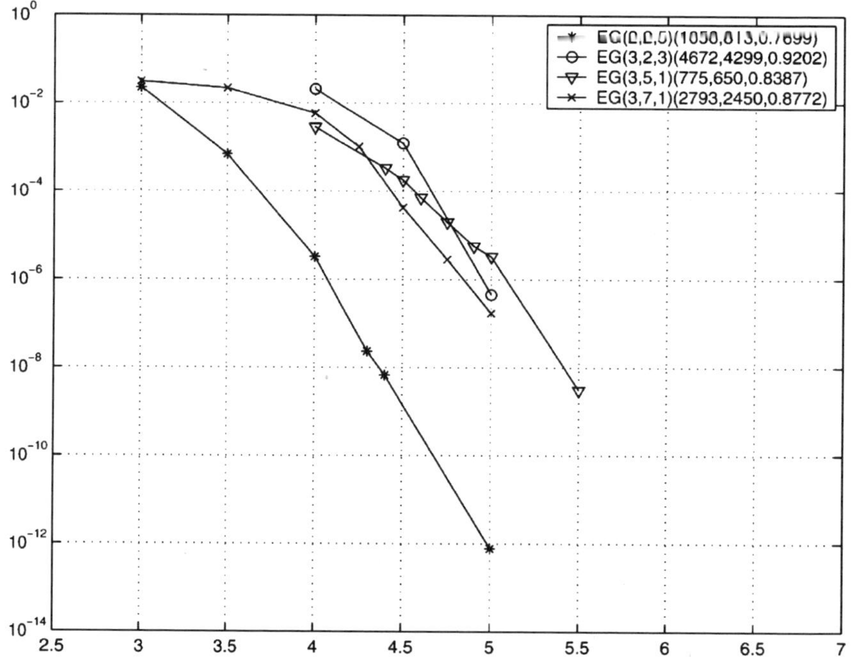

Figure 5.15 Euclidean-geometry codes in AWGN channel

We consider the codes based on Euclidean geometries $EG(2, 2^5)$, $EG(3, 2^3)$, $EG(3, 5)$, $EG(3, 7)$. Minimum distances of these codes are, correspondingly, 33, 9, 10, and 14. As can be seen from the plots, the code with minimum distance 33 shows significantly better performance.

5.6.4 Shortened EG-Codes

We consider two methods of Euclidean-geometry code shortening: the first method means shortening on columns correspondent to parallel classes of geometries $EG(2, p^s)$ (EG-planes) and $EG(3, p^s)$ (EG-spaces). In this case the regularity of construction is retained, i.e., equality of row and column weights.

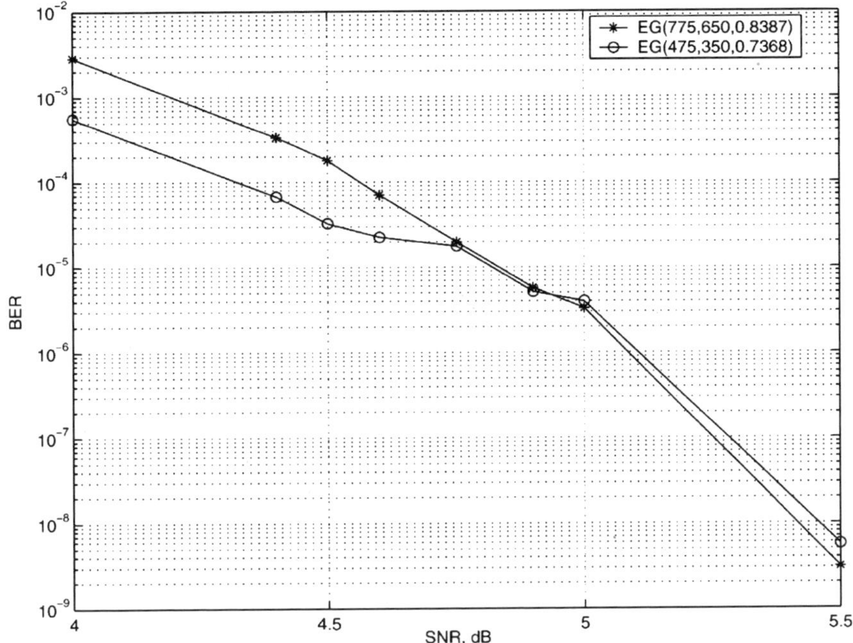

Figure 5.16   Shortening code $EG(3,5)$

The second method is applied to EG-spaces, and removes the lines, corresponding to parallel classes in planes, containing in particular space. In general, this leads to breaking regularity. The intention of this shortening method is to improve spectral properties of codes, since according to Theorem 5.4 words of weight $\gamma + 1$ are removed from planes.

In Figure 5.16 the results of such shortenings are shown for codes based on $EG(3,5)$. The obtained code gives gain on low SNRs, where the spectral characteristics are important. However, this shortening method gives no increase of minimum distance of the code, and with SNR increasing the performance is practically the same as for the original code.

In Figure 5.17 the results are shown for the same shortening, but with geometry $EG(3,2^3)$ (EGsh3 code). The results of shortening this code on 18 and 41 parallel classes are also shown (correspondingly sh1 and sh2). Shortening sh1 gives code with the same parameters as sh3, and its performance in AWGN is practically the same.

In Figure 5.18 the code from $EG(2,2^5)$ is presented, and shortenings on 5 and 10 parallel classes (sh5 and sh10, correspondingly). Shortenings allow codes with new parameters to be obtained, and better (lower) error probability.

In Figure 5.19 the results of shortening for codes $EG(2,2^4)$ are presented. The obtained code has practically the same performance as the original one. Using shortening in this case is the method of obtaining Euclidean-geometry codes at different lengths and rates.

5.6.5   Block-structured LDPC-codes

Here we consider the general construction with parity-check matrix consisting blocks — powers of cyclic permutation matrix (5.20). This matrix is constructed using the Vandermonde matrix (5.21) as a basis.

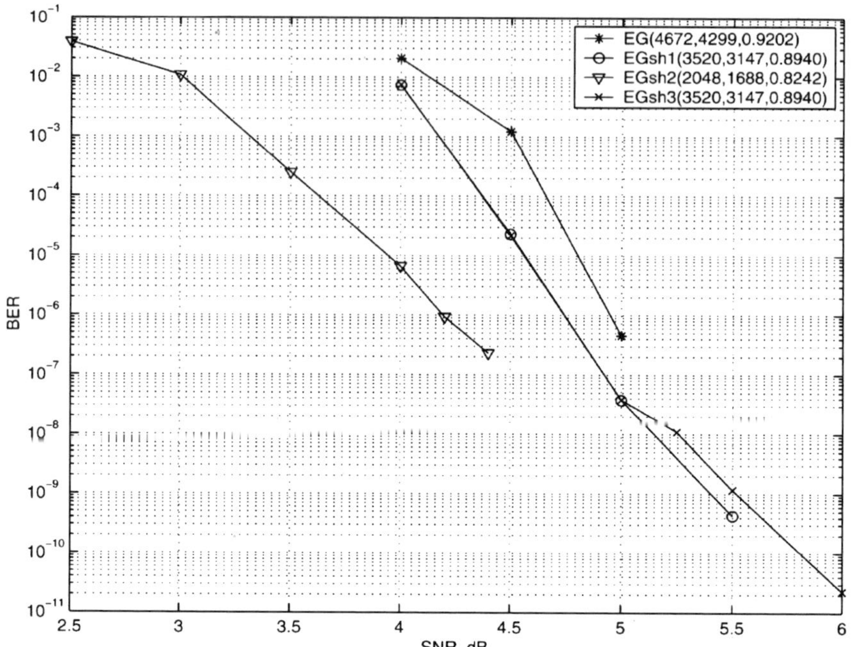

Figure 5.17 Shortening code $EG(3, 2^3)$

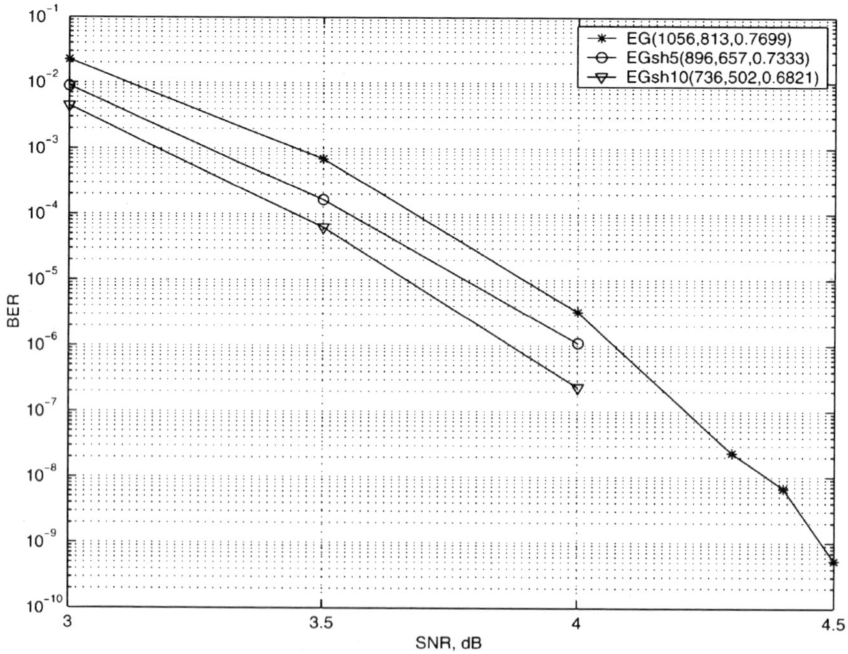

Figure 5.18 Shortening code $EG(2, 2^5)$

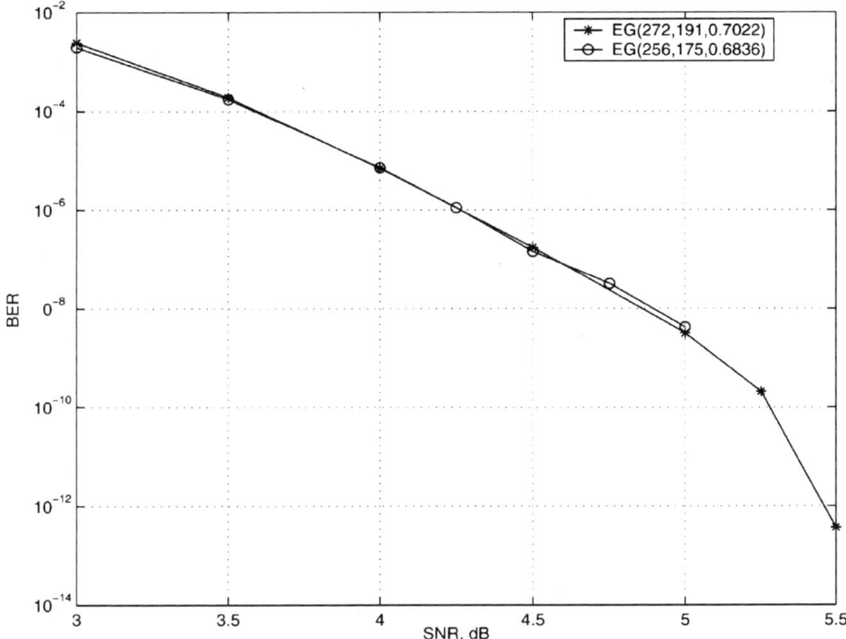

Figure 5.19   Shortening code $EG(2, 2^4)$

The simulation results for these codes on lengths about 600 and rates about 0.74 are presented in Figure 5.20. The construction is defined by the parameters (m, ρ, γ), and curves show that better quality is obtained using large m and less ρ. This can be explained by the fact that the functionality of the decoder is sensitive to the number of ones in the parity-check matrix. For instance, for code (25,24,7) the ratio of ones in parity-check matrix is 0.016, while for the code (40,15,4) it is 0.009, i.e. almost half, with the same size of parity-check matrix.

The simulation results for the codes with lengths 1024 and 2048 are presented in Figure 5.21. Here we can make similar conclusions about block size m and the row weight ρ and column weight γ: the code (32,32,15) shows significantly worse results than the code (64,16,4).

5.6.6   Constructions Comparison

Here we give the comparison plots of constructions considered above, on lengths of about 1024 and 2056.

In Figure 5.22 the results of different LDPC-constructions of length 1024 are shown. On such lengths and high (about 0.76) rate the PEG construction does not show good performance. Other constructions show near the same performance, slightly better is RS-LDPC construction.

In Figure 5.23 the simulation results for LDPC-codes on lengths about 2048 are presented. On low SNRs the better results show Euclidean-geometry code and PEG, however, with SNR increasing PEG construction becomes worse than any other. In conclusion we note that

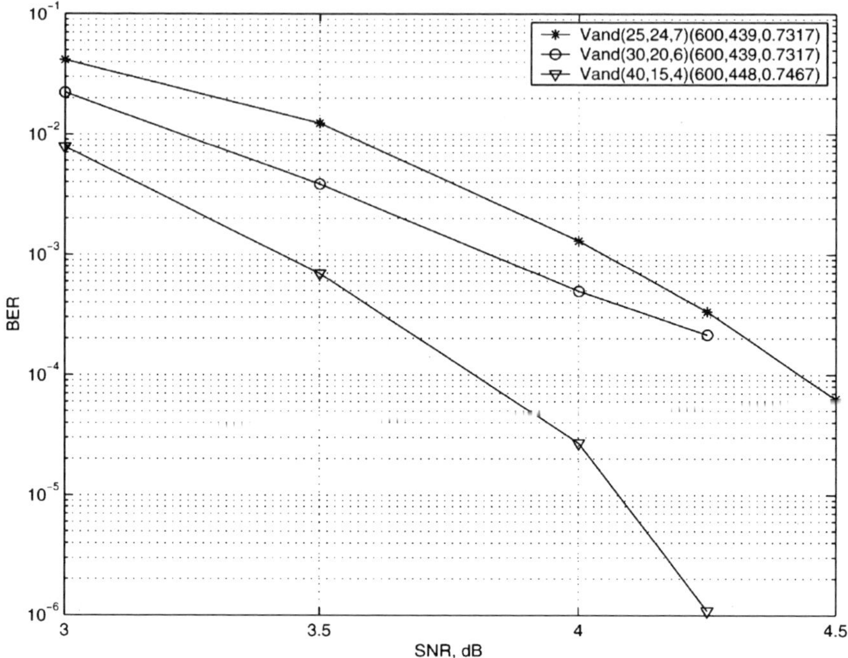

Figure 5.20   Codes based on Vandermonde matrix, in AWGN channel

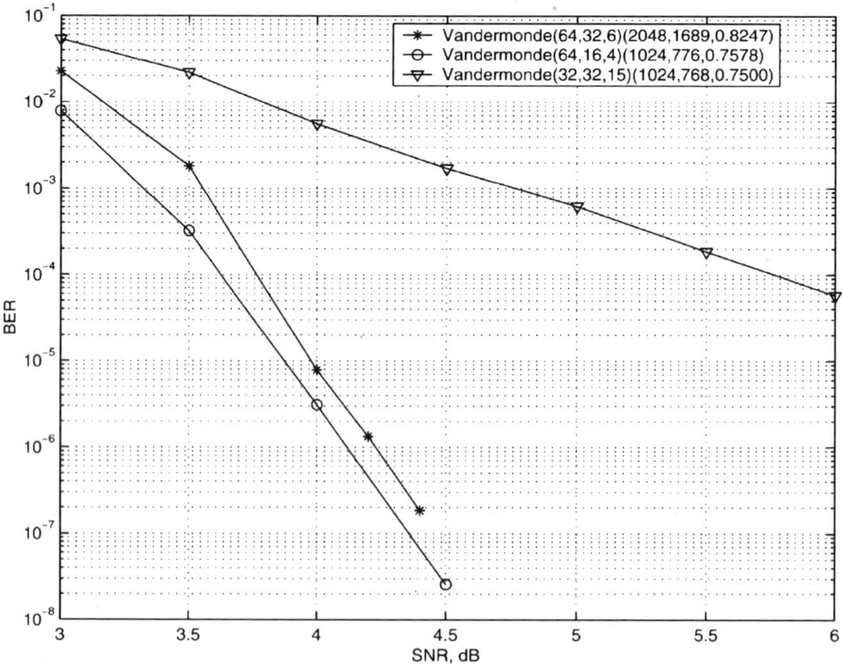

Figure 5.21   Codes based on Vandermonde matrix, in AWGN channel

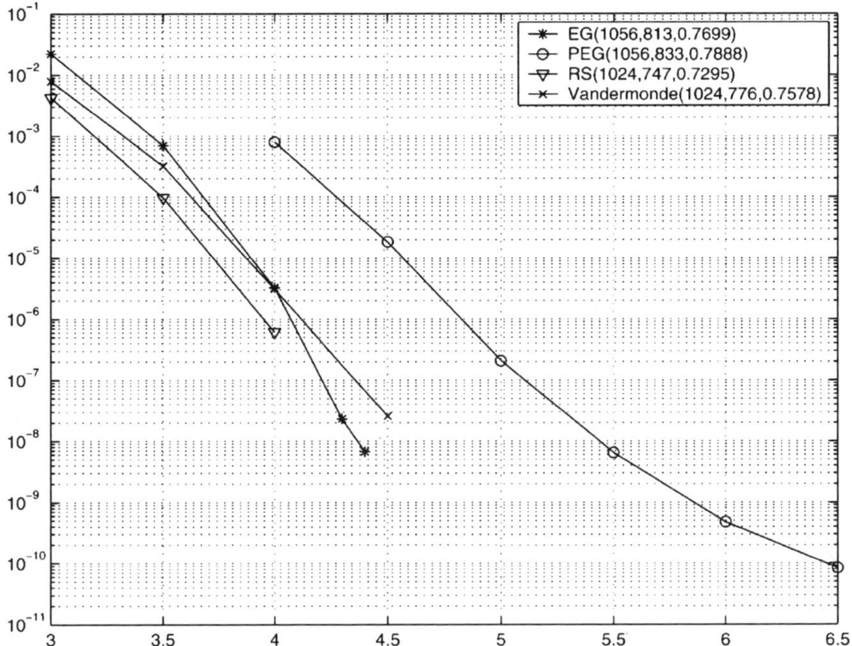

Figure 5.22   LDPC-Codes comparison with code length 1024

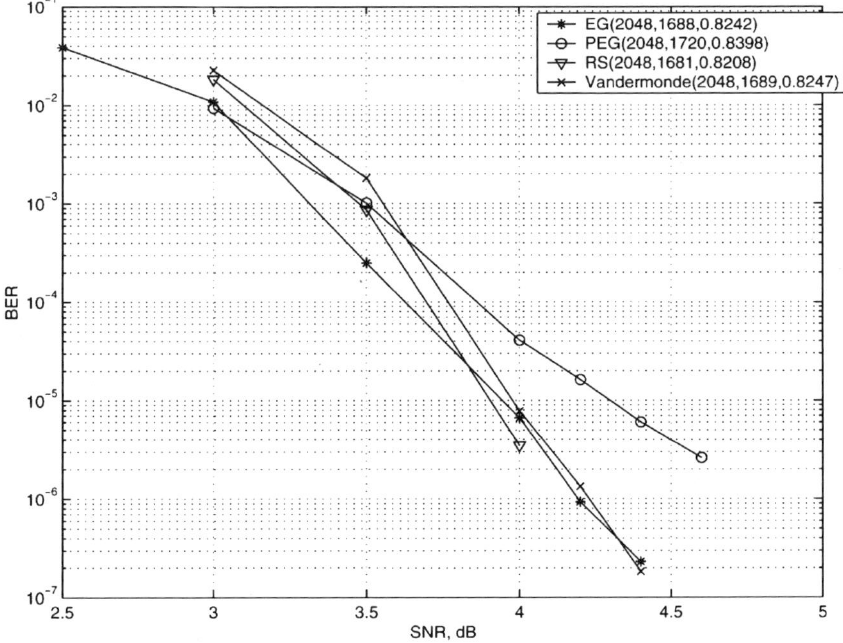

Figure 5.23   LDPC-Codes comparison with code length 2048

the construction based on Vandermonde matrix shows the performance, comparable to EG-codes, but it has more simple decoding procedure due to less number of ones in parity-check matrix.

APPENDIX 5.A   EUCLIDEAN GEOMETRIES

Euclidean geometry *EG* [12,32] is the set of objects: *points* and *lines*, satisfying the following axioms:

1. exactly one line can be passed through every two points;

2. for any line L and any point p not on L, there is a line passing through p that does not cross L (i.e. parallel to L);

3. each line contains at least three points.

It should be noted that these axioms are not the only ones that can define Euclidean geometry. However, any other axiom can be reduced to these, and vice versa.

One of the most used and practically important forms of Euclidean geometry is the representation using finite fields. Euclidean geometry $EG(m, q)$ is defined by finite field $GF(q^m)$ (extension of the field $GF(q)$) as follows: points of Euclidean geometry are elements $\alpha^j \in GF(q^m)$, $j = -\infty, 0, 1, \ldots, q^m - 2$ (α is the primitive element of $GF(q^m)$). Note that the set of Euclidean geometry points contains the zero point, i.e., the zero element $\alpha^{-\infty}$ of $GF(q^m)$. Then the line passing through the zero point and some non-zero point α^j is defined by the equation

$$L(0, \alpha^j) = \{\beta\alpha^j\} = \{\beta\alpha^j : \beta \in GF(q), \alpha^j \in GF(q^m), \alpha^j \neq 0\} \qquad (5.A1)$$

i.e., passing through points $\alpha^{-\infty} = 0$ and $\alpha^j \neq 0$ and containing the field elements, obtained from α^j by multiplying it by all elements β (including zero) of sub-field $GF(q)$. If some elements α^i and α^j are linearly independent (i.e., α^i does not belong to the line $L(0, \alpha^j)$), then, in accordance with the axiom 2, it is possible to draw the line parallel to $L(0, \alpha^j)$ and passing through α^i:

$$L(\alpha^i, \alpha^j) = \{\alpha^i + \beta\alpha^j\} = \{\alpha^i + \beta\alpha^j : \beta \in GF(q)\} \qquad (5.A2)$$

Consider the fields of characteristics p, where p is prime, i.e., Euclidean geometries $EG(m, q)$, $q = p^s$. Since element β in (5.A1), (5.A2) takes $q = p^s$ different values, each line in Euclidean geometry contains

$$\rho = q \qquad (5.A3)$$

points. There are in total

$$|L| = q^{(m-1)}(q^m - 1)/(q - 1) \qquad (5.A4)$$

lines in $EG(m,q)$. Each line has $q^{(m-1)} - 1$ lines parallel to it, there are

$$\gamma = (q^m - 1)/(q - 1) \tag{5.A5}$$

lines passing through each point (in other words, γ lines intersect in one point).

Euclidean geometry $EG(2,q)$ is called *plane*. The points of the plane can be obtained as linear combinations of three points α^i, α^j, α^v, not belonging the same line:

$$\{\alpha^i + \mu\alpha^j + \eta\alpha^v\} \quad \mu, \eta \in GF(q) \tag{5.A6}$$

It follows from (5.A4) and (5.A6) that a plane contains q^2 points and $q(q + 1)$ lines.

REFERENCES

1. Gallager, R.G. (1963). *Low-density parity-check codes*, M.I.T. Press, Cambridge, MA: USA.
2. Gallager, R.G. (1962). *Low density parity check codes*. IRE Transactions on Information Theory.
3. MacKay, D. (1999). *Good error correcting codes based on very sparse matrices*. IEEE Transactions on Information Theory, **45**.
4. MacKay, D. and Neal, R. (2001). *Near Shannon Limit Performance of Low-Density Parity-Check Codes*. IEEE Transactions on Information Theory, **47**(2).
5. Richardson, T.J. and Urbanke, R.L. (2001). *The Capacity of low-Density Parity-Check Codes Under Message-Passing Decoding. IEEE Transactions on Information Theory*, **47**(2).
6. Zyablov, V.V. and Pinsker, M.S. (1975). *Estimation of the error-correction complexity for Gallager low-density codes*. Problemy Peredachi Informatsii, **11**(1).
7. Tanner, M. (1981). *A Recursive Approach to Low Complexity Codes*. IEEE Transactions on Information Theory, IT, **27**.
8. Richardson, T.J. Urbanke, R.L., and Shokrollahi, M.A. (2001). *Design of Capacity-Approaching Irregular low-Density Parity-Check Codes*. IEEE Transactions on Information Theory, **47**(2).
9. Richardson, T.J. Urbanke, R.L., and Chung, S.-Y. (2001). *Analysis of Sum-Product Decoding of low-Density Parity-Check Codes Using a Gaussian Approximation*. IEEE Transactions on Information Theory, **47**(2).
10. Hu, X.-Y., Eleftheriou, E., and Arnold, D.-M. (2003). *Regular and Irregular Progressive Edge-Growth Tanner Graphs*. IBM Research, Zurich Research Laboratory.
11. Blahut, R. (1984). *Theory and Practice of Error Control Codes*. Addison-Wesley, Reading, MA, USA.
12. Kasami, T., Tokura, N. Ivadari, E., and Inaghaki, Y. (1978). *Coding Theory* (in Russian). Mir, Russia.
13. Kolesnik, V.D. (1971). *Probability decoding of majority codes, Prob. Peredachi Inform.*, **7**, pp. 3–12.
14. Kou, Y., Lin, S. and Fossorier, M.P.C. (2000). *Construction of low-density parity-check codes: A geometric approach*. In *Proc. 2nd Int. Symp. Turbo Codes and Related Topics*, pp. 137–140, Brest, France.
15. Kou, Y., Lin, S., and Fossorier, M.P.C. (2001). *Low-Density Parity-Check Codes Based on Finite Geometries: A Rediscovery and New Results, IEEE Trans Inf. Theor.*, **47**, (7).
16. Chen, C.-L. (1971). *On Majority-Logic Decoding of Finite Geometry Codes. IEEE Transactions on Information Theory, IT*, **17**(3).
17. Kasami, T. and Lin, S. (1971). *On Majority-Logic Decoding for Duals of Primitive Polynomial Codes. IEEE Transactions on Information Theory*, IT **17**(3).
18. Kolesnik, V. and Mironchikov, E. (1968). *Decoding of cyclic codes* (In Russian). Svyaz, Russia.
19. Peterson, W.W. and Weldon, Jr. E.J. (1972). *Error-Correcting Codes*. MIT Press, Cambridge, Mass, USA.

20. Lin, S. (1972). *On the Number of Information Symbols in Polynomial Codes. IEEE Transactions on Information Theory*, **18**(6).

21. Djurdjevic, I., Xu, J. Abdel-Ghaffar, K., and Lin, S. (2003). *A Class of Low-Density Parity-Check Codes Constructed Based on Reed-Solomon Codes With Two Information Symbols, IEEE Communication Letters*, **7**(7).

22. Gilbert, E.N. (1960). *A problem in binary encoding.* In *Proceedings of the Symposium in Applied Mathematics*, **10**, pp. 291.

23. Arazi B. (1978). *The Optimal Burst Error-Correcting Capability of the Codes Generated by* $f(x) = (x^p + 1)(x^q + 1)/(x + 1)$. *Inform. and Contr.*, **39**(3).

24. Krouk, E.A. and Semenov, S.V. (1990). *Low-density parity-check burst error-correcting codes.* In 2nd International Workshop 'Algebraic and combinatorial coding theory,' pp. 121–4.

25. Zhang, W. and Wolf, J. (1988). *A Class of Binary Burst Error-Correcting Quasi-Cyclic Codes. IEEE Transactions on Information Theory*, **34**; pp. 463–79.

26. Tian, T., Jones, C. Villasenor, J., and Wesel, R. (2003). *Construction of Irregular LDPC Codes with Low Error Floors. In Proceedings of ICC'2003*, Anchorage, Alaska. pp. 3125–9.

27. Lin, S. (1972). *Shortened Finite Geometry Codes. IEEE Transactions on Information Theory*, **18**(5).

28. Ovchinnikov, A. (1999). *About one class of burst error-correcting codes.* In *Proc. of Second Int. school-seminar* BICAMP'99, SPb.

29. Ovchinnikov, A. (1999). *A class of binary burst error-correcting codes.* In Volume 2, EuroXChange.

30. Ovchinnikov, A. (2000). *About modification of one class of burst error-correcting codes.* In volume 3, EuroXChange.

31. Fossorier, M.P.C. Mihaljevic, M., and Imai, H. (1999). *Reduced Complexity Iterative Decoding of Low-Density Parity-Check Codes Based on Belief Propagation. IEEE Transactions on Communications*, **47**(5).

32. MacWilliams, F. and Sloane, N. (1977). *The Theory of Error-Correcting Codes.* North-Holland, Amsterdam, The Netherlands.

33. Lechner, G. (2003). *Convergence of Sum-Product Algorithm for Finite Length Low-Density Parity-Check Codes. Winter School on Coding and Information Theory*, Monte Verita, Switzerland.

34. Lucas, R., Fossorier, M.P.C., Kou, Y., and Lin, S. (2000). *Iterative decoding of one-step majority logic decodable codes based on belief propagation. IEEE Trans. Commun.*, **48**(6).

35. Ryan, W.E. (2004). *An Introduction to LDPC Codes.* In (B. Vasic, ed.) *CRC Handbook for Coding and Signal Processing for Recording Systems* CRC Press, to be published in 2004.

36. Snyders, J. (1991). *Reduced lists of error patterns for maximum likelihood soft decoding*, IEEE Trans on Information Theory, Vol. **IT-37** pp. 1134–1200.

37. Yeo, E. Nikolic, B. and Anatharam, V. (2002). *Architectures and Implementations of Low-Density Parity Check Decoding Algorithms*, Department of Electrical Engineering and Computer Sciences, University of California, Berkley.

6

Convolutional Codes and Turbo-Codes

6.1 CONVOLUTIONAL CODES REPRESENTATION AND ENCODING

The general structure of the convolutional encoder is represented in Figure 6.1. Every moment a block of k symbols is fed to the input of the encoder. The encoder has memory, which keeps the values of $\nu - 1$ previous input blocks ($k \cdot (\nu - 1)$ symbols). The encoder forms n output symbols at a time. The block of n output symbols depends on $k \cdot \nu$ input symbols and each output symbol is the linear combination of information symbols. The number k is called the *number of information symbols*, the number n is called the *number of encoded symbols* and the number ν is known as the *constraint length* of the code. The value k/n is called the *code rate*. The corresponding convolutional code is usually denoted as (n, k, ν) code. Note that in some literature the notation used is $(n, k, \nu - 1)$.

The encoder works as follows: The input data block of k symbols is fed to k shift registers (each symbol to its own register) with the help of demultiplexer. Each shift register consists of no more than $\nu - 1$ delay elements. The outputs of the delay elements are multiplied by some fixed coefficients $g_{ij}^{(l)}$ $(i = 0, \ldots, n - 1; j = 0, \ldots, k - 1; l = 0, \ldots, \nu - 1)$. These weighted outputs are distributed by n sets and each set generates the output symbol just by adding the members of the set. All multiplications and additions are executed on field K. Thus, the output block of n encoded symbols corresponds to input data block of k symbols. After encoding, the n output symbols are multiplexed into a single sequence. Note that each encoded symbol is obtained with the help of a linear combination of the outputs of the delay elements.

It is obvious that the output symbol depends on the input symbol and on the contents of the shift registers. In this sense a convolutional code is close to a block code. However, unlike a block code that has a fixed word length, a convolutional code has no particular word length. We can say that a word of a convolutional code is a semi-infinite sequence.

We can describe the encoder of the convolutional code with the help of its *impulse response*, that is the output of the encoder obtained after feeding the '1' to the encoder input and assuming that the encoder starts from the zero state. It is easy to verify that if the '1' is fed to the j th input then the output sequence (or impulse response) from the i th output will be represented by the sequence of coefficients $\mathbf{g}_{ij} = g_{ij}^{(l)}$, $(l = 0, \ldots, \nu - 1)$.

Error Correcting Coding and Security for Data Networks   G. Kabatiansky, E. Krouk and S. Semenov
© 2005 John Wiley & Sons, Ltd   ISBN: 0-470-86754-X

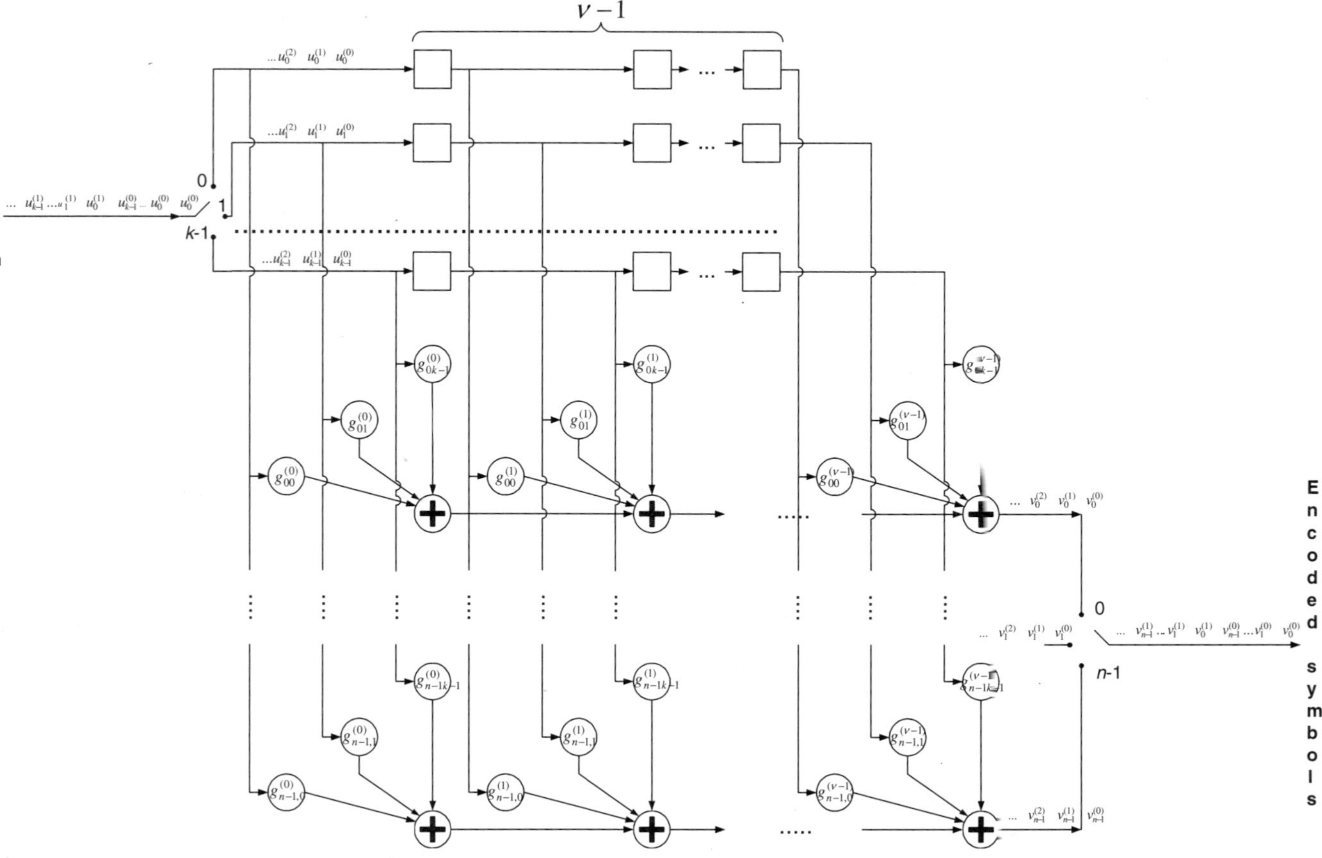

Figure 6.1 A general convolutional encoder

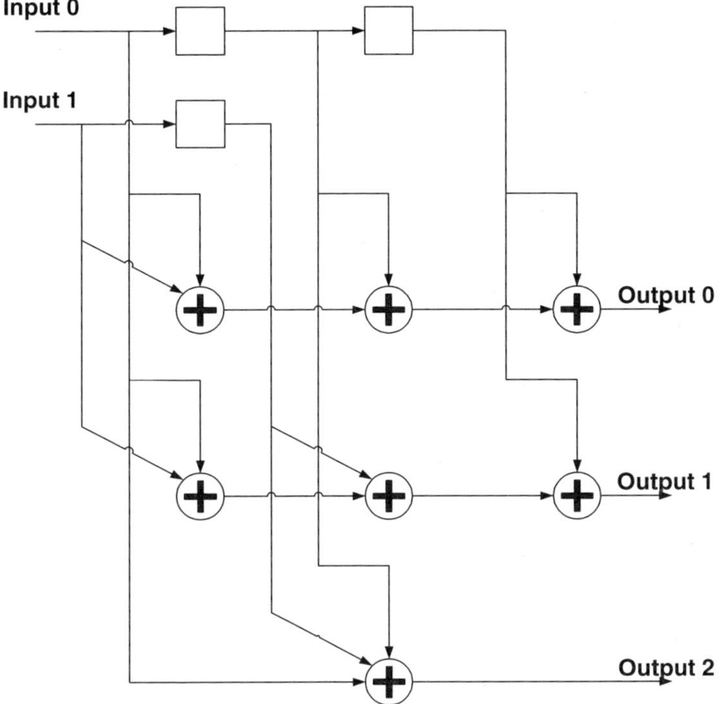

Figure 6.2   A (3,2,3) binary convolutional encoder

Example 6.1   Consider a binary (3,2,3) convolutional code. The encoder of this code is shown in Figure 6.2. Note that the constraint length of the code is 3. That means the length of the impulse response is also equal to 3. The impulse responses of this encoder can be written as follows:

$\mathbf{g}_{00} = (1,1,1)$; Output sequence from the output 0, assuming that the input sequence at the input 0 is (1 0 0 ...).

$\mathbf{g}_{01} = (1,0,0)$; Output sequence from the output 0, assuming that the input sequence at the input 1 is (1 0 0 ...).

$\mathbf{g}_{10} = (1,0,1)$; Output sequence from the output 1, assuming that the input sequence at the input 0 is (1 0 0 ...).

$\mathbf{g}_{11} = (1,1,0)$; Output sequence from the output 1, assuming that the input sequence at the input 1 is (1 0 0 ...).

$\mathbf{g}_{20} = (1,1,0)$; Output sequence from the output 2, assuming that the input sequence at the input 0 is (1 0 0 ...).

$\mathbf{g}_{21} = (0,1,0)$; Output sequence from the output 2, assuming that the input sequence at the input 1 is (1 0 0 ...).

Let us denote the input symbols, which after the demultiplexer are fed to the j th input of the encoder as u_j, and the encoded symbols at the i th output we denote as v_i. Then the input sequence can be written as $\mathbf{u} = \left(u_0^{(0)}, u_1^{(0)}, \ldots, u_{k-1}^{(0)}, u_0^{(1)}, u_1^{(1)}, \ldots, u_{k-1}^{(1)}, u_0^{(2)}, u_1^{(2)}, \ldots, \right.$

$u_{k-1}^{(2)}, \ldots)$ and the corresponding output sequence after multiplexing as $\mathbf{v} = (v_0^{(0)}, v_1^{(0)}, \ldots,$ $v_{n-1}^{(0)}, v_0^{(1)}, v_1^{(1)}, \ldots, v_{n-1}^{(1)}, v_0^{(2)}, v_1^{(2)}, \ldots, v_{n-1}^{(2)}, \ldots)$. Now let us consider k input sequences (after the demultiplexer):

$$\mathbf{u}_0 = (u_0^{(0)}, u_0^{(1)}, u_0^{(2)}, \ldots);$$

$$\mathbf{u}_1 = (u_1^{(0)}, u_1^{(1)}, u_1^{(2)}, \ldots);$$

$$\cdots\cdots\cdots\cdots\cdots\cdots\cdots\cdots$$

$$\mathbf{u}_{k-1} = (u_{k-1}^{(0)}, u_{k-1}^{(1)}, u_{k-1}^{(2)}, \ldots)$$

and n output sequences (before the multiplexer):

$$\mathbf{v}_0 = (v_0^{(0)}, v_0^{(1)}, v_0^{(2)}, \ldots);$$

$$\mathbf{v}_1 = (v_1^{(0)}, v_1^{(1)}, v_1^{(2)}, \ldots)$$

$$\cdots\cdots\cdots\cdots\cdots\cdots\cdots\cdots$$

$$\mathbf{v}_{n-1} = (v_{n-1}^{(0)}, v_{n-1}^{(1)}, v_{n-1}^{(2)}, \ldots).$$

Then it is easy to verify that the output sequences can be written as the sum of the convolutions of the corresponding input sequence and the impulse response:

$$\mathbf{v}_0 = \sum_{j=0}^{k-1} \mathbf{u}_j * \mathbf{g}_{0j};$$

$$\mathbf{v}_1 = \sum_{j=0}^{k-1} \mathbf{u}_j * \mathbf{g}_{1j};$$

$$\cdots\cdots\cdots\cdots\cdots$$

$$\mathbf{v}_{n-1} = \sum_{j=0}^{k-1} \mathbf{u}_j * \mathbf{g}_{n-1, j};$$

$$(6.1)$$

where $*$ denotes discrete convolution. This feature explains the name convolutional for the codes.

The equations (6.1) can be rewritten as

$$v_i^{(f)} = \sum_{j=0}^{k-1} \sum_{l=0}^{\nu-1} u_j^{(f-l)} \cdot g_{ij}^{(l)}, \quad i = 0, \ldots, n-1, \tag{6.2}$$

where $u_j^{(f-l)} = 0$ if $f < l$.

These equations can also be written in the form of a matrix multiplication:

$$\mathbf{v} = \mathbf{u} \cdot \mathbf{G}, \tag{6.3}$$

where $\mathbf{v}$ and $\mathbf{u}$ are the semi-infinite sequences and $\mathbf{G}$ is a semi-infinite matrix:

$$\mathbf{G} = \begin{bmatrix} \mathbf{G}^{(0)} & \mathbf{G}^{(1)} & \cdots & \mathbf{G}^{(\nu-1)} & & \cdots \\ & \mathbf{G}^{(0)} & \cdots & \mathbf{G}^{(\nu-2)} & \mathbf{G}^{(\nu-1)} & \cdots \\ & & \cdots & & & \cdots \end{bmatrix}, \tag{6.4}$$

and submatrix $\mathbf{G}^{(l)}$ is a $k \times n$ matrix, which can be written as follows:

$$\mathbf{G}^{(l)} = \begin{bmatrix} g_{00}^{(l)} & g_{10}^{(l)} & \cdots & g_{n-1,0}^{(l)} \\ g_{01}^{(l)} & g_{11}^{(l)} & \cdots & g_{n-1,1}^{(l)} \\ \cdots & \cdots & \cdots & \cdots \\ g_{0,k-1}^{(l)} & g_{1,k-1}^{(l)} & \cdots & g_{n-1,k-1}^{(l)} \end{bmatrix}, \, l = 0, \ldots, \nu - 1. \tag{6.5}$$

Matrix $\mathbf{G}$ is called the *generator matrix* of the convolutional code. It is easy to see that each output (encoded) sequence $\mathbf{v}$ can be obtained as a linear combination of rows of the matrix $\mathbf{G}$. Hence the convolutional code defined by the generator matrix $\mathbf{G}$ is the *linear* code.

Example 6.2   Consider the same binary (3,2,3) convolutional code as in previous example. The generator matrix $\mathbf{G}$ of this code can be written as:

$$\mathbf{G} = \begin{bmatrix} 1 & 1 & 1 & 1 & 0 & 1 & 1 & 1 & 0 & & & \\ 1 & 1 & 0 & 0 & 1 & 1 & 0 & 0 & 0 & & & \\ & & 1 & 1 & 1 & 1 & 0 & 1 & 1 & 1 & 0 \\ & & 1 & 1 & 0 & 0 & 1 & 1 & 0 & 0 & 0 \\ & & & & 1 & 1 & 1 & 1 & 0 & 1 \\ & & & & 1 & 1 & 0 & 0 & 1 & 1 \end{bmatrix}.$$

Let the input sequence $\mathbf{u} = (u_0^{(0)}, u_1^{(0)}, u_0^{(1)}, u_1^{(1)}, u_0^{(2)}, u_1^{(2)}, \ldots)$ be $\mathbf{u} = (1, 0, 0, 1, 1, 1, \ldots)$. Then it can be easily verified that in accordance with (6.1)–(6.5) the output sequence $\mathbf{v} = (v_0^{(0)}, v_1^{(0)}, v_2^{(0)}, v_0^{(1)}, v_1^{(1)}, v_2^{(1)}, v_0^{(2)}, v_1^{(2)}, v_2^{(2)}, \ldots) = (1, 1, 1, 0, 1, 1, 1, 0, 0, \ldots)$.

The convolutional encoder can be described as a device that may take up a finite number of states. The state of the encoder is defined by the contents of the shift registers. As was mentioned above the n output symbols are defined by the present k input symbols and by the encoder state. And the input symbols change the encoder state. This kind of device is known as a *finite state machine* (FSM). It is convenient to describe the operation of FSM with the help of a *state diagram.* The state diagram is a directed graph. The nodes represent the possible states and the branches represent the allowed transitions between states. The branches are labelled with input symbols that cause this transition and with the output symbols, which are emanated during the transition. The total number of encoder states is equal to $M^{k(\nu-1)}$, where M is the cardinal number of the input alphabet. It is obvious that this method of code representation is convenient only for the case of an encoder with a small number of states.

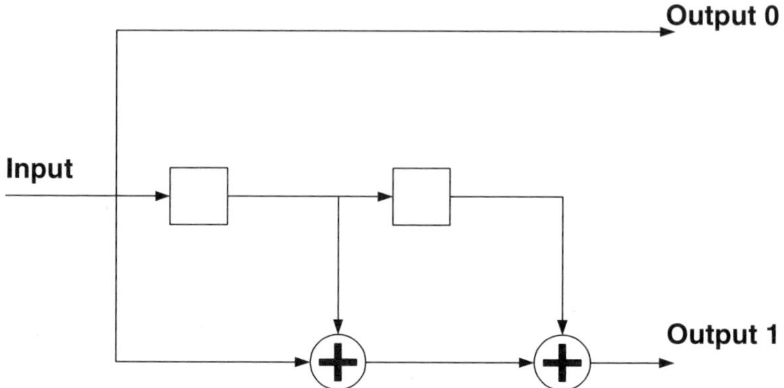

Figure 6.3 A (2,1,3) binary systematic convolutional encoder

Example 6.3 Consider a binary (2,1,3) convolutional code. The encoder of this code is shown in Figure 6.3. The state diagram for this code is represented in Figure 6.4. Note that output 0 of this encoder represents just the input symbol. That means the output (encoded) sequence contains the unchanged input (information) sequence. This kind of code is called a *systematic* convolutional code. This code has only four states: 00, 01, 10 and 11. The branches are labelled with the following notation: inS/outS0outS1, where inS is the input symbols that cause this transition and outS0, outS1 are the output symbols at the corresponding outputs. Another form of the

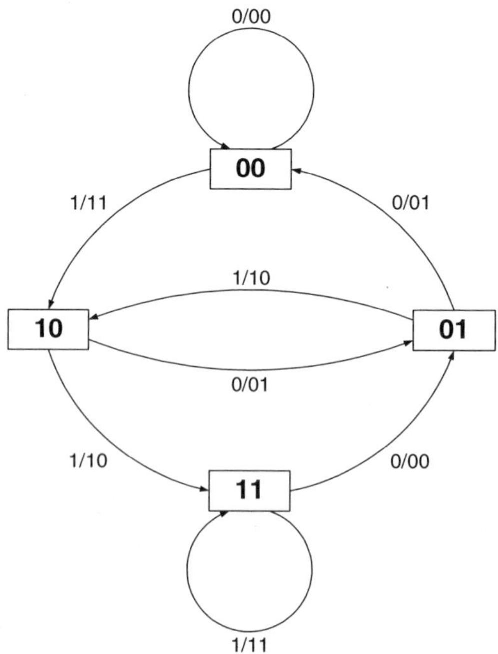

Figure 6.4 A state diagram of the (2,1,3) binary systematic convolutional code

convolutional code representation is the *tree diagram*. The tree diagram represents the encoding process as a tree with the branches corresponding to the transition of the encoder from one possible state to another at a given moment of time. Each node of the tree corresponds to the possible state of the encoder. The number of branches stemming from each node is equal to the number of possible combinations of input symbols, i.e. each branch starting from the given node corresponds to one of the possible combinations of symbols at the input of the encoder. For example, if the cardinal number of the input alphabet is M and the number of the encoder inputs is k, then the number of branches stemming from each node is equal to M^k.

The tree diagram for the code of Example 6.3 for four input bit intervals is shown in Figure 6.5. The upper branch (marked with solid line) from each node corresponds to input data '0', the lower branch (marked with dashed line) to '1'. The labels on the branches show the corresponding data at the encoder output. The tree diagram helps in visualisation of the

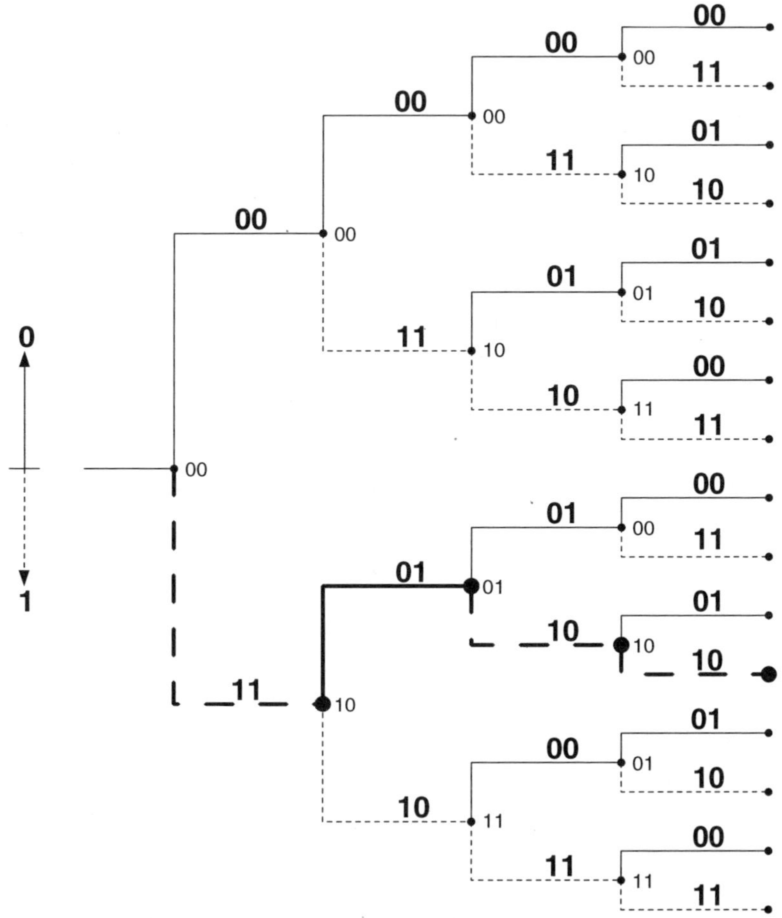

Figure 6.5   A tree diagram of the (2,1,3) binary systematic convolutional code

encoding process since it adds the dimension of time to the state diagram (i.e. it represents each moment of time with a separate state diagram). Now the encoding procedure can be described by traversing the tree diagram from left to right in accordance with the input symbols. An input information sequence defines a specific path through the tree diagram, and an output sequence corresponding to the input information sequence can easily be read from the branch labels of this path. For example, it is easy to see from Figure 6.5 that the input sequence 1011 corresponds to the output sequence 11 01 10 10. Thus any code sequence is represented as a path through the tree.

However, the number of branches in the tree increases exponentially with the length of the input sequence and it is a very hard work to draw the tree diagram for a long input sequence. On the other hand it is easy to see that the tree diagram contains a lot of redundant elements. It is enough to compare parts of the diagram starting at the same state to understand that they are identical. This means the tree diagram can be simplified.

Let us merge all the parts of the tree diagram where the encoder takes the same state at the same time. Now we obtain a diagram in which the number of nodes at any time moment is no more than the number of states. This kind of diagram is called a *trellis diagram*. The trellis diagram provides a more manageable encoder description than does the tree diagram. This compact representation is very helpful in describing the decoding of the convolutional codes, as will be discussed later. The trellis diagram for the code of Example 6.3 is shown in Figure 6.6. Here we use the same convention as for the tree diagram. A solid line denotes the branch generated by an input data '0', and a dashed line denotes the branch generated by an input data '1'. The branches are labelled with the output data. As one can see from Figure 6.6 the structure of the diagram becomes fixed after trellis depth 3 is reached. Generally the

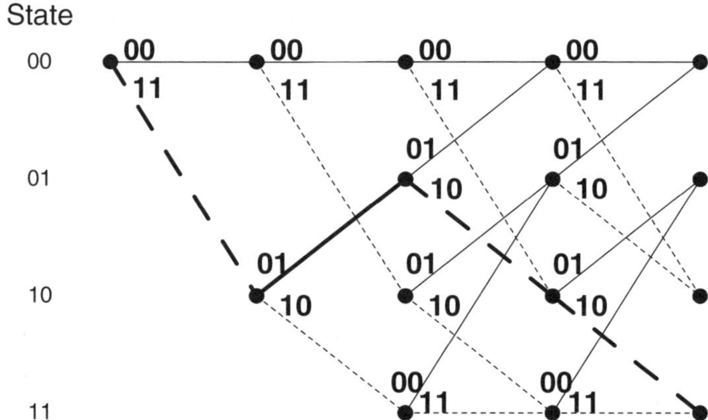

Figure 6.6   A trellis diagram of the (2,1,3) binary systematic convolutional code

structure of the trellis becomes fixed after depth ν is reached. With the help of a trellis diagram we can see the correspondence between paths through the diagram and code sequences, while the complexity of the diagram no longer grows exponentially. The bold lines in the trellis in Figure 6.6 show the path corresponding to the input sequence 1011.

With the help of the trellis diagram we can define the most important measure of performance of a convolutional code, which is called *free distance*. It is the analogue of the minimum distance of block code. As we cannot divide the code sequences into code words of the same length, we have to consider the distance between the complete code sequences. The free distance of a convolutional code d_{free} can be defined as

$$d_{free} = \min\{d(\mathbf{v}, \mathbf{v}') : \mathbf{u} \neq \mathbf{u}'\}, \tag{6.6}$$

where $\mathbf{v}$, $\mathbf{v}'$ are the code sequences corresponding to the information sequences $\mathbf{u}$, $\mathbf{u}'$, respectively. It is assumed that if sequences $\mathbf{u}$, $\mathbf{u}'$ are of different length, the necessary number of zeros is appended to the shorter sequence. In other words, free distance is the smallest Hamming distance between any pair of code sequences. Because a convolutional code is a linear code, there is no loss in generality in finding the minimum distance between each code sequence and the all-zero sequence. Assuming that the all-zero code sequence can be generated only by the all-zero information sequence (on condition that the initial state of encoder is all-zero), the paths of interest (to compare with all-zero code sequence) are those that diverge from and remerge with the all-zero state. These are the closest sequences that could be confused by the decoder. For example, the free distance of the code of Example 6.3 can be find from the trellis in Figure 6.6 by computing the distances between the all-zero path and the paths starting from the left-hand node and returning to the all-zero state later. It is easy to verify that the free distance of this code is 4. For calculating the error-correcting capability of the code we can use the following equation

$$t = \left\lfloor \frac{d_{free} - 1}{2} \right\rfloor, \tag{6.7}$$

where t is the maximum number of errors that the code is capable of correcting, $\lfloor x \rfloor$ means the largest integer not to exceed x. In accordance with (6.7) the code of Example 6.3 can correct only 1 error. It is possible to build a nonsystematic convolutional code with the same parameters (2,1,3), which is capable of correcting 2 errors. In general, making the convolutional code systematic reduces the maximum possible free distance for a given constraint length and rate. This means that unlike the block codes a convolutional nonsystematic code cannot be transformed into a systematic code with the same parameters and error-correcting capability. In Table 6.1 the free distances of systematic and nonsystematic codes of rate $\frac{1}{2}$ are compared [1].

Some convolutional codes cause an infinite number of errors in the output sequence after decoding when only a finite number of errors occur during the transmission of a code

Table 6.1   Comparison of free distance for systematic and nonsystematic convolutional codes. Rate $= 1/2$

Constraint length	d_{free} Systematic	d_{free} Nonsystematic
2	3	3
3	4	5
4	4	6
5	5	7
6	6	8
7	6	10
8	7	10

sequence over the channel. This event is called the *catastrophic error propagation* and this kind of code is called a *catastrophic convolutional code*. This type of code needs to be avoided and can be identified by the state diagram. A state diagram having a loop in which a nonzero information sequence corresponds to all-zero output sequence identifies a catastrophic convolutional code. The examples of such loops are shown in Figure 6.7.

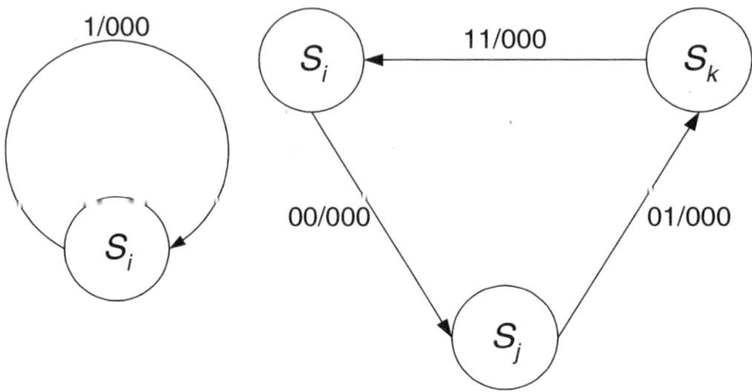

Figure 6.7   Examples of catastrophic convolutional code

6.2 VITERBI DECODING ALGORITHM

The best known algorithm of decoding of the convolutional codes was introduced by A.Viterbi in 1967 [2]. The code sequence $\mathbf{v}$ which is transmitted over the channel and the received sequence $\mathbf{r}$ can be written as

$$\mathbf{r} = \mathbf{v} + \mathbf{e}, \tag{6.8}$$

where $\mathbf{e}$ is the error sequence. The Viterbi algorithm finds a code sequence $\mathbf{y}$ such that it maximises the probability $P(\mathbf{r}|\mathbf{y})$ that sequence $\mathbf{r}$ is received in condition that sequence $\mathbf{y}$ is transmitted. Usually it is more convenient to maximise the logarithm of the probability $P(\mathbf{r}|\mathbf{y})$ rather than $P(\mathbf{r}|\mathbf{y})$ itself. However, since the logarithms are the monotonic functions the code sequence $\mathbf{y}$ that maximises value $\log P(\mathbf{r}|\mathbf{y})$ maximises $P(\mathbf{r}|\mathbf{y})$ also. In the case where the channel is BSC the Viterbi algorithm finds the code sequence $\mathbf{y}$ that is closest to the received sequence $\mathbf{r}$ in the sense of minimum Hamming distance. That means that the Viterbi algorithm is a maximum likelihood sequence detection algorithm. The advantage of the Viterbi algorithm compared with e.g. brute-force maximum likelihood decoding, is that the complexity of the Viterbi algorithm does not depend exponentially on the number of symbols in the code sequence.

As was discussed earlier the Viterbi algorithm selects the code sequence $\mathbf{y}$ such that it maximises the probability $P(\mathbf{r}|\mathbf{y})$ that sequence $\mathbf{r}$ is received on condition that sequence $\mathbf{y}$ is transmitted. This probability is called the *likelihood function*. The channel is assumed to be memoryless, and thus the noise process affects each received symbol independently of the all the other received symbols. From the probability theory it is known that the probability of joint, independent events is equivalent to the product of the probabilities of the individual events. Thus,

$$P(\mathbf{r}|\mathbf{y}) = \prod_i P(r_i|y_i), \qquad (6.9)$$

where $P(r_i|y_i)$ is a channel transition probability. It is more convenient to use log-likelihood function $\log P(\mathbf{r}|\mathbf{y})$ rather than the likelihood function itself. It follows from (6.9) that

$$\log P(\mathbf{r}|\mathbf{y}) = \sum_i \log P(r_i|y_i). \qquad (6.10)$$

To simplify the manipulation of the summations over the log function, a *symbol metric* $M(r_i|y_i)$ is defined as

$$M(r_i|y_i) = c_1 \cdot (\log P(r_i|y_i) + c_2), \qquad (6.11)$$

where c_1 and c_2 are chosen such that the symbol metric can be well approximated by integers.

From the symbol metric a *path metric* and a *branch metric* are defined as follows:

$$M(\mathbf{r}|\mathbf{y}) = \sum_{j=0}^{L-1} M(\mathbf{r}_j|\mathbf{y}_j) = \sum_{j=0}^{L-1} \left(\sum_{i=0}^{n-1} M(r_i|y_i) \right), \qquad (6.12)$$

where $M(\mathbf{r}|\mathbf{y})$ is the path metric, $M(\mathbf{r}_j|\mathbf{y}_j)$ is the branch metric and L is the number of blocks of n symbols in the sequence. In the same manner we can define the *partial path metric* $M^j(\mathbf{r}|\mathbf{y})$ as

$$M^i(\mathbf{r}|\mathbf{y}) = \sum_{j=0}^{i} M(\mathbf{r}_j|\mathbf{y}_j). \qquad (6.13)$$

The symbol metric shows the cost of choosing symbol y_i as the estimate of the corresponding symbol r_i. The branch metric indicates the cost of choosing the branch from the trellis, the partial path metric $M^j(\mathbf{r}|\mathbf{y})$ corresponds to the cost of choosing given path $\mathbf{y}$ up to time index j as a part of decoded sequence and finally the path metric shows the total cost of estimating the received sequence $\mathbf{r}$ with the sequence $\mathbf{y}$.

6.2.1 Hard Decision Viterbi Algorithm

Let us consider for simplicity the Viterbi algorithm for BSC (i.e. the hard-decision Viterbi algorithm) first. If the symbols are transmitted over the BSC with the crossover probability p the likelihood function for the received sequence $\mathbf{r}$ of length N can be written as

$$P(\mathbf{r}|\mathbf{y}) = (1-p)^{N-d(\mathbf{r},\mathbf{y})} \cdot p^{d(\mathbf{r},\mathbf{y})} = \prod_{i=0}^{N-1} (1-p) \cdot \left(\frac{p}{1-p} \right)^{d(r_i,y_i)}, \qquad (6.14)$$

where $d(\bullet, \bullet)$ is the Hamming distance. Then the symbol metric in accordance with (6.11) and (6.14) can be written in the form of

$$M(r_i|\,y_i) = c_1 \cdot \left(\log\left((1-p) \cdot \left(\frac{p}{1-p} \right)^{d(r_i, y_i)} \right) + c_2 \right), \qquad (6.15)$$

where coefficients c_1 and c_2 can be chosen as follows:

$$c_1 = \left(\log \frac{p}{1-p} \right)^{-1}, \qquad (6.16)$$

$$c_2 = -\log(1-p). \qquad (6.17)$$

Then the symbol metric becomes the Hamming metric and can be written as

$$M(r_i|\,y_i) = 1 - d(r_i,\, y_i). \qquad (6.18)$$

In this case the problem of finding the sequence $\mathbf{y}$ that maximises the probability $P(\mathbf{r}|\,\mathbf{y})$ can be formulated as the search of optimum path $\mathbf{y}$ (with the minimum Hamming distance between $\mathbf{y}$ and $\mathbf{r}$) through the trellis. This is equivalent to a dynamic programming problem of finding the path with minimum weight through a weighted graph [3]. The Viterbi algorithm is based on the *principle of optimality*. The principle of optimality states that if any two paths in the trellis merge to the same state, one of them can always be discarded in the search for an optimum path, because the path with more weight could not turn out to be the prefix of the optimum path through the trellis. This statement can be illustrated by the picture in Figure 6.8.

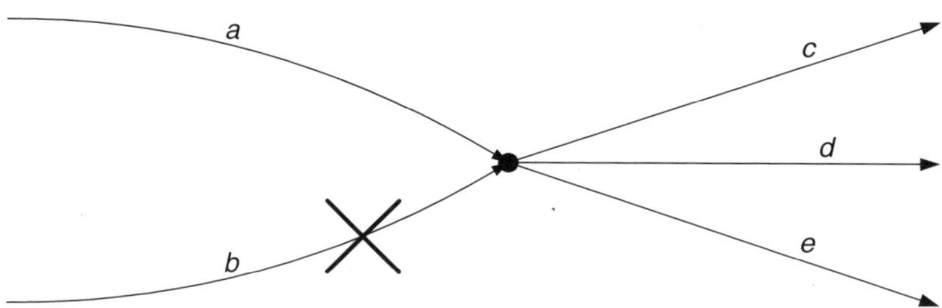

Figure 6.8 Eliminating one of the merged paths in trellis in accordance with the principle of optimality

The weights of branches a, b, c, d, e correspond to the Hamming distance between parts of two sequences; this means all the weights are nonzero values. Let $a < b$, then the weights of paths generated by path with weight a, i.e. $a+c, a+d, a+e$ will be less than $b+c, b+d, b+e$ for all possible values of c, d, e. In this case we can eliminate the path with weight b in the search for an optimum path with minimum weight. This principle allows us to consider only the constant number of paths on the each stage of the decoding procedure.

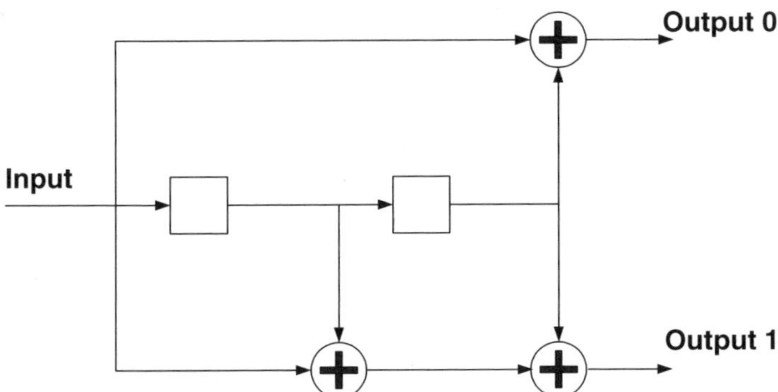

Figure 6.9   A (2,1,3) binary nonsystematic convolutional encoder

Example 6.4   Let us consider the hard-decision Viterbi algorithm on the example of decoding of a nonsystematic (2,1,3) binary convolutional code. The encoder of this code is shown in Figure 6.9. The corresponding trellis diagram is represented in Figure 6.10. As one can see from the trellis diagram the free distance of this code is 5. That means the code is capable of correcting 2 errors. The information sequence is $\mathbf{u} = 1100011111$. This information sequence generates the code sequence $\mathbf{v} =$ 11 10 10 11 00 11 10 01 01 01.

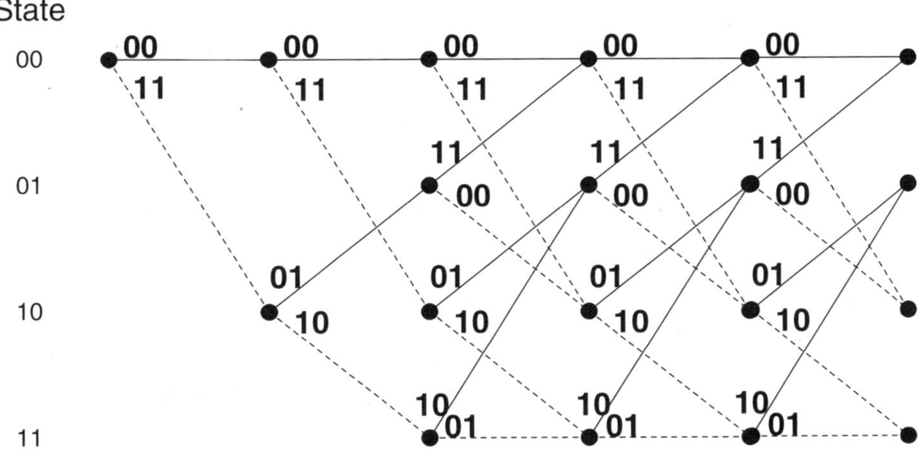

Figure 6.10   A trellis diagram of the (2,1,3) binary nonsystematic convolutional code

Assume that the received sequence is $\mathbf{r} = 10\ 00\ 10\ 11\ 00\ 11\ 10\ 01\ 01\ 01$, i.e. 2 errors occur during the transmission of the code sequence over the channel. Let us go through the trellis from left to right and search the optimum path that is closest to the received sequence $\mathbf{r}$ in the sense of Hamming distance assuming that the initial state of the encoder is all-zero (at the beginning of encoding of each sequence the encoder should be flushed). The process of decoding is shown in Figure 6.11. At the first stage (Figure 6.11.a) we have two partial

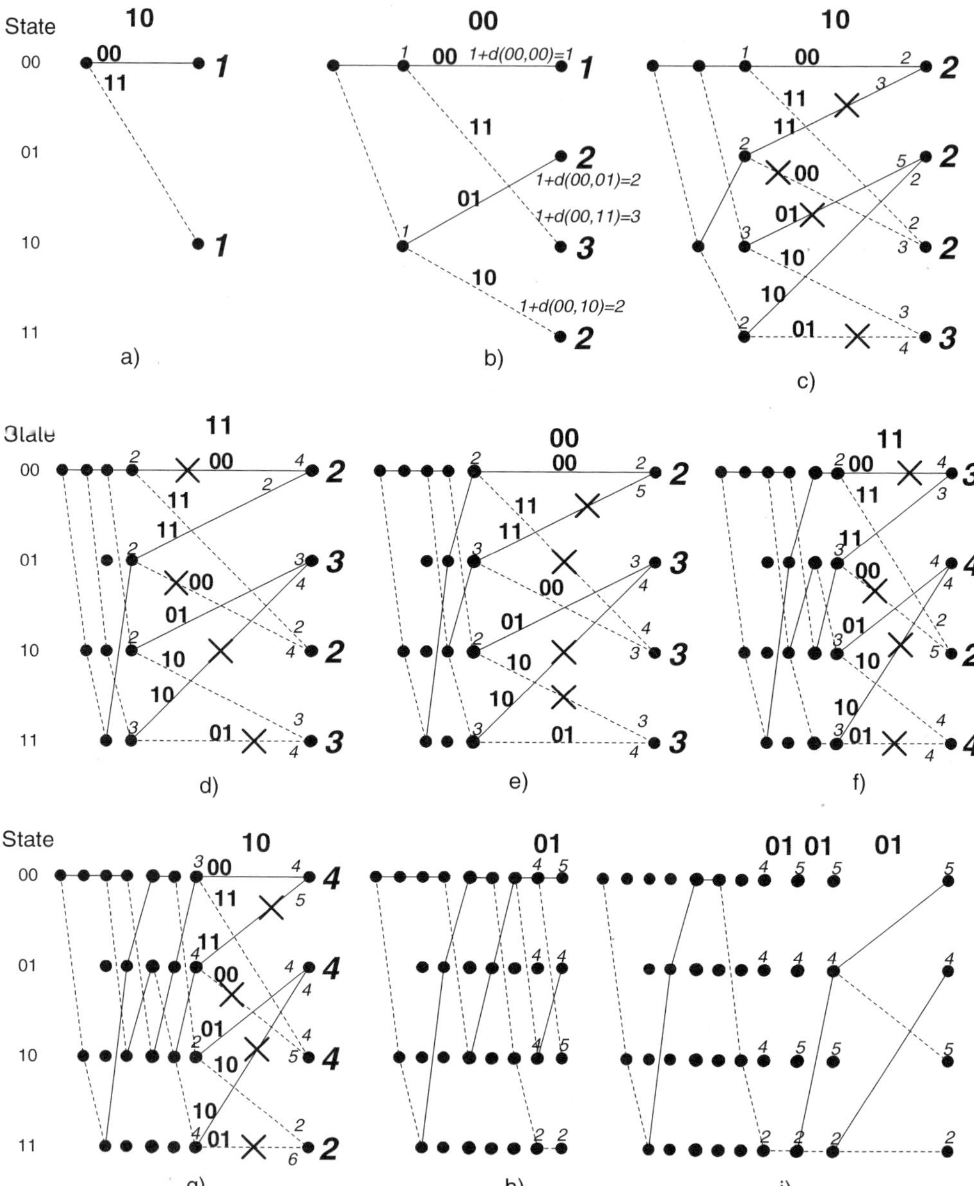

Figure 6.11 The stages of Viterbi decoding of the received sequence $r = 10\,00\,10\,11$ $00\,11\,10\,01\,01\,01$ in trellis diagram of the (2,1,3) binary nonsystematic convolutional code

paths in the trellis: 00 and 11. The partial path is defined as the path from state $S = 0$ at time $j = 0$ to a particular state $S = i$ at time $j \geq 0$. Both of these paths differ from the received symbols 10 in one position. Hence, both of these paths have weight 1. There are no paths merging at the same state at this stage, so we keep all the paths, because we have no choice yet. At the second stage (Figure 6.11.b) the number of partial paths is 4. The weight of each partial path (*path metric*) is obtained as the sum of weight of the previous partial path and the weight of the corresponding branch (*branch metric*). The corresponding symbols of the received sequence **r** are 00. Then the branch metric of the branch from state 00 to state 00 is equal to $d(00, 00) = 0$, the branch metric of the branch from state 00 to state 10 is $d(00, 11) = 2$, the branch metric of the branch from state 10 to state 01 is $d(00, 01) = 1$, the branch metric of the branch from state 10 to state 11 is $d(00, 10) = 1$. Summing up the branch metric and the path metric of the previous partial path gives the following path metrics of the corresponding paths: 1, 3, 2, 2. Here we also have no merging paths, so we keep all 4 paths. At the third stage (Figure 6.11.c) we have 8 paths. We can calculate the path metrics as in previous stages and then we have 2 paths merging to each state. Now in accordance with the principle of optimality we can eliminate the path with greater weight. For example, the path merging to state 00 from state 00 has the path metric 2 and the path merging to the same state 00 from state 01 has the path metric 3 that means the last path should be discarded. The path merging to state 00 from state 00 is called a *survivor path*. After this procedure we only have to keep again 4 paths. Hereafter at each stage we only have to keep 4 survivor paths.

In the case of two paths with the same weight merged to a single node, an arbitrary decision about discarding one of these two paths can be made (e.g. at the sixth stage in Figure 6.11.f two paths with weight 4 merge to the state 01 and two paths with weight 4 merge to the state 11; in this example we eliminate the lower of the two paths with the same weight). As can be seen from Figure 6.11.e already at the fifth stage the path metric of the partial path corresponding to the sequence 11 10 10 11 00 (which is the part of the transmitted sequence **v**) is the best among the 4 metrics of the survivor paths. Since in our example there are no new errors in the latter symbols it is obvious that the correct path will be found at the end. On the other hand we can see from Figure 6.11 that the survivor paths can differ from each other over a long time. It is only at stage 10 as can be seen from Figure 6.11.i that the first 8 branches of the survivor paths coincide. At this time we can make a decision about the first 8 transmitted symbols since the survivor paths merge. The depth of this merge is an arbitrary value and it depends only on errors in the channel during the transmission. In any case we can see that the decoding introduces a severe delay, which is much more than one stage period. In practice it is impossible to wait until the survivor paths merge. Usually the fixed depth of the decoding is defined for the decoder. After reaching this *decoding depth* (the certain number of stages) the decision about the first symbols (in accordance with path with best metric) is made. This of course leads to some degradation in performance and the algorithm is no longer optimal but now suboptimal. On the other hand if the depth of making a decision is five or seven times the constraint length, the degradation of performance is negligible [4].

Now we can formulate the hard decision Viterbi algorithm as follows [5]:

Hard-Decision Viterbi Decoding:

$S_{i,j}$ is the state in the trellis diagram that corresponds to the state S_i at time j. Every state in the trellis is assigned a value denoted $V(S_{i,j})$. L is the decoding depth (or as it is often called the *truncation window length*).

1.
   a. Initialise time $j = 0$.
   b. Initialise $V(S_{0,0}) = 0$ and all other $V(S_{i,j}) = \infty$.

2.

   a. Set time $j = j + 1$.

   b. For all i compute the partial path metrics for all paths going to state S_i at time j. To do this: first, calculate the branch metric, and then add the branch metric to $V(S_{i,j-1})$.

3.

   a. For all i set $V(S_{i,j})$ to the best partial path metric going to state S_i at time j.

   b. If there is a tie for the best partial path metrics, then any one of the tied partial path metric may be chosen.

4. If $j < L$ go to the step 2.

5.

   a. Start trace-back through the trellis by following the branches of the best survivor path.

   b. Store the associated survivor k symbols. These are currently decoded k information symbols.

   c. Set time $j = 0$; go to step 2. Here is the start of the new truncation window.

Usually it is more convenient to use code words of fixed length rather than the semi-infinite code sequences. In this case it is possible to add $k \cdot (\nu - 1)$ dummy zeros (so called *tail symbols*) to the end of information sequence of fixed length before encoding, which forces the encoder to return to the all-zero state and terminates the trellis. This simplifies the work of the decoder, because now only the survivor that ends at the all-zero state needs to be checked. Obviously by using this technique the convolutional code becomes the block code.

6.2.2 Soft Decision Viterbi Algorithm

The soft-decision Viterbi algorithm exploits the additional information, which is provided by the soft-decision demodulator and this additional information allows the performance to be increased. The algorithm itself is the same as for the hard decisions. The only difference is that the Hamming distance is not used as a metric. Generally speaking, the metric used in the algorithm should be defined by the channel. For example, the Euclidean distance is the optimal metric for the Gaussian channel. Let us consider the example of the soft-decision Viterbi decoding for the *discrete memoryless channel* (DMC).

Example 6.5 Let us consider the binary input, 8-ary output DMC represented in Figure 6.12. The transition probabilities $P(r|y)$ of this channel are shown in the following table:

$P(r\|y)$	0_4	0_3	0_2	0_1	1_1	1_2	1_3	1_4
0	0.439	0.2	0.17	0.1	0.06	0.025	0.005	0.001
1	0.001	0.005	0.025	0.06	0.1	0.17	0.2	0.439

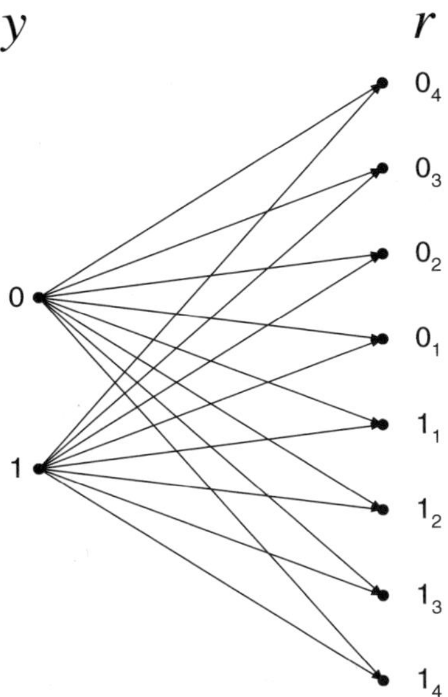

Figure 6.12   Binary input, 8-ary output DMC

Taking the logarithms obtain the log-likelihood values $\log P(r|y)$:

$\log P(r\mid y)$	0_4	0_3	0_2	0_1	1_1	1_2	1_3	1_4
0	-0.82	-1.61	-1.77	-2.3	-2.81	-3.69	-5.3	-6.91
1	-6.91	-5.3	-3.69	-2.81	-2.3	-1.77	-1.61	-0.82

Let us choose the coefficient $c_2 = -\min_y P(r|y)$ and the coefficient $c_1 = 1.35$. Then the symbol metric values $M(r|y)$ in accordance with (6.11) can be written as follows:

$M(r\mid y)$	0_4	0_3	0_2	0_1	1_1	1_2	1_3	1_4
0	8	5	3	1	0	0	0	0
1	0	0	0	0	1	3	5	8

Let the information and encoded sequence be the same as in previous example $\mathbf{u} = 1100011111$, $\mathbf{v} = 11\ 10\ 10\ 11\ 001110010101$. Assume that the received sequence is $\mathbf{r} = 1_4 0_3\ 0_2 0_2\ 0_1 0_4\ 1_1 1_1\ 0_2 0_3\ 1_2 1_3\ 1_2 0_3\ 0_4 1_1\ 0_2 1_1\ 0_1 1_2$. The decoding process for this received sequence is shown in Figure 6.13.

As one can see from Figure 6.13 the first 8 symbols are decoded correctly at stage 10. Notice that if we merge the soft decisions outputs $0_1\ 0_2\ 0_3\ 0_4$ into the hard decision output 0 and

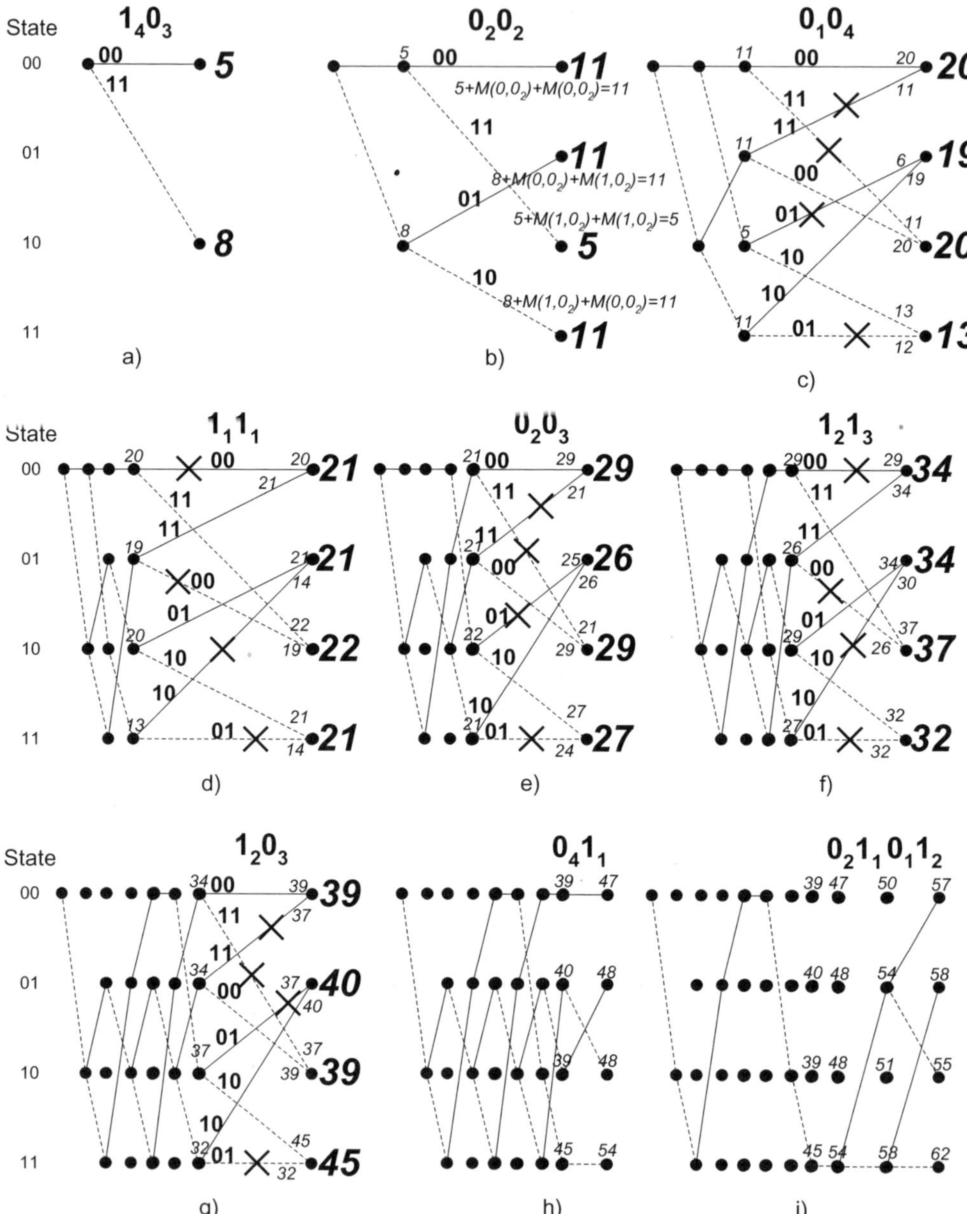

Figure 6.13 The stages of soft-decision Viterbi decoding of the received sequence $r = 1_4 0_3\, 0_2 0_2$ $1_1 0_4\, 1_1 1_1\, 0_2 0_3\, 1_2 1_3\, 1_2 0_3\, 0_4 1_1\, 0_2 1_1\, 0_1 1_2$ in trellis diagram of the $(2,1,3)$ binary nonsystematic convolutional code

the outputs $1_1\, 1_2\, 1_3\, 1_4$ into the hard decision output 1, the hard decision received sequence becomes $\mathbf{r} = 10\,00\,00\,11\,00\,11\,10\,01\,01\,01$. The hard-decision decoding of this sequence is represented in Figure 6.14. As expected, in this case the decoder chooses the path that does not coincide with the transmitted sequence (Figure 6.14.f).

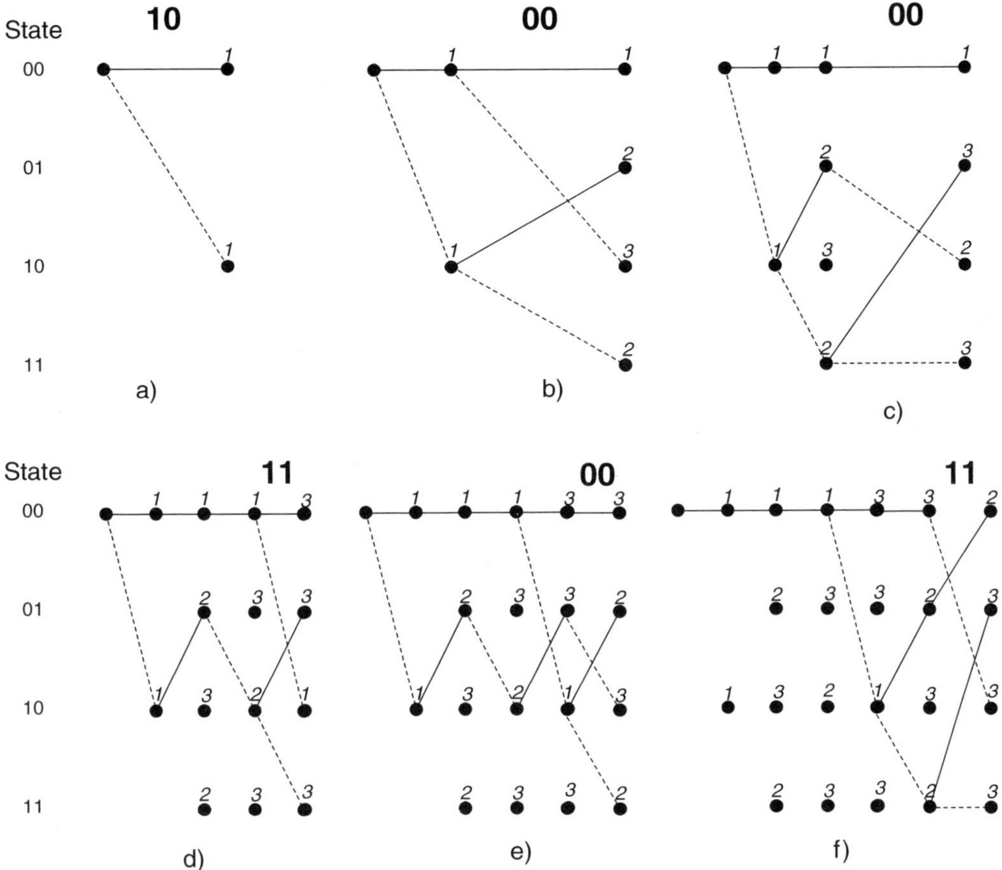

Figure 6.14 The stages of hard-decision Viterbi decoding of the received sequence $r = 10\,00\,10\,11$ $00\,11\,10\,01\,01\,01$ in a trellis diagram of the (2,1,3) binary nonsystematic convolutional code

Now we can write the soft-decision Viterbi algorithm as follows:

Soft-Decision Viterbi Decoding:

1.

    a. Initialise time $j = 0$.

    b. Initialise $V(S_{0,0}) = 0$ and all other $V(S_{i,j}) = -\infty$.

2.

    a. Set time $j = j + 1$.

    b. For all i compute the partial path metrics for all paths going to state S_i at time j. To do this: first, calculate the branch metric $M(\mathbf{r}_j \mid \mathbf{y}_j)$ in accordance with (6.12), and then compute the j-th partial path metric $M^j(\mathbf{r} \mid \mathbf{y}) = V(S_{i,j-1}) + M(\mathbf{r}_j \mid \mathbf{y}_j)$.

3.

   a. For all i set $V(S_{i,j})$ to the 'best' partial path metric going to state S_i at time j.

   b. If there is a tie for the best partial path metrics, then any one of the tied partial path metric may be chosen.

4. If $j < L$ go to step 2.

5.

   a. Start trace-back through the trellis by following the branches of the best survivor path.

   b. Store the associated survivor k symbols. These are currently decoded k information symbols.

   c. Set time $j = 0$, go to step 2. Here is the start of the new truncation window.

As can be seen the soft-decision algorithm differs from the hard-decision algorithm only by the used metric. The Example 6.5 demonstrates the gain which can be obtained by using the same Viterbi algorithm exploiting the additional information from the soft-decision demodulator. Usually soft-decision decoding increases the coding gain of a convolutional code by about 2 dB.

As was discussed earlier the number of nodes at each trellis stage is equal to $M^{k(\nu-1)}$, where M is the cardinal number of the input alphabet. At each node of the trellis M^k calculations are needed to perform the Viterbi algorithm. Hence, the complexity of the Viterbi algorithm is of the order of $O(M^{k(\nu-1)} \cdot M^k \cdot L)$. This value is significantly less than the complexity of brute-force ML decoding, which can be estimated as $O(M^{kL})$. However, the increase in the number of information symbols k or the constraint length ν leads to exponential growth of the Viterbi algorithm complexity.

6.3   LIST DECODING

List decoding is a suboptimal non-backtracking algorithm, which consists in choosing the best partial paths at the each stage of the decoding process. These partial paths form a list of size L. Unlike the Viterbi algorithm the list decoder considers the extensions only of these best partial paths from the list, not all partial paths. The list decoding algorithm belongs to the class of breadth-first algorithms, as does the Viterbi algorithm. Of course, the list size L should be less than the number of states $M^{\nu-1}$. Obviously the complexity of the list decoding is less than the complexity of the Viterbi algorithm but because some partial paths are not considered, list decoding is not the optimal algorithm.

> ***Example 6.6*** Let us consider the hard-decision list decoding of the received sequence $\mathbf{r} = 10\ 00\ 10\ 11\ 00\ 11\ 10\ 01\ 01\ 01$ of Example 6.4. The size of the list is 3, i.e. we will find the extensions of only 3 best paths at each decoding stage. The decoding process is shown in Figure 6.15. As can be seen from Figure 6.15 the received sequence is successfully decoded and already at stage 9 (Figure 6.15.h) the first 7 branches of the survivor paths coincide.

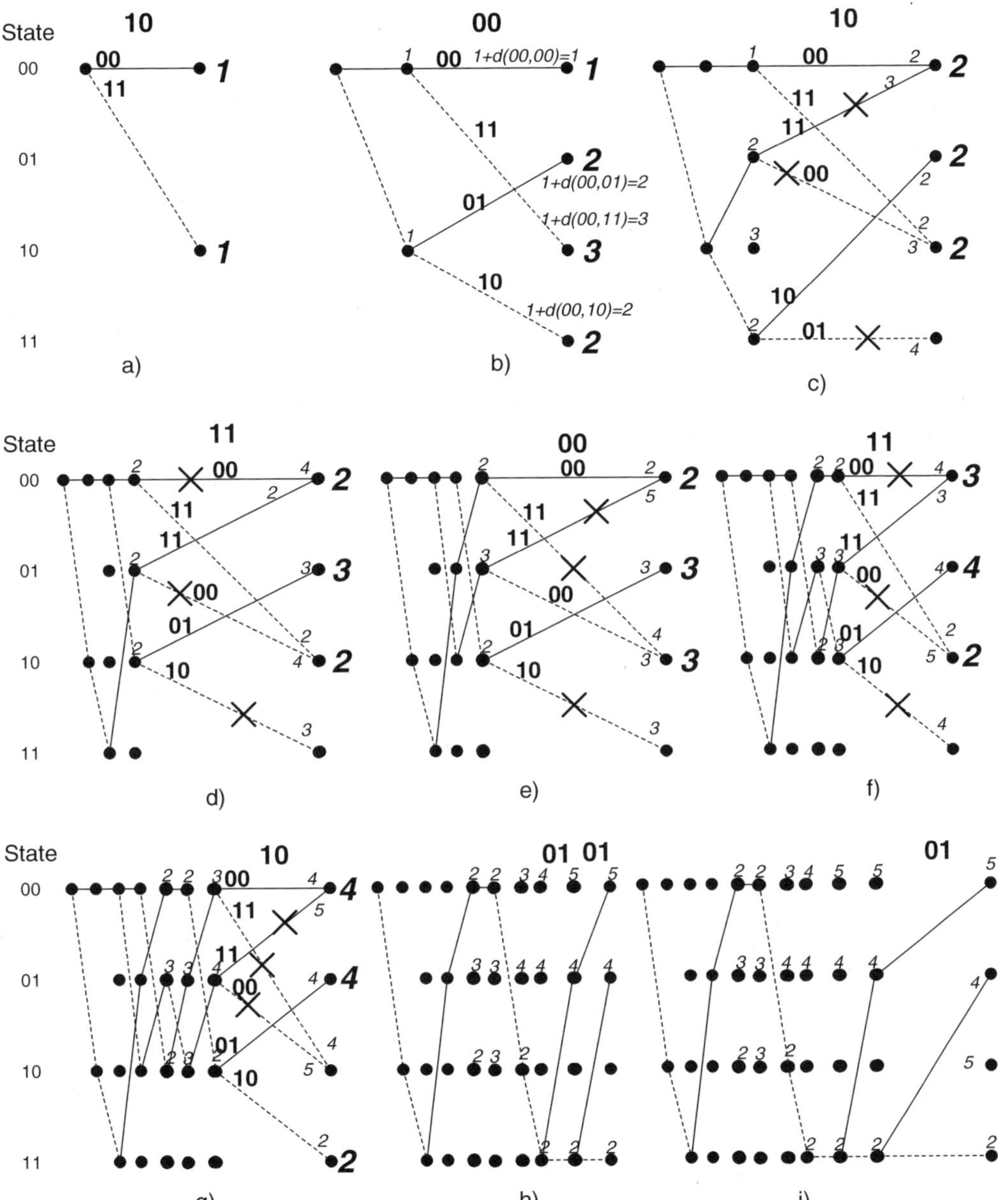

Figure 6.15 The stages of hard-decision List decoding of the received sequence $r = 10\,00\,10\,11$ $00\,11\,10\,01\,01\,01$ in trellis diagram of the $(2, 1, 3)$ binary nonsystematic convolutional code

So in this particular case we have managed to decode correctly the received sequence with less complexity than the Viterbi algorithm requires. Unfortunately the list algorithm as was mentioned above is not the optimal one and in some cases it is possible to miss the correct path. The greater the list size, the less is the probability of missing the correct path.

6.4 SEQUENTIAL DECODING

Sequential decoding algorithms were invented by Wozencraft and then by Fano before the discovery of the Viterbi algorithm. Due to the exponential growth of the Viterbi algorithm complexity with the growth of constraint length of the code, application of this algorithm is limited to the case of relatively small values of the constraint lengths. Unlike the Viterbi algorithm the complexity of sequential decoding is essentially independent of constraint length. Sequential decoding algorithms belong to the class of backtracking algorithms known as *depth first*, because they go forward in the depth of the code tree tracing a given path as long as the metric indicates that the choice is likely, otherwise they return and start tracing a new path. The Viterbi algorithm on the contrary belongs to the class of algorithms known as *breadth first*, because it explores all the paths on the given stage of trellis before considering the next stage. It is possible that a sequential decoding algorithm misses the best path because it does not explore the whole trellis. Hence, the sequential decoding algorithms are the sub-optimal algorithms.

In sequential decoding we have to compare paths of different lengths. To do this usually the *Fano metric* is used. The hard-decision Fano metric for the path $\mathbf{y}$ can be represented as

$$M_F(\mathbf{r}, \mathbf{y}) = \sum_{j=0}^{L-1} M_F(\mathbf{r}_j, \mathbf{y}_j), \tag{6.19}$$

where L is the path length in n-tuples (branches) and $M_F(\mathbf{r}_j, \mathbf{y}_j)$ is the *Fano branch metric*. The Fano branch metric can be written as

$$M_F(\mathbf{r}_j, \mathbf{y}_j) = (n - d(\mathbf{r}_j, \mathbf{y}_j)) \cdot a + d(\mathbf{r}_j, \mathbf{y}_j) \cdot b, \tag{6.20}$$

where

$$\begin{aligned} a &= \log(1 - p) + 1 - R \\ b &= \log p + 1 - R \end{aligned}, \tag{6.21}$$

p is crossover probability for BSC and $R = \dfrac{k}{n}$ is the rate of a convolutional code. The Fano branch metric can be expressed as the sum of *Fano symbol metrics*:

$$M_F(\mathbf{r}_j, \mathbf{y}_j) = \sum_{i=0}^{n-1} M_F(r_i, y_i), \tag{6.22}$$

where the Fano symbol metric $M_F(r_i, y_i)$ is

$$M_F(r_i, y_i) = \begin{cases} a, & \text{if } r_i = y_i \\ b, & \text{if } r_i \neq y_i \end{cases}, \tag{6.23}$$

where a and b are represented in (6.21). The Fano metric increases along the correct path and decreases along the incorrect path.

6.4.1 Stack Algorithm

The idea of the stack algorithm is very simple. The decoder creates the list (stack) of the most likely candidates to be the correct path. The list is sorted in such a way that the more likely candidate (with greatest Fano metric) is always on the top of the stack. So, when the exploration of the code tree is finished the top path is assumed to be the decoded sequence. The algorithm can be formulated as follows [6]:

Stack Decoding Algorithm:

1. Load the stack with the root and the metric zero.

2. Remove the top node and place its successors in the stack according to their metrics.

3. If the top path leads to the end of the tree, then stop and choose the top path to be the decoded sequence; otherwise go to step 2.

Example 6.7 Consider the (2, 1, 3) binary convolutional code of Example 6.4. Let the channel be BSC with the crossover probability $p = 0.05$. Let the information sequence be $\mathbf{u} = 01000$ and the corresponding encoded sequence $\mathbf{v} = 00\,11\,01\,11\,00$. Assume that 2 errors occur during the transmission and the received sequence is $\mathbf{r} = 10\,01\,01\,11\,00$. Let us decode the received sequence $\mathbf{r}$ with the help of a stack-algorithm using the Fano metric (for hard-decisions). First of all, let us calculate the symbol Fano metric for given convolutional code and BSC:

$$a = \log(1 - 0.05) + 1 - \frac{1}{2} = 0.449$$
$$b = \log 0.05 + 1 - \frac{1}{2} = -2.496$$

$$(6.24)$$

For convenience we will use more rough values than in (6.24):

$$a = 0.5 \quad \text{if } y_i = r_i;$$
$$b = -2.5 \quad \text{if } y_i \neq r_i.$$

Then following the steps of stack-algorithm let us find the decoded sequence by exploring the code tree and putting the obtained values of the Fano path metric in the stack.

1. The initial metric value for the root node is 0.

2. Explore paths 00 and 11. Compare the corresponding part of the received sequence 10 with branches (paths) 00 and 11. Both branches differ from the received sequence in 1 bit. This means that the Fano metric for each branch is equal to $a + b = 0.5 - 2.5 = -2$. Both paths have metric -2. Path 00 is on the top of the stack.

Path	Path metric
00	-2
11	-2

3. Explore branches 00 and 11, which are the successors of the path 00. The corresponding part of the received sequence is 01. Add corresponding branch metrics (-2 and -2) to the path metric. Put the obtained path metrics (-4 and -4) and the corresponding paths in the stack. Now the path 11 with metric -2 is on the top of the stack.

Path	Path metric
11	-2
00 00	-4
00 11	-4

4. Explore branches 01 and 10, which are the successors of the path 11. The corresponding part of the received sequence is 01. Add corresponding branch metrics ($+1$ and 5) to the path metric. Put the obtained path metrics (-1 and -7) and the corresponding paths in the stack. Now path 11 01 with metric -1 is on the top of the stack.

Path	Path metric
11 01	-1
00 00	-4
00 11	-4
11 10	-7

5. Explore branches 11 and 00, which are the successors of path 11 01. The corresponding part of the received sequence is 01. Add corresponding branch metrics (-2 and -2) to the path metric. Put the obtained path metrics (-3 and -3) and the corresponding paths in the stack. Now path 11 01 11 with metric -3 is on the top of the stack.

Path	Path metric
11 01 11	-3
11 01 00	-3
00 00	-4
00 11	-4
11 10	-7

6. Explore branches 00 and 11, which are the successors of path 11 01 11. The corresponding part of the received sequence is 11. Add corresponding branch metrics (-5 and $+1$) to the path metric. Put the obtained path metrics (-8 and -2) and the corresponding paths in the stack. Now path 11 01 11 11 with metric -2 is on the top of the stack.

Path	Path metric
11 01 11 11	-2
11 01 00	-3
00 00	-4
00 11	-4
11 10	-7
11 01 11 00	-8

7. Explore the successors of path 11 01 11 11. The corresponding part of the received sequence is 00.

Path	Path metric
11 01 00	−3
00 00	−4
00 11	−4
11 01 11 11 01	−4
11 01 11 11 10	−4
11 10	−7
11 01 11 00	−8

8. Explore the successors of path 11 01 00. The corresponding part of the received sequence is 11.

Path	Path metric
00 00	−4
00 11	−4
11 01 11 11 01	−4
11 01 11 11 10	−4
11 01 00 01	−5
11 01 00 01	−5
11 10	−7
11 01 11 00	−8

9. Explore the successors of path 00 00. The corresponding part of the received sequence is 01.

Path	Path metric
00 11	−4
11 01 11 11 01	−4
11 01 11 11 10	−4
11 01 00 01	−5
11 01 00 01	−5
00 00 00	−6
00 00 11	−6
11 10	−7
11 01 11 00	−8

10. Explore the successors of path 00 11. The corresponding part of the received sequence is 01.

Path	Path metric
00 11 01	−3
11 01 11 11 01	−4
11 01 11 11 10	−4
11 01 00 01	−5
11 01 00 01	−5
00 00 00	−6
00 00 11	−6
11 10	−7
11 01 11 00	−8
00 11 10	−9

11. Explore the successors of path 00 11 01. The corresponding part of the received sequence is 11

Path	Path metric
00 11 01 11	-2
11 01 11 11 01	-4
11 01 11 11 10	-4
11 01 00 01	-5
11 01 00 01	-5
00 00 00	-6
00 00 11	-6
11 10	-7
00 11 01 00	-8
11 01 11 00	-8
00 11 10	-9

12. Explore the successors of path 00 11 01 11. The corresponding part of the received sequence is 00. The path on top of the stack reached the end of the code tree. The decoded path is 00 11 01 11 00, which coincides with the transmitted sequence **v**.

Path	Path metric
00 11 01 11 00	-1
11 01 11 11 01	-4
11 01 11 11 10	-4
11 01 00 01	-5
11 01 00 01	-5
00 00 00	-6
00 00 11	-6
11 10	-7
00 11 01 11 11	-7
00 11 01 00	-8
11 01 11 00	-8
00 11 10	-9

The corresponding partially explored code tree is represented in Figure 6.16.

As can be seen from Figure 6.16 we have to explore about 1/3 of the code tree in this example to find the decoded sequence. The Fano metric can also easily be used for soft-decisions and the stack-algorithm does not need any changes except the new metric for soft-decision decoding. The serious drawback of the stack-algorithm is the necessity to keep the long list (stack) of path-candidates and to sort the stack at the each stage.

6.4.2 Fano Algorithm

The Fano algorithm differs from the stack algorithm in such a way that it explores only the immediate successors or predecessors of the current path. It never 'jumps' as the stack-algorithm. It moves along a certain path while the metric exceeds some threshold. In another case it returns back and explores the next successor of the previous partial path.

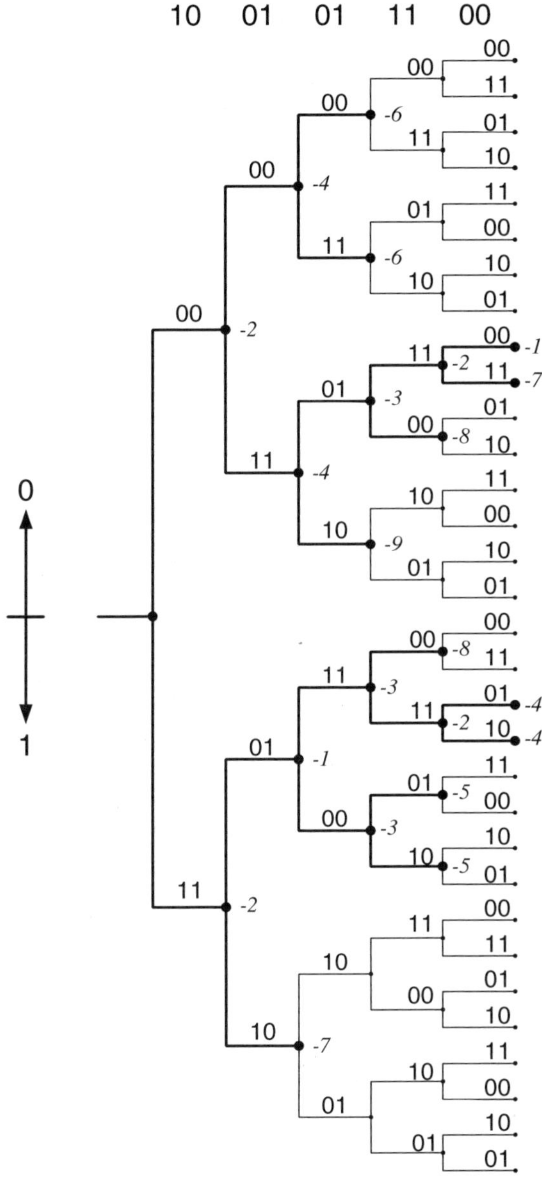

Figure 6.16 The partially explored code tree of Example 6.7. Stack-algorithm

Example 6.8 Consider the decoding of the same received sequence $\mathbf{r} =$ 10 01 01 11 00 as in Example 6.7 with the help of the Fano algorithm. The same Fano symbol metric $a = 0.5$; $b = -2.5$ is used. Let the threshold be -5.

1. Compare branches 00 and 11 with the received symbols 10. Both branches have the same metric -2. The decoder can arbitorily choose any of them. Let us in this example always

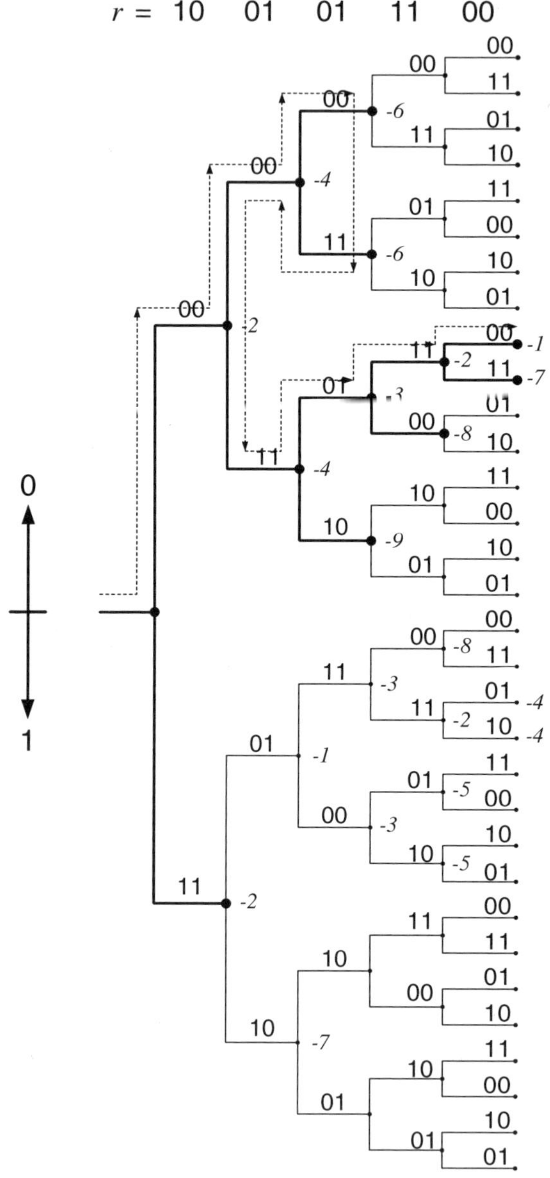

Figure 6.17 The partially explored code tree of Example 6.8. Fano algorithm

choose moving to the upper part of the code tree if the alternatives are equal. Then the decoder choice is the one moving along the path 00. The current path metric is -2.

2. Compare the successors of path 00 branches 00 and 11 with the received symbols 01. Again the branch metrics are equal and the decoder chooses path 00 00. The current path metric is -4.

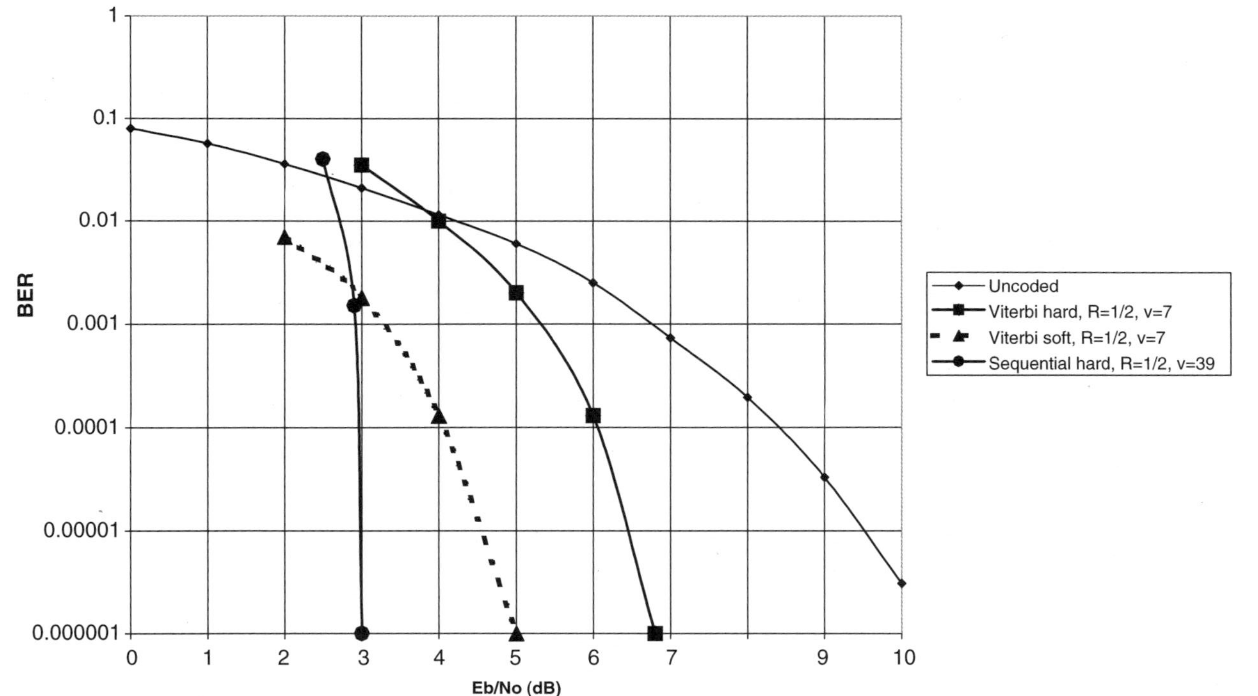

Figure 6.18 Performance comparison of different decoding algorithms for rate 1/2 convolutional codes. BPSK over AWGN channel

3. Compare the successors of path 00 00 branches 00 and 11 with the received symbols 01. The branches again have equal metrics but now the path metric -6 is less than the threshold. That means the decoder returns one step back to path 00.

4. Explore the next successor of path 00, i.e. the branch 11. Now the current path is 00 11 and the current metric is -4.

5. Compare the successors of path 00 11 branches 01 and 10 with the received symbols 01. Branch 01 has the better metric and the new chosen path is 00 11 01 with metric -3.

The decoder continues in this manner and in two steps it finds the decoded sequence 00 11 01 11 00, which is the correct answer. The corresponding code tree is represented in Figure 6.17. The dashed line shows the moving of the decoder along the tree. In fact, only the idea of the Fano algorithm is represented here. The algorithm itself is more complicated and the threshold needs changing whilst moving along the code tree.

The performance of different decoding algorithms is compared in Figure 6.18 (bit error probability vs. SNR). As one can see from the curves in Figure 6.18 the soft decision Viterbi decoding outperforms the hard-decision Viterbi algorithm by 2 dB. The Viterbi decoding of code with the constraint length 7 is compared with the sequential decoding of code with $\nu = 39$ because of the significant difference in the decoding complexity of these methods. As can be seen from Figure 6.18 the comparably complex Viterbi and sequential decoding algorithms provide significantly different performances. Unfortunately, in real life there is a serious limitation on the use of sequential decoding because of the necessity to buffer the input sequence while the algorithm is exploring the code tree. If the input symbol arrival rate exceeds the decoding rate, the buffer will overflow and the data will be lost. Also, the buffer overflow threshold is a very sensitive function of SNR [5]. This fact restricts the use of sequential decoding algorithms.

6.5 PARALLEL-CONCATENATED CONVOLUTIONAL CODES AND SOFT INPUT SOFT OUTPUT DECODING

Parallel-concatenated convolutional codes were introduced by Berrou, Glavieux and Thitimajshima [7]. The encoder of the parallel-concatenated convolutional code (or *turbo-code*) consists of two *recursive systematic convolutional* (RSC) code encoders concatenated in such a way that one of the encoders is fed via an interleaver. The codes are concatenated to employ the principle of iterative decoding. This principle will be discussed later. As mentioned above at high values of SNR the systematic convolutional codes usually show worse performance than nonsystematic convolutional codes, but at low SNR values the opposite is true [7].

The structure of the RSC encoder is shown in Figure 6.19. As can be seen from Figure 6.19 the RSC encoder can be represented as infinite impulse response (IIR) filter. In fact, any nonrecursive nonsystematic convolutional code is equivalent to some RSC code in the sense that it generates the same set of code sequences. The main difference in a nonrecursive nonsystematic and a RSC form of the encoder is that for the RSC encoder the low weight input sequence could generate an infinite weight output sequence. For example, the unit weight input sequence 00 ... 00100 ... will always generate a low weight sequence at the output of the nonrecursive nonsystematic convolutional encoder, which is not the case for

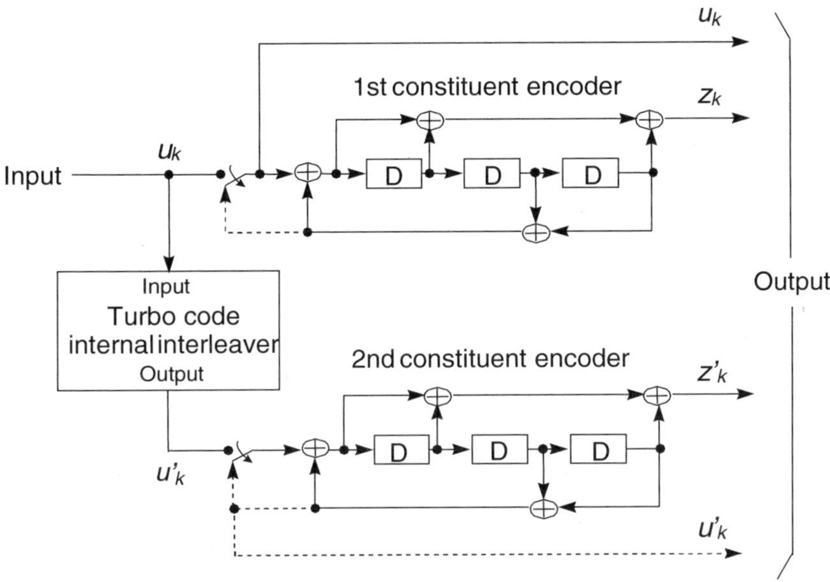

Figure 6.19 Structure of rate 1/3 Turbo encoder (dotted lines apply for trellis termination only)

the RSC encoder. This is a very important feature for the design of the parallel-concatenated codes. The other important issue in the design of the parallel-concatenated codes is the use of the interleaver. Let us consider the encoder of the turbo code used in a 3GPP mobile radio system. The structure of this encoder is shown in Figure 6.19. The output code sequence of the encoder is the concatenation of the output sequences of the constituent codes. It is obvious that the weight distribution of the code sequences of the turbo code depends on the way that the code sequences of the constituent encoders are combined. For example, the pairing of low weight code sequences of the constituent codes leads to poor performance of the result code. This kind of pairing can be avoided with the help of the interleaver. For the encoder of Figure 6.19 the minimum weight sequences of the constituent encoder is generated by the information sequence ...0 0 1 0 0 0 1 1 0 0.... The weight of this code sequence is 6. Another information sequence that generates the code sequence of the same weight 6 is ...0 0 1 1 0 1 0 0.... Due to the interleaver it is unlikely that one of these sequences appears at the input of the second constituent encoder if it has already appeared at the input of the first encoder. That means the interleaver makes it possible to decrease the probability of combining two code sequences of minimum weight. As was mentioned above, the interleaver cannot help in the case where the low weight code sequence is generated by the information sequence of weight 1, but because of the recursive structure of the constituent encoder of Figure 6.19 the information sequence ...0 0 1 0 0... generates the code sequence of weight 13. Thus the effect of the recursive structure of the constituent encoder combined with the effect of the interleaver leads to the rise in the distant structure of the turbo code. Usually the turbo codes are used for transmission of the finite length codewords rather than the half-infinite sequences. In this case some zero tail bits are added at the end of each information sequence, which leads to a flush of the constituent encoders. For the encoder of Figure 6.19 6 tail bits are padded after the encoding of information bits.

The first 3 tail bits are used to terminate the first constituent encoder and the second 3 tail bits are used to terminate the second constituent encoder. For the encoder of Figure 6.19 the flushing is performed by taking the tail bits from the shift register feedback after all the information bits are encoded, which corresponds to the lower position of the switches.

An encoder of the type shown in Figure 6.19 is usually employed together with *iterative decoding*. Each iteration of iterative decoding is executed in two phases. Each phase corresponds to the decoding of the codeword of one of two constituent codes. The idea is that additional information about the reliability of the information symbols obtained during decoding of one constituent code should be used in the following phase.

The calculation of likelihood functions used in iterative decoding is based on Bayes' rule:

$$P(u = i\mid r) = \frac{p(r\mid u = i) \cdot P(u = i)}{p(r)}, \quad i = 0, \ldots, M - 1$$

$$p(r) = \sum_{i=1}^{M-1} p(r\mid u = i) \cdot P(u = i) \tag{6.25}$$

where u is the symbol transmitted over the channel, M is the cardinal number of the alphabet, r is a random variable at the channel output, $P(u = i\mid r)$ is the a posteriori probability (APP) of the decision that transmitted symbol $u = i$ conditioned on r, $p(r\mid u = i)$ is the probability density function (pdf) of the random variable r conditioned on the transmitted symbol $u = i$, $P(u = i)$ is the *a priori probability* of occurrence symbol i at the channel input.

We will consider the binary case, i.e. $M = 2$. Let the binary symbols 0 and 1 be represented by the voltages -1 and $+1$ respectively. Then the *maximum a posteriori* (MAP) rule states that the decision $(u = +1)$ should be chosen in case $P(u = +1\mid r) > P(u = -1\mid r)$ and otherwise the decision $(u = -1)$ should be chosen in case $P(u = +1\mid r) < P(u = -1\mid r)$. The MAP rule provides the minimum probability of error. MAP conditions can be written in terms of likelihood ratios:

$$\frac{P(u = +1\mid r)}{P(u = -1\mid r)} > 1, \quad \text{decision } (u = +1)$$

$$\frac{P(u = +1\mid r)}{P(u = -1\mid r)} < 1, \quad \text{decision } (u = -1). \tag{6.26}$$

Using the Bayes' rule the *likelihood ratio* $\dfrac{P(u = +1\mid r)}{P(u = -1\mid r)}$ can be written as follows:

$$\frac{P(u = +1\mid r)}{P(u = -1\mid r)} = \frac{p(r\mid u = +1) \cdot P(u = +1)}{p(r\mid u = -1) \cdot P(u = -1)} \tag{6.27}$$

In practice the metric called *log-likelihood ratio* (LLR) is more useful. It can be obtained by taking the logarithm of the likelihood ratio and is denoted by $L(u\mid r)$. Using (6.27) we obtain

$$L(u\mid r) = \ln\left(\frac{p(r\mid u = +1) \cdot P(u = +1)}{p(r\mid u = -1) \cdot P(u = -1)}\right) = \ln\left(\frac{p(r\mid u = +1)}{p(r\mid u = -1)}\right) + \ln\left(\frac{P(u = +1)}{P(u = -1)}\right)$$

$$\tag{6.28}$$

or

$$L(u|\,r) = L_c(r) + L(u). \tag{6.29}$$

The value $L(u|\,r)$ can be interpreted as the soft decision output of the demodulator; the value $L_c(r)$ as the reliability of the detected symbol, which can be obtained by measurement of the channel at the receiver input; and $L(u)$ is the a priori LLR of the transmitted symbol (bit). For the decoder of a systematic code the information is available both from the received information symbols and from the redundant symbols. The information available from the redundant symbols is called the *extrinsic information*. The information from code stream $L(u)$ is called the *intrinsic information*. It was shown in [7] that for systematic codes the extrinsic information does not depend on decoder input. Then it is possible to obtain the soft output of the decoder in the following form:

$$L(\hat{u}) = L(u|\,r) + L_{ext}(\hat{u}), \tag{6.30}$$

where $L_{ext}(\hat{u})$ denotes the extrinsic information. Substituting (6.29) to (6.30) obtain

$$L(\hat{u}) = L_c(r) + L(u) + L_{ext}(\hat{u}), \tag{6.31}$$

The decoder soft output $L(\hat{u})$ represents both the hard decision itself and the reliability of that hard decision. The hard decision is defined by the sign of $L(\hat{u})$ and the magnitude of $L(\hat{u})$ defines the reliability of the hard decision. We can regard the extrinsic information $L_{ext}(\hat{u})$ as the improvement of the reliability of the received information symbol. The idea of the iterative decoding is to forward to the next phase of decoding only the extrinsic information $L_{ext}(\hat{u})$, since the information from the information symbols is already available to it.

A constituent decoder that accepts a priori information at its input and produces a posteriori information at its output is called a soft-input/soft-output (SISO) decoder. The constituent SISO decoder is shown in Figure 6.20. The input of the SISO decoder is the information LLRs $L_c\!\left(r_k^{(i)}\right)$ that agrees with the transmitted information bits u_k, the parity

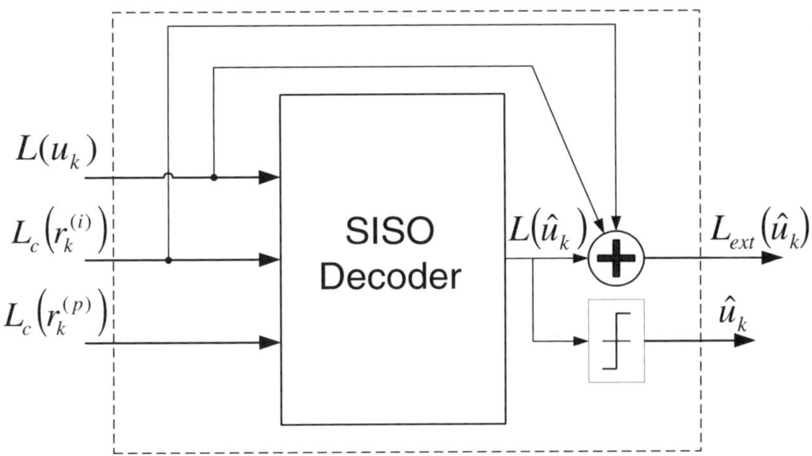

Figure 6.20   A constituent Soft Input/Soft Output decoder

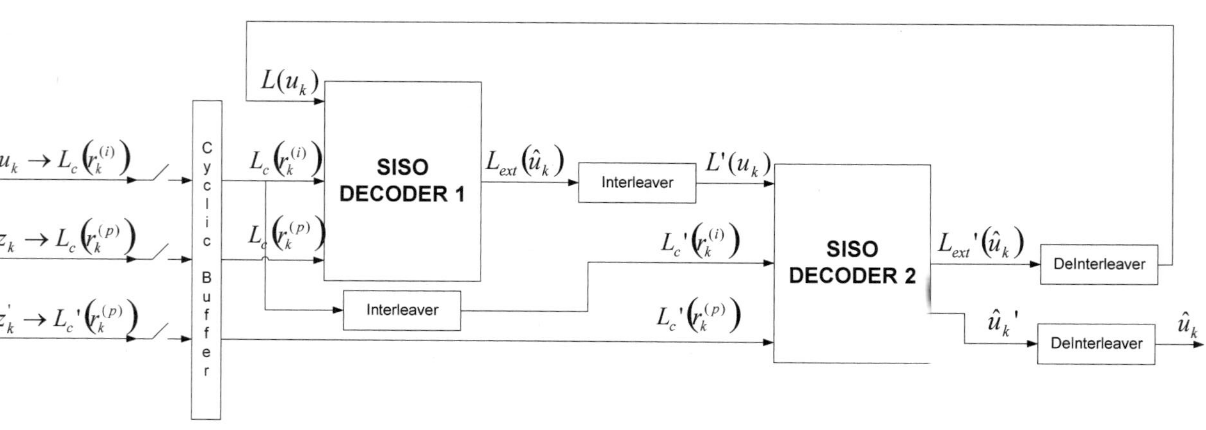

Figure 6.21   A standard turbo decoder

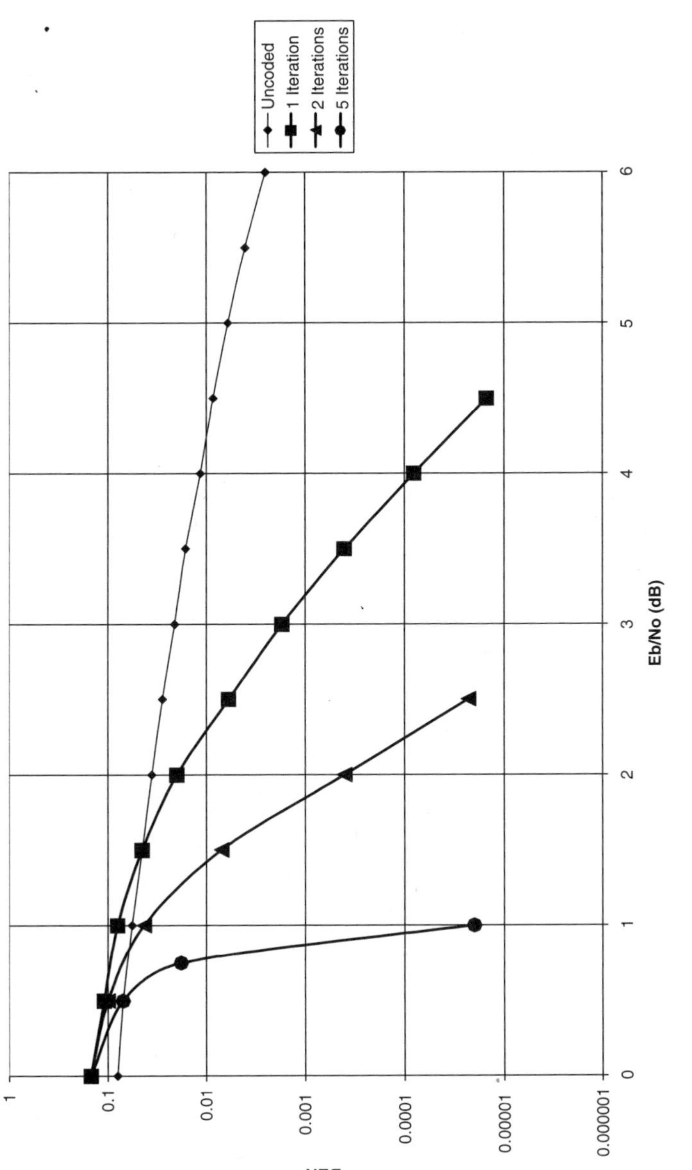

Figure 6.22 Performance of turbo decoder with different number of iterations. BPSK over AWGN, MAP algorithm

LLRs $L_c\left(r_k^{(p)}\right)$ that agrees with the transmitted parity bits z_k or z_k', and the a priori information $L(u_k)$. The decoder output is $L(\hat{u}_k)$. In accordance with (6.31) we can obtain the extrinsic information by subtracting from the output the inputs $L_c\left(r_k^{(i)}\right)$ and $L(u_k)$. Also the sign of the output LLR $L(\hat{u}_k)$ gives us the hard decision, i.e. the decoded information bits $\hat{u}_k$. Regarding these operations of subtraction and comparison with the threshold as the inner operations of the decoder we can say that the outputs of the SISO constituent decoder are the extrinsic information $L_{ext}(\hat{u}_k)$ and the decoded information bits $\hat{u}_k$. This form of the decoder corresponds to the dashed line in Figure 6.20.

The scheme of a standard turbo decoder is shown in Figure 6.21. The first SISO decoder receives the information LLRs $L_c\left(r_k^{(i)}\right)$ that agrees with the transmitted information bits u_k, the parity LLRs $L_c\left(r_k^{(p)}\right)$ that agree with the transmitted parity bits z_k, and the a priori information $L(u_k)$. For the first decoding iteration it is assumed that the a priori information $L(u_k) = 0$, i.e. the information bits assumed to be equally likely. The output of the first SISO decoder is the extrinsic information $L_{ext}(\hat{u}_k)$. The second SISO decoder receives the interleaved information LLRs $L_c'\left(r_k^{(i)}\right)$ that agree with the interleaved information bits u_k' and the parity LLRs $L_c'\left(r_k^{(p)}\right)$ that agree with the transmitted parity bits z_k'. The interleaved extrinsic information obtained from the output of the first SISO decoder is fed to the input of the second SISO decoder and is used as the a priori information. The extrinsic information produced by the second SISO decoder in turn is used after deinterleaving as the input a priori information for the first SISO decoder in the next decoding iteration. It is extrinsic information that is passed between constituent SISO decoders, rather than the decoded data. As was mentioned above, usually the constituent encoders of turbo encoder are terminated by the tail bits. Similarly, the constituent decoders of the turbo decoder work on a block-by-block principle. The LLRs corresponding to the block of received symbols (bits) are stored in the cyclic buffers and are fed to the inputs of the constituent SISO decoders at each iteration. Usually several iterations are needed to achieve the required performance. After the last iteration the decoded information bits (hard decisions) $\hat{u}_k$ can be obtained from the output of the second SISO decoder.

In Figure 6.22 the curves of bit error probability vs. SNR for the decoder with different number of iterations are represented. As one can see, the performance of the turbo decoder increases with the increase in the number of iterations.

6.6 SISO DECODING ALGORITHMS

The main part of the turbo decoder is the SISO constituent decoder. Several different algorithms can be used to implement the SISO decoder. These algorithms can be split into two groups. One group represents the algorithms derived from the Viterbi algorithm and the second group includes algorithms based on the *Maximum A Posteriori* (MAP) *algorithm*. In this section we will follow mostly the study [8].

6.6.1 MAP Algorithm and Its Variants

The symbol-by-symbol MAP algorithm was introduced in 1974 by Bahl, Cocke, Jelinek and Raviv [9] for decoding the convolutional codes. Very often the MAP algorithm is called the BCJR algorithm. Unlike the Viterbi algorithm that minimises the probability of error per sequence, the MAP algorithm minimises the probability of error per symbol. The implementation of the MAP algorithm is close to the implementation of the Viterbi algorithm performing in forward-backward directions over a block of code symbols. The MAP algorithm finds for each decoded bit u_k the a posteriori LLR $L(u_k|\,\mathbf{r})$, where $\mathbf{r}$ is the received sequence. This LLR corresponds to $L(\hat{u})$ in (6.31). We will denote it as $L(\hat{u}_k)$. Let us consider one trellis section of a terminated trellis shown in Figure 6.23. If the previous state $S_{k-1} = i$ and the present state $S_k = j$ are known in a trellis then the input bit u_k, which

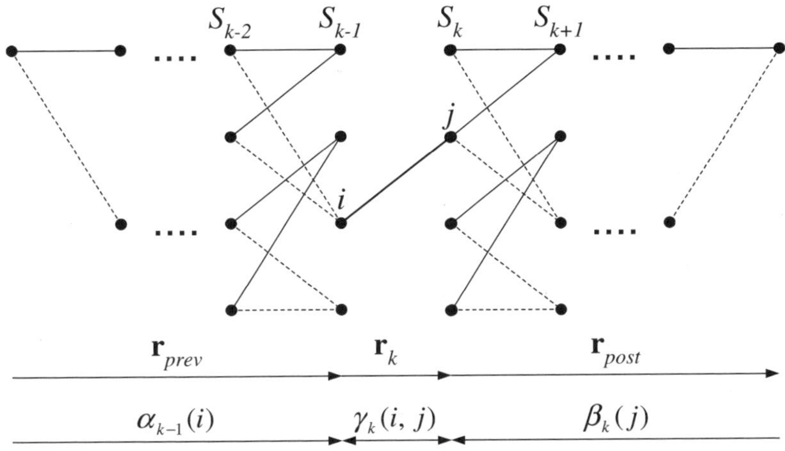

Figure 6.23 A section of MAP decoder trellis

caused the transition between these two states is also known. Then using the Bayes' rule we can write $L(u_k|\,\mathbf{r})$ as follows

$$L(\hat{u}_k) = L(u_k|\,r) = \ln\left(\frac{\sum\limits_{(i,\,j)\,\Rightarrow\,u_k\,=\,+1} P(S_{k-1} = i, S_k = j, \mathbf{r})}{\sum\limits_{(i,j)\,\Rightarrow\,u_k\,=\,-1} P(S_{k-1} = i, S_k = j, \mathbf{r})}\right), \tag{6.32}$$

where $(i,\,j) \Rightarrow u_k = +1$ is the set of transitions from the state $S_{k-1} = i$ to the state $S_k = j$ that can occur if the input bit $u_k = +1$, and similarly for $(i,j) \Rightarrow u_k = -1$. The received sequence $\mathbf{r}$ can be split into three sections: the received symbols associated with the current transition $\mathbf{r}_k$, the received sequence prior the current transition $\mathbf{r}_{prev}$ and the received sequence after the current transition $\mathbf{r}_{post}$ as shown in Figure 6.23.

Assuming that the channel is memoryless the properties of the Markov process can be used to write the probability $P(S_{k-1} = i, S_k = j, \mathbf{r})$ as follows

$$\begin{aligned}
P(S_{k-1} = i, S_k = j, \mathbf{r}) &= P\big(S_{k-1} = i, S_k = j, \mathbf{r}_{prev}, \mathbf{r}_k, \mathbf{r}_{post}\big) \\
&= P\big(\mathbf{r}_{post}|\,S_k = j\big) \cdot P\big(S_{k-1} = i, S_k = j, \mathbf{r}_{prev}, \mathbf{r}_k\big) \\
&= P\big(\mathbf{r}_{post}|\,S_k = j\big) \cdot P(\{S_k = j, \mathbf{r}_k\}|\,S_{k-1} = i) \cdot P\big(S_{k-1} = i, \mathbf{r}_{prev}\big).
\end{aligned} \tag{6.33}$$

Let us denote the probability that the trellis is in state i at time moment $k - 1$ and the received sequence prior to time moment k is $\mathbf{r}_{prev}$ as $\alpha_{k-1}(i)$:

$$\alpha_{k-1}(i) = P\big(S_{k-1} = i, \mathbf{r}_{prev}\big), \tag{6.34}$$

the probability that the received sequence after time moment k is $\mathbf{r}_{post}$, conditioned on the trellis is in state j at time moment k as $\beta_k(j)$:

$$\beta_k(j) = P\left(\mathbf{r}_{post}\middle|\, S_k = j\right) \tag{6.35}$$

and the probability of transition from state i at the time moment $k-1$ to state j at time moment k caused by the received symbols $\mathbf{r}_k$ as $\gamma_k(i, j)$:

$$\gamma_k(i, j) = P(\{S_k = j, \mathbf{r}_k\}|\, S_{k-1} = i). \tag{6.36}$$

The value $\gamma_k(i, j)$ is called the branch metric associated with the transition $i \rightarrow j$. Substituting (6.33), (6.34), (6.35) and (6.36) to (6.32) obtain

$$L(\hat{u}_k) = L(u_k|\, \mathbf{r}) = \ln\left(\frac{\sum\limits_{(i,\, j)\, \Rightarrow\, u_k = +1} \alpha_{k-1}(i) \cdot \beta_k(j) \cdot \gamma_k(i, j)}{\sum\limits_{(i,\, j)\, \Rightarrow\, u_k = -1} \alpha_{k-1}(i) \cdot \beta_k(j) \cdot \gamma_k(i, j)}\right). \tag{6.37}$$

The MAP algorithm finds probabilities $\alpha_k(j)$ and $\beta_k(j)$ for all possible states j throughout the trellis, i.e. for all values of k: and the branch metric $\gamma_k(i, j)$ for all branches in the trellis.

Let us consider the calculation of $\alpha_k(i)$, $\beta_k(i)$ and $\gamma_k(i, j)$.

Consider the definition of the branch metric $\gamma_k(i, j)$ in (6.36). Using the Bayes' rule we can write (6.36) as follows

$$\gamma_k(i, j) = P(\{S_k = j, \mathbf{r}_k\}|\, S_{k-1} = i) = P(\mathbf{r}_k|\, \{S_{k-1} = i, S_k = j\}) \cdot P(S_k = j|\, S_{k-1} = i)$$
$$= P(\mathbf{r}_k|\, \{S_{k-1} = i, S_k = j\}) \cdot P(u_k) = P(\mathbf{r}_k|\, \mathbf{x}_k) \cdot P(u_k), \tag{6.38}$$

where u_k is the input information bit necessary to cause the transition from state $S_{k-1} = i$ to state $S_k = j$; $P(u_k)$ is a priori probability of bit u_k.

Obviously the event $(S_{k-1} = i, S_k = j)$ coincides with the event that bit vector $\mathbf{x}_k$ generated by the transition $(S_{k-1} = i) \rightarrow (S_k = j)$ was transmitted over the channel. Then (6.38) can be written as

$$\gamma_k(i, j) = P(\mathbf{r}_k|\, \{S_{k-1} = i, S_k = j\}) \cdot P(u_k) = P(\mathbf{r}_k|\, \mathbf{x}_k) \cdot P(u_k). \tag{6.39}$$

If the encoder forms n output symbols (bits) during each transition the probability $P(\mathbf{r}_k|\, \mathbf{x}_k)$ can be written in following way

$$P(\mathbf{r}_k|\, \mathbf{x}_k) = \prod_{s=0}^{n-1} P\left(r_{k,s}\middle|\, x_{k,s}\right), \tag{6.40}$$

where $x_{k,s}$ is the transmitted bit and $r_{k,s}$ is the corresponding received symbol. Let us assume the channel model to be flat fading with Gaussian noise. Then the probability that the received symbol is r conditioned on the transmitted bit x is

$$P(r|x) = \frac{1}{\sqrt{\pi \cdot N_o}} \cdot e^{-\frac{E_b}{N_0}(r-ax)^2}, \tag{6.41}$$

where $\dfrac{E_b}{N_0}$ is the signal to noise ratio per bit; a is the fading amplitude (for nonfading Gaussian channel $a = 1$). Then the branch metric can be calculated in the following way

$$\gamma_k(i, j) = P(u_k) \cdot \prod_{s=0}^{n} \frac{1}{\sqrt{\pi \cdot N_o}} \cdot e^{-\frac{E_b}{N_0}\left(r_{k,s} - a x_{k,s}\right)^2}. \tag{6.42}$$

The a priori probability $P(u_k)$ can be derived from the input a priori LLR $L(u_k)$

$$L(u_k) = \ln\left(\frac{P(u_k = +1)}{P(u_k = -1)}\right). \tag{6.43}$$

Solving the equation (6.43) for $P(u_k = +1)$ or for $P(u_k = -1)$, we obtain

$$P(u_k = +1) = \frac{e^{L(u_k)}}{1 + e^{L(u_k)}},$$

$$P(u_k = -1) = \frac{1}{1 + e^{L(u_k)}}. \tag{6.44}$$

Taking in account that for calculation of numerator in (6.37) we need in $P(u_k = +1)$ rather than $P(u_k)$ and for calculation of denominator we need in $P(u_k = -1)$ rather than $P(u_k)$, the equations (6.44) can be used for calculation of the branch metrics in accordance with (6.39) or (6.42).

Consider $\alpha_k(j)$. From the definition of $\alpha_{k-1}(i)$ in (6.34) we can write

$$\alpha_k(j) = P(S_k = j, \mathbf{r}_{prev}, \mathbf{r}_k) = \sum_i P(S_{k-1} = i, S_k = j, \mathbf{r}_{prev}, \mathbf{r}_k), \tag{6.45}$$

Using the Bayes' rule and the properties of Markov process we can write (6.45) as follows

$$\begin{aligned}
\alpha_k(j) &= \sum_i P(S_{k-1} = i, S_k = j, \mathbf{r}_{prev}, \mathbf{r}_k) \\
&= \sum_i P(\{S_k = j, \mathbf{r}_k\} | \{S_{k-1} = i, \mathbf{r}_{prev}\}) \cdot P(S_{k-1} = i, \mathbf{r}_{prev}) \\
&= \sum_i P(\{S_k = j, \mathbf{r}_k\} | S_{k-1} = i) \cdot P(S_{k-1} = i, \mathbf{r}_{prev}) = \sum_i \alpha_{k-1}(i) \cdot \gamma_k(i, j).
\end{aligned} \tag{6.46}$$

That means the probabilities $\alpha_k(j)$ can be calculated recursively. Assuming that the trellis has the initial state $S_0 = 0$, the initial conditions for this recursion are

$$\begin{aligned}
\alpha_0(S_0 = 0) &= 1 \\
\alpha_0(S_0 = i) &= 0, \quad i \neq 0.
\end{aligned} \tag{6.47}$$

This is the forward recursion. Using the same technique it can be shown that the probabilities $\beta_k(i)$ can be calculated with the help of backward recursion:

$$\beta_k(i) = \sum_j \beta_{k+1}(j) \cdot \gamma_{k+1}(i, j). \tag{6.48}$$

Assuming the length of information sequence is equal to K and that the tail bits put the encoder in the zero state, the conditions for backward recursion can be written as follows

$$\beta_K(S_K = 0) = 1$$
$$\beta_K(S_K = i) = 0, \quad i \neq 0. \tag{6.49}$$

The process of calculating forward and backward recursion in MAP algorithm is illustrated in Figure 6.24.

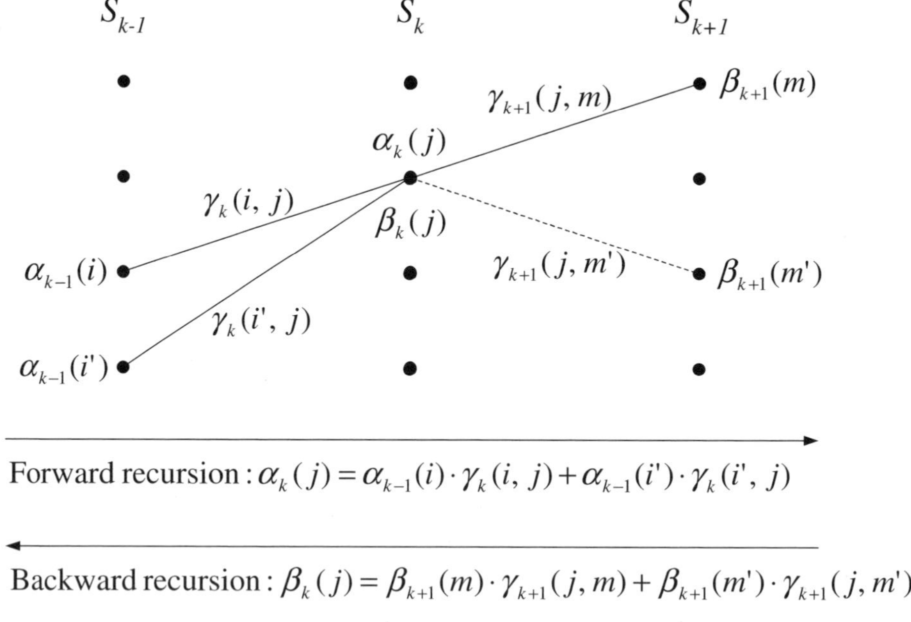

Figure 6.24 Calculation of forward and backward recursion in MAP algorithm

From the description given above, we see that the MAP decoding of a received sequence $\mathbf{r}$ to give the a posteriori LLRs $L(u_k| \mathbf{r})$ can be carried out as follows. As the channel values $r_{k,s}$ are received, they and the a priori LLRs $L(u_k)$ are used to calculate $\gamma_k(i, j)$ according to (6.39) or (6.42). Then the forward recursion in accordance with (6.46) and (6.47) can be used to calculate $\alpha_{k-1}(i)$. When all channel values have been received the backward recursion in accordance with (6.48) and (6.49) can be used to calculate $\beta_k(j)$. Finally, all the calculated values $\alpha_{k-1}(i)$, $\beta_k(j)$ and $\gamma_k(i,j)$ are used to calculate a posteriori LLRs $L(u_k|\mathbf{r})$ in accordance with (6.37).

In accordance with (6.31) the a posteriori LLR $L(u_k|\mathbf{r})$ can be represented as follows

$$L(u_k|\mathbf{r}) = L_c\left(r_k^{(i)}\right) + L(u_k) + L_{ext}(\hat{u}_k), \tag{6.50}$$

where $L_c\left(r_k^{(i)}\right)$ is the information LLR that agrees with the transmitted information bit u_k and $L_{ext}(\hat{u}_k)$ is the extrinsic information obtained as a result of decoding. As was mentioned above we are interested in passing to the next decoder the extrinsic information rather than the a posteriori LLR $L(u_k|\mathbf{r})$. To obtain extrinsic information $L_{ext}(\hat{u}_k)$ we should subtract values of $L_c\left(r_k^{(i)}\right)$ and $L(u_k)$ from the a posteriori LLR $L(u_k|\mathbf{r})$. In accordance with (6.28) the information LLR $L_c\left(r_k^{(i)}\right)$ can be represented as

$$L_c\left(r_k^{(i)}\right) = \ln\left(\frac{p\left(r_k^{(i)}\Big|u_k = +1\right)}{p\left(r_k^{(i)}\Big|u_k = -1\right)}\right). \tag{6.51}$$

Assuming the channel model to be flat fading with Gaussian noise, equation (6.51) can be written as

$$L_c\left(r_k^{(i)}\right) = 4\frac{E_b}{N_0} \cdot a \cdot r_k^{(i)}, \tag{6.51}$$

where $r_k^{(i)}$ is the received symbol corresponding to the transmitted information bit u_k and a is the fading amplitude. For simplicity (6.51) can be rewritten as

$$L_c\left(r_k^{(i)}\right) = L_c \cdot r_k^{(i)}, \tag{6.52}$$

where

$$L_c = 4\frac{E_b}{N_0} \cdot a. \tag{6.53}$$

The MAP algorithm is, in the form described above, extremely complex. However, much work was done to reduce its complexity. Initially the Max-Log-MAP algorithm was proposed by Koch and Bayer [10] and Efranian *et al* [11]. This technique simplifies the MAP algorithm by transferring the recursions into the log domain and invoking an approximation to dramatically reduce the complexity. Because of this approximation its performance is suboptimal compared to that of the MAP algorithm. Later Robertson [12] proposed the Log-MAP algorithm, which corrected the approximation used in the Max-Log-MAP algorithm and hence gave a performance almost identical to that of the MAP algorithm, but at a fraction of its complexity.

Max-Log-MAP Algorithm

The Max-Log-MAP algorithm simplifies the calculations of MAP algorithm by transferring
the equations into the log arithmetic domain and then using the approximation

$$\ln\left(\sum_i e^{x_i}\right) \approx \max_i(x_i). \tag{6.54}$$

Then, with $A_k(i)$, $B_k(i)$ and $\Gamma_k(i,j)$ defined as follows:

$$\begin{aligned}
A_k(i) &= \ln(\alpha_k(i)), \\
B_k(i) &= \ln(\beta_k(i)), \\
\Gamma_k(i,j) &= \ln(\gamma_k(i,j))
\end{aligned} \tag{6.55}$$

we can rewrite (6.46) as

$$A_k(j) = \ln(\alpha_k(j)) = \ln\left(\sum_i \alpha_{k-1}(i)\cdot\gamma_k(i,\,j)\right) = \ln\left(\sum_i e^{A_{k-1}(i)+\Gamma_k(i,j)}\right) \approx$$
$$\max_i(A_{k-1}(i) + \Gamma_k(i,\,j)). \tag{6.56}$$

Equation (6.56) implies that for each path in Figure 6.23 from the previous stage in the trellis
to the state $S_k = j$ at the present stage, the algorithm adds a branch metric term $\Gamma_k(i,j)$ to the
previous value $A_{k-1}(i)$ to find a new value $A_k(j)$ for that path. The new value of $A_k(j)$
according to (6.56) is then the maximum of the values of the various paths reaching the state
$S_k = j$. This can be thought of as selecting one path as the 'survivor' and discarding any
other paths reaching the state. The value of $A_k(j)$ should give the natural logarithm of the
probability that the trellis is in state $S_k = j$ at stage k, given that the received channel
sequence up to this point has been $\mathbf{r}_{prev}$. However, because of the approximation of (6.54)
used to derive (6.56), only the Maximum Likelihood (ML) path through the state $S_k = j$ is
considered when calculating this probability. Thus, the value of $A_k(j)$ in the Max-Log-MAP
algorithm actually gives the probability of the most likely path through the trellis to the state
$S_k = j$, rather than the probability of *any* path through the trellis to state $S_k = j$. This
approximation is one of the reasons for the sub-optimal performance of the Max-Log-MAP
algorithm compared to the MAP algorithm.

We see from (6.56) that in the Max-Log-MAP algorithm the forward recursion used to
calculate $A_k(j)$ is exactly the same as the forward recursion in the Viterbi algorithm – for
each pair of merging paths the survivor is found using two additions and one comparison.
Notice that for binary trellises the summation, and maximisation, over all previous states
$S_{k-1} = i$ in (6.56) will in fact be over only two states, because there will be only two
previous states $S_{k-1} = i$ with paths to the present state $S_k = j$. For all other values of S_{k-1} we
will have $\gamma_k(i,j) = 0$.

Similarly to (6.56) for the forward recursion used to calculate the $A_k(j)$, we can rewrite
(6.48) as

$$B_k(i) = \ln(\beta_k(i)) \approx \max_j(B_{k+1}(j) + \Gamma_{k+1}(i,\,j)) \tag{6.57}$$

giving the backward recursion used to calculate the $B_k(i)$ values. Again, this is equivalent to the recursion used in the Viterbi algorithm except that it proceeds backward rather than forward through the trellis.

Using (6.39) and (6.42), we can write the branch metrics in the above recursive equations (6.56) and (6.57) as

$$\Gamma_k(i,j) = \ln(\gamma_k(i,j)) = C + \frac{1}{2} u_k L(u_k) + \frac{L_c}{2} \sum_{s=0}^{n-1} r_{k,s} \cdot x_{k,s} \tag{6.58}$$

where C does not depend on u_k or on the transmitted bits sequence $\mathbf{x}_k$ and so can be considered a constant and omitted. Hence, the branch metric is equivalent to that used in the Viterbi algorithm, with the addition of the a-priori LLR term $u_k L(u_k)$. Furthermore, the correlation term $\sum_{s=0}^{n-1} r_{k,s} \cdot x_{k,s}$ is weighted by the channel reliability value L_c of (6.53).

Finally, from (6.37), we can write for the a-posteriori LLRs $L(\hat{u}_k) = L(u_k|\mathbf{r})$ that the Max-Log-MAP algorithm calculates

$$L(\hat{u}_k) = L(u_k|\mathbf{r}) = \ln\left(\frac{\sum\limits_{(i,j)\,\Rightarrow\,u_k=+1} \alpha_{k-1}(i) \cdot \beta_k(j) \cdot \gamma_k(i,j)}{\sum\limits_{(i,j)\,\Rightarrow\,u_k=-1} \alpha_{k-1}(i) \cdot \beta_k(j) \cdot \gamma_k(i,j)}\right)$$

$$\approx \max_{(i,j)\,\Rightarrow\,u_k=+1} \left(A_{k-1}(i) + B_k(j) + \Gamma_k(i,j)\right) - \max_{(i,j)\,\Rightarrow\,u_k=-1} \left(A_{k-1}(i) + B_k(j) + \Gamma_k(i,j)\right).$$

$$\tag{6.59}$$

This means that in the Max-Log-MAP algorithm for each bit u_k the a-posteriori LLR $L(\hat{u}_k) = L(u_k|\mathbf{r})$ is calculated by considering every transition from the trellis stage S_{k-1} to the stage S_k. These transitions are grouped into those that might have occurred if $u_k = +1$, and those that might have occurred if $u_k = -1$. For both of these groups the transition giving the maximum value of $A_{k-1}(i) + B_k(j) + \Gamma_k(i,j)$ is found, and the a-posteriori LLR is calculated based on only these two 'best' transitions.

Log-MAP Algorithm

Due to the approximation (6.54) used in the Max-Log-MAP algorithm its performance is worse in comparison with MAP algorithm. Robertson *et al* [12] proposed using instead of approximation (6.54) the exact formula

$$\ln(e^{x_1} + e^{x_2}) = \max(x_1, x_2) + \ln(1 + e^{-|x_1 - x_2|}), \tag{6.60}$$

where $\ln(1 + e^{-|x_1 - x_2|})$ can be regarded as a correction factor that will tend to zero as the difference between the arguments increases. Similarly to the Max-Log-MAP algorithm, values for $A_k(j) = \ln(\alpha_k(j))$ and $B_k(j) = \ln(\beta_k(j))$ are calculated using a forward and a backward recursion. However, the maximisation in (6.56) and (6.57) is complemented by the correction factor in (6.60). This means that the exact rather than approximate values of $A_{k-1}(i)$ and $B_k(j)$ are calculated. The correction factor can be stored in a look-up table. This means that the Log-MAP algorithm is only slightly more complex than the Max-Log-MAP algorithm, but it gives almost the same performance as the MAP algorithm. Depending on the size of look-up table used the performance of the Log-MAP algorithm can achieve the performance of MAP algorithm.

6.6.2 Soft-In/Soft-Out Viterbi Algorithm (SOVA)

The Soft-In/Soft-Out Viterbi Algorithm (SOVA) was proposed by Hagenauer [13]. The SOVA operates similarly to the Viterbi algorithm except that the ML sequence is found with the help of a modified metric. The path metrics used in SOVA are modified to take account of a priori information when selecting the ML path through the trellis. As it was shown above the Max-Log-MAP algorithm also outputs the ML sequence over the whole trellis. Moreover, the recursion defined by (6.56) selects the metric corresponding to the ML path to node j, which in terms of the Viterbi algorithm is the survivor path. Thus the forward recursion performs the same operations as the Viterbi algorithm [14]. Due to this fact the Viterbi algorithm can be modified so that it provides a soft output in the form of the a posteriori LLR $L(\hat{u}_k) = L(u_k|\, \mathbf{r})$ for each decoded bit.

Consider the state sequence $\mathbf{S}_k^{(m)}$, which gives the states along the surviving path m at stage k in the trellis. The Viterbi algorithm searches for the state sequence $\mathbf{S}^{(m)}$ that maximises the a posteriori probability $P\left(\mathbf{S}^{(m)}|\, \mathbf{r}\right)$. By using Bayes' rule the a posteriori probability can be written as

$$P\left(\mathbf{S}^{(m)}|\, \mathbf{r}\right) = p\left(\mathbf{r}|\, \mathbf{S}^{(m)}\right) \cdot \frac{P\left(\mathbf{S}^{(m)}\right)}{p(\mathbf{r})}. \tag{6.61}$$

Since the received sequence $\mathbf{r}$ is fixed and does not depend on m it can be discarded. Then we can equivalently maximise

$$\max_m p\left(\mathbf{r}|\, \mathbf{S}^{(m)}\right) \cdot P\left(\mathbf{S}^{(m)}\right). \tag{6.62}$$

This maximisation is realised in the code trellis, when for each state sequence $\mathbf{S}^{(m)}$ and each stage k, the path with the largest probability $P\left(\mathbf{S}_k^{(m)}, \mathbf{r}_k\right)$ is selected. This probability can be calculated by multiplying the branch transition probabilities associated to path m. They are $\gamma_l\left(i^{(m)}, j^{(m)}\right)$ for $1 \le l \le k$ and defined in (6.39). The maximum is not changed if we take the logarithm, and hence the metric computations are the same as described for the forward recursion of the Max-Log-MAP algorithm. The values $A_k(i)$ and $B_k(i)$ from (6.56) and (6.57) are additive and the same for all paths m and therefore are irrelevant for the maximisation. Let us denote the path m entering state $S_k = j^{(m)}$ at the stage k by $\mathbf{S}_k^{(j^{(m)})}$. Then if the path $\mathbf{S}_k^{(j^{(m)})}$ at the kth stage has the path $\mathbf{S}_{k-1}^{(i^{(m)})}$ as its prefix and assuming the memoryless channel we can choose the following metric $M\left(\mathbf{S}_k^{(j^{(m)})}\right)$ for the path $\mathbf{S}_k^{(j^{(m)})}$

$$M\left(\mathbf{S}_k^{(j^{(m)})}\right) = \ln\left(P\left(\mathbf{S}_k^{(j^{(m)})}, \mathbf{r}_k\right)\right) = M\left(\mathbf{S}_{k-1}^{(i^{(m)})}\right) + \ln\left(\gamma_k\left(i^{(m)}, j^{(m)}\right)\right). \tag{6.63}$$

Using (6.58) and omitting the constant C we obtain

$$M\left(\mathbf{S}_k^{(j^{(m)})}\right) = M\left(\mathbf{S}_{k-1}^{(i^{(m)})}\right) + \frac{1}{2}u_k \cdot L(u_k) + \frac{1}{2}\sum_{l=0}^{n-1}L_c \cdot r_{k,l} \cdot x_{k,l}, \tag{6.64}$$

where u_k is the information bit and $x_{k,l}$ are the coded bits of path m at stage k.

This slight modification of the metric of the Viterbi Algorithm in (6.64) with the additional term $u_k \cdot L(u_k)$ included, incorporates the a priori information about the probability of the information bits. The balance between a priori information $L(u)$ and the channel reliability L_c is very important for the SOVA metric. If the channel is very good, $|L_c \cdot r|$ will be larger than $|L(u)|$, and decoding relies on the received channel values. If the channel is bad, such as during deep fade, decoding relies on the a priori information $L(u)$. In iterative decoding this is the extrinsic information from the previous decoding step [15]. If this balance is not achieved, catastrophic effects may result in the degradation of the decoder performance.

Let us now discuss the second modification of the algorithm required, i.e. to give soft output.

The modified Viterbi algorithm proceeds in the usual way by calculating the path metrics using (6.64). If the two paths $\mathbf{S}_k^{(j^{(m)})}$ and $\mathbf{S}_k^{(j^{(l)})}$ reaching state $S_k = j$ have metrics $M\left(\mathbf{S}_k^{(j^{(m)})}\right)$ and $M\left(\mathbf{S}_k^{(j^{(l)})}\right)$ respectively, and the path $\mathbf{S}_k^{(j^{(m)})}$ is selected as the survivor because of higher metric, then we can define the metric difference Δ_k^j as

$$\Delta_k^j = M\left(\mathbf{S}_k^{(j^{(m)})}\right) - M\left(\mathbf{S}_k^{(j^{(l)})}\right) \geq 0. \tag{6.65}$$

The probability that the decision is correct $P(\text{correct decision at } S_k = j)$ can be written as follows:

$$P(\text{correct decision at } S_k = j) = \frac{P\left(S_k^{(j^{(m)})}\right)}{P\left(S_k^{(j^{(m)})}\right) + P\left(S_k^{(j^{(l)})}\right)}. \tag{6.66}$$

Using the metric definition (6.63) we can rewrite (6.66) as

$$P(\text{correct decision at } S_k = j) = \frac{e^{M\left(S_k^{(j^{(m)})}\right)}}{e^{M\left(S_k^{(j^{(m)})}\right)} + e^{M\left(S_k^{(j^{(l)})}\right)}} = \frac{e^{\Delta_k^j}}{1 + e^{\Delta_k^j}}. \tag{6.67}$$

Therefore the LLR that the decision is correct or 'soft' value of this binary path decision is Δ_k^j because

$$\ln\frac{P(\text{correct decision at } S_k = j)}{1 - P(\text{correct decision at } S_k = j)} = \Delta_k^j. \tag{6.68}$$

The examples of metric differences in trellis are shown in Figure 6.25. Along the ML path several nonsurviving paths were discarded. Usually all the surviving paths at some stage in

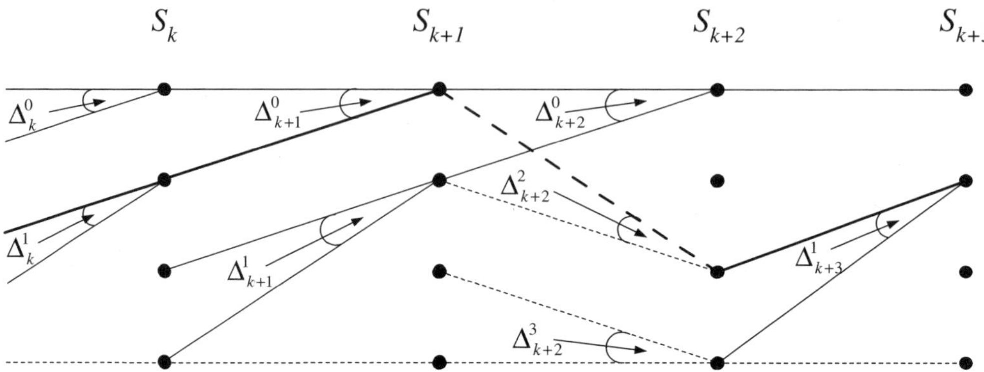

Figure 6.25 A section of SOVA decoder trellis

the trellis have come from the same path at some point at most δ transitions before the given stage in the trellis. The value δ is set to be five or six times the constraint length of the convolutional code. If the value of the bit u_k associated with the transition from state $S_{k-1} = i$ to state $S_k = j$ on the ML path differs from the value of the corresponding bit of the competing path which merged with the ML path at stage $k + \delta$ and if this competing path was chosen by the decoder, then there is a bit error. Thus, when calculating the LLR of the bit u_k, the SOVA must take into account the probability that the paths merging with the ML path from stage k to stage $k + \delta$ in the trellis were incorrectly discarded. This is done by considering the values of the metric difference $\Delta_i^{s_i}$ for all states s_i along the ML path from stage $i = k$ to stage $i = k + \delta$. It is shown in [16] that this LLR can be approximated by

$$L(u_k \mid \mathbf{r}) \approx u_k \cdot \min_{\substack{i = k...k + \delta, \\ u_k \neq u_k^i}} \Delta_i^{s_i}, \tag{6.69}$$

where u_k is the value of the bit given by the ML path and u_k^i is the value of the corresponding bit in the competing path that merged with the ML path and was discarded at stage i. The minimum is only over those nonsurviving paths, which would have led to different values of the bit u_k. Consider the section of trellis shown in Figure 6.25. In this figure solid lines represent transitions taken when the input bit is -1, and dashed lines represent transitions taken when the input bit is $+1$. The bold line marks the ML path. As can be seen from Figure 6.25 the ML path gives a value of -1 for u_k. Assume that $\delta = 3$. Then we can make decision about u_k at stage S_{k+3}. The other paths merge with ML path at stages S_k, S_{k+1}, S_{k+2}, S_{k+3}, which gives us the metric differences Δ_k^1, Δ_{k+1}^0, Δ_{k+2}^2, Δ_{k+3}^1. However, only the competing paths merging with the ML path at stages S_{k+2} and S_{k+3} give the opposite value $+1$ for bit u_k. That means only the minimum of values Δ_{k+2}^2 and Δ_{k+3}^1 should be taken into account during the calculation of soft output for u_k.

The SOVA can be implemented as follows:

Soft Input Soft Output Viterbi Decoding:

1. a. Initialise stage number $k = 0$.

    b. Initialise $M\left(\mathbf{S}_0^{(j)}\right) = 0$ for $j = 0$, $M\left(\mathbf{S}_0^{(j)}\right) = -\infty$ for all $j \neq 0$

2. a. Set $k = k + 1$.

   b. Compute the metric $M\left(\mathbf{S}_k^{(j^{(m)})}\right) = M\left(\mathbf{S}_{k-1}^{(i^{(m)})}\right) + \frac{1}{2}u_k \cdot L(u_k) + \frac{1}{2}\sum_{l=0}^{n-1} L_c \cdot r_{k,l} \cdot x_{k,l}$ for each state in the trellis, where

   m denotes the path (branch) number;

   u_k is the information bit of the path m;

   $x_{k,l}$ is the lth bit of n bits for stage k associated with the branch m;

   $r_{k,l}$ is the received value from the channel corresponding to $x_{k,l}$;

   $L_c = 4\dfrac{E_b}{N_0} \cdot a$ is the channel reliability value;

   $L(u_k)$ is the a priori information. This value is obtained from the previous decoding step.

   If there was no previous decoding step this value is set to zero.

3. Find $M_k^{(j)} = \max\limits_{m} M\left(\mathbf{S}_k^{(j^{(m)})}\right)$ for each state j.

4. Store $M_k^{(j)}$ and its associated survivor bit and state paths.

5. Compute $\Delta_k^j = M_k^j - M\left(\mathbf{S}_k^{(j^{(l)})}\right) \geq 0$ for each state j, where $M\left(\mathbf{S}_k^{(j^{(l)})}\right)$ is the metric of the discarded path.

6. Update $\Delta_k^{j\min} = \min\limits_{\substack{i=k-\delta...k,\\ u_k \neq u_k^i}} \Delta_i^{S_i}$ by choosing the minimum metric difference.

7. Go to step 2 until the end of the received sequence.

8. Output the estimated bit sequence $\mathbf{u} = \{u_k\}$ and the corresponding LLRs $\{L(\hat{u}_k) = L(u_k|\,\mathbf{r}) = u_k \cdot \Delta_k^{j\min}\}$.

As was already mentioned earlier the recursion used to find the metric in SOVA is identical to the forward recursion (6.56) in the Max-Log-MAP algorithm. Using the notation associated with the Max-Log-MAP algorithm, once a path merges with the ML path, it will have the same value of $B_k(j)$ as the ML path. Hence, as the metric in the SOVA is identical to

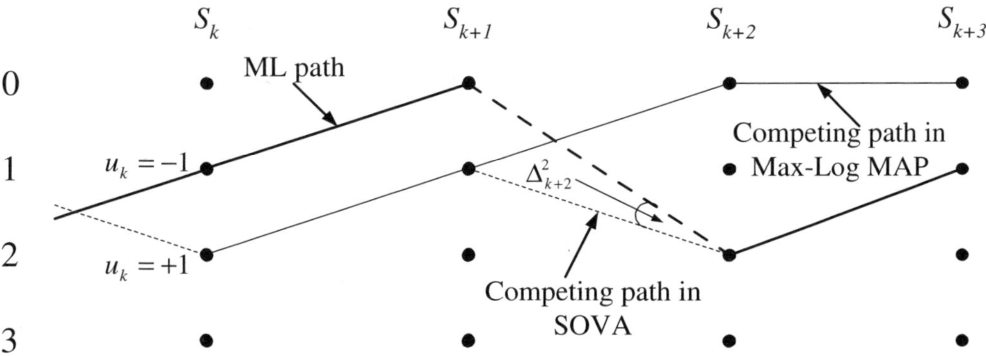

Figure 6.26 Different competing paths in Max-Log-MAP and SOVA algorithm

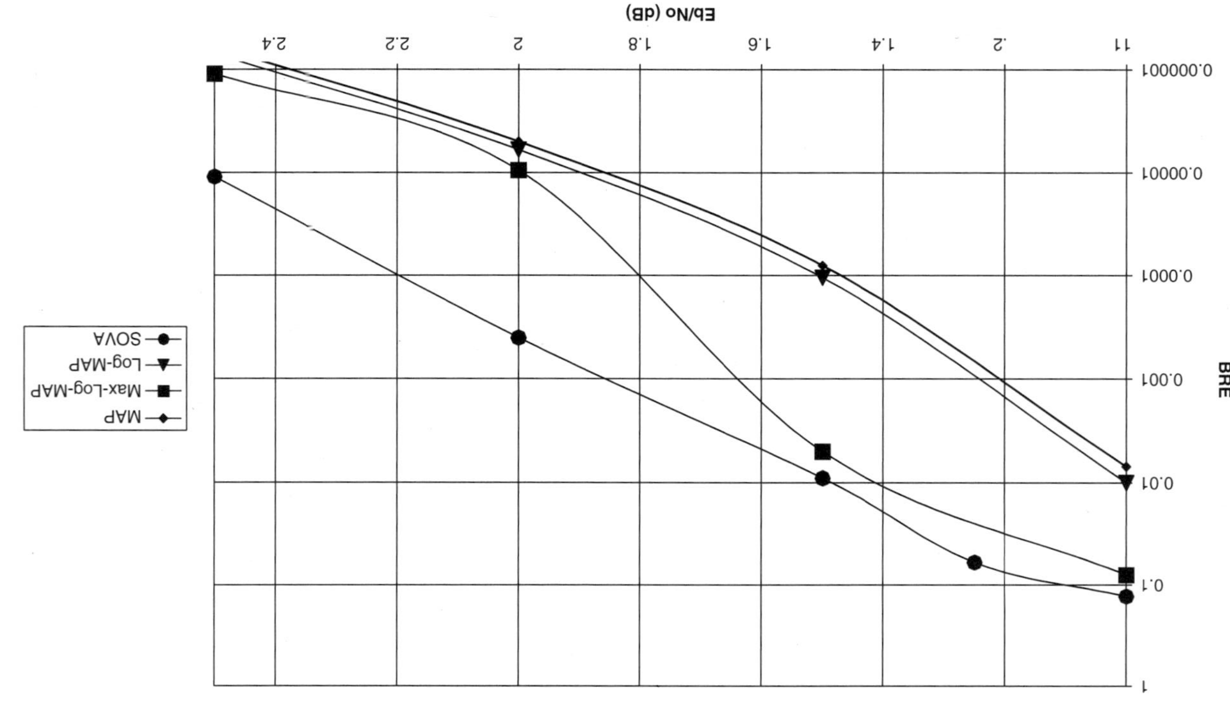

Figure 6.27 Comparison of different SISO decoding algorithms

the $A_k(j)$ values in the Max-Log-MAP, taking the difference between the metrics of the two merging paths in the SOVA algorithm to be equivalent to taking the difference between two values of $A_{k-1}(i) + B_k(j) + \Gamma_k(i, j)$ in the Max-Log-MAP algorithm, as in (6.59). The only difference is that in the Max-Log-MAP algorithm one path will be the ML path, and the other will be the most likely path to give a different hard decision for u_k. In the SOVA algorithm again one path will be the ML path, but the other may not be the most likely path to give a different hard decision for u_k. Instead, it will be the most likely path to give a different hard decision for u_k and survive to merge with the ML path. Other, more likely paths, which give a different hard decision for the bit u_k to the ML path, may have been discarded before merging with the ML path, as shown in Figure 6.26. Due to this fact the performance of the SOVA algorithm is slightly worse compared to the performance of the Max-Log-MAP algorithm. However, as pointed out in [12] the SOVA and Max-Log-MAP algorithms will always give the same hard decisions, as in both algorithms these hard decisions are determined by the ML path, which is calculated using the same metric in both algorithms.

In Figure 6.27 the performance of the different SISO decoding algorithms is compared. The simulation results were obtained for QPSK modulation over flat fading channel, the number of iterations in the turbo-decoder was set to 8. As can be seen from Figure 6.27 the Log-MAP algorithm shows almost the same performance as MAP algorithm. The performance of Max-Log-MAP and SOVA appears to be quite close, especially in the range of low SNRs, and in the range of high SNRs the Max-Log-MAP outperforms SOVA by about 0.5 dB for the reasons discussed above.

REFERENCES

1. Viterbi, A. J. and Omura, J. K. (1979). *Principles of Digital Communication and Coding*, McGraw-Hill Book Company, N.Y., USA.
2. Viterbi, A. J. (1967). Error Bounds for Convolutional Codes and an Asymptotically Optimum Decoding Algorithm, *IEEE Trans. Inf. Theory*, **IT-13**, pp. 260–9.
3. Omura, J. K. (1969). On the Viterbi Decoding Algorithm, *IEEE Trans. Inf. Theory*, **IT-15**, pp. 177–9.
4. Proakis, J. G. (1995). *Digital Communications*, 3rd ed. McGraw-Hill, New York, USA.
5. Rappaport, T. S. (1996). *Wireless Communications Principles and Practice*, Prentice-Hall, New Jersey, USA.
6. Johannesson, R. and Zigangirov, K. (1999). *Fundamentals of Convolutional Coding* John Wiley & Sons IEEE Press, Chichester, UK.
7. Berrou, C., Glavieux, A., and Thitimajshima, P. (1993). Near Shannon limit error-correcting coding: turbo codes. *Proceedings IEEE International Conference on Communications, Geneva, Switzerland*, pp. 1064–70.
8. Woodard, J. P. and Hanzo, L. (2000). Comparative Study of Turbo Decoding Techniques: An Overview. *IEEE Transactions on Vehicular Technology*, **49**, (6) pp. 2208–33.
9. Bahl, L., Cocke, J., Jelinek, F., and Raviv, J. (1974). Optimal decoding of linear codes for minimizing symbol error rate. *IEEE Transactions on Information Theory*, **IT-20**, pp. 284–87.
10. Koch, W. and Baier, A. (1990). Optimum and sub-optimum detection of coded data disturbed by time-varying inter-symbol interference, *IEEE Globecom*, pp. 1679-84.
11. Erfanian, J. A., Pasupathy, S., and Gulak, G. (1994). Reduced complexity symbol detectors with parallel structures for ISI channels, *IEEE Trans. Commun.*, **42**, pp. 1661–71.

12. Robertson, P. Villebrun, E., and Hoeher, P. (1995). *A comparison of optimal and sub-optimal MAP decoding algorithms operating in the log domain*, Proc. Int. Conf. Communications, pp. 1009–13.

13. Hagenauer, J. and Hoeher, P. (1989). *A Viterbi Algorithm with Soft-DecisionOutputs and Its Applications*, GLOBECOM 1989, Dallas, Texas, pp. 1680–86.

14. Burr, A. (2001). *Modulation and Coding for Wireless Communications*. Prentice Hall, NJ, USA.

15. Hagenauer, J., Offer, E., and Papke, L. (1996). Iterative Decoding of Binary Block and Convolutional Codes, *IEEE Transactions of Information Theory*, **42**, (2), pp. 429–45.

16. Hagenauer, J. (1995). Source-Controlled Channel Decoding, *IEEE Transactions on Communications*, **43**, (9), pp. 2449–57.

7

Coding of Messages at the Transport Layer of the Data Network

In this chapter we consider the applications of error correcting codes to the data network. It is well known that using error controlling codes adapted to typical errors in a defended system is the universal method of error protection. However, in modern data networks error correcting (or controlling) codes are used only as a means of increasing the reliability of information during the data transmission over different channels; no correlation between coding and other network procedures is considered. The application of coding not only to the physical layer but also to procedures at higher layers (e.g. transport layer) gives us some unexpected results, indicating that coding in a network helps not only to increase the reliability of the transmitted information but can also be used to improve such important characteristics of a network as a mean message delay. In this chapter we consider mostly the packet switching network with datagram routing (or in datagram mode). *Packet switching* is switching in which messages are broken into packets and one packet at a time is transmitted on a communication link. Thus, when a packet arrives at a switching node on its path to the destination site, it waits in a queue for its turn to be transmitted to the next link in its path. *Datagram routing* is packet switching in which each packet finds its own path through the network according to the current information available at the nodes visited [1]. To be more precise, there is only one restriction on the considered network model, which is essential for the exposition of this chapter. This is the possibility of getting the packets out of order at the destination node. It is shown in [1] that not only the datagram networks but also the virtual circuit networks have this feature as well. However, for simplicity we assume that we are dealing with a datagram network and that packets can get out of order arbitrarily on the network.

7.1 DECREASING THE MESSAGE DELAY WITH THE HELP OF TRANSPORT CODING

One of the most important measures of the effectiveness of a data network is the information delay. The mean packet delay has been subject to many studies, for example [2], [3], [4]. However, in a packet-switching network, the parameter of interest is not the delay of a separate packet but the delay of a message as a whole. And the mean message delay can differ from the mean packet delay, as the assembly of a message at a destination node can be

Error Correcting Coding and Security for Data Networks   G. Kabatiansky, E. Krouk and S. Semenov
© 2005 John Wiley & Sons, Ltd   ISBN: 0-470-86754-X

delayed due to the absence of a small number of packets (for example, one). This section deals with an analysis of the method of decreasing the mean message delay with the help of error-correcting code at the transport level of network. This method was suggested in [5]. The possibility of using error-correcting code in a bipolar network was described in [6].

Let us consider a model of a network having M channels, in which the capacity of the ith channel is C_i. We will consider separately a case where all the channels are assumed to be reliable and another where they are unreliable. We will assume that the channels in the network are reliable if the following inequality is always satisfied:

$$P_{err.} < P_{err.a.}, \tag{7.1}$$

where $P_{err.}$ is the probability of obtaining a distorted message at a destination node, and $P_{err.a.}$ is the acceptable error probability of the message. The time taken to transmit a packet over a channel has an exponential distribution with the expectation $1/\mu$. When the servicing device is busy, the packet may be placed in a queue. Each message, arriving in the network, is divided into K packets of the same length. The length of each packet is s bits. The traffic arriving in the network from external sources forms a Poisson process with an intensity γ (packets per second). We will denote the mean number of packets passing through the ith channel per second by λ_i. The total network traffic is then

$$\lambda = \sum_{i=1}^{M} \lambda_i. \tag{7.2}$$

If the packets arrive at a node via different routes, we can assume that the dependence between packet delays is negligible. Hence, the model of the network turns out to be close to the Kleinrock model, for which the Kleinrock 'assumption of independence' holds [2], [3]. According to this assumption, the packet delays can be regarded as independent random variables. This statement was proved in [7] for some network types. Then the ith channel can be represented in the form of a queuing system with a Poisson flow of intensity λ_i at the input and an exponential servicing time with mean $\dfrac{1}{\mu \cdot C_i}$. In this case we can assume that the packet delays in the network have an exponential distribution with the expectation $\bar{t}(\lambda, \mu)$, where

$$\bar{t}(\lambda, \mu) = \sum_{i=1}^{M} \frac{\lambda_i}{\gamma} \cdot \frac{1}{\mu C_i - \lambda_i}. \tag{7.3}$$

If we consider a case where all M channels have the same carrying capacity while the external traffic is uniformly distributed between the channels (so that the intensity of the packet flow for all channels is the same), expression (7.3) can be written as follows:

$$\bar{t}(\lambda, \mu) = \frac{\bar{l}}{\mu C} \cdot \frac{1}{1 - \rho}, \tag{7.4}$$

where $\bar{l} = \dfrac{\lambda}{\gamma} = \dfrac{M \cdot \lambda_i}{\gamma}$ is the mean path length traversed by a packet along the network, $\rho = \dfrac{\lambda}{\mu \cdot C}$ is the network load, and $C = \sum\limits_{i=1}^{M} C_i$ is the overall capacity of the network channels.

The value of the network load in this case is identical with the $\rho_i = \dfrac{\lambda_i}{\mu \cdot C_i}$, the load of a single channel. In fact, as will be shown later, all needed assumptions are as follows: the packet delays are independent random variables with exponential distribution and with expectation of form $\dfrac{a}{1-\rho}$, where ρ is the network load and a is the constant for the given network.

Let us consider a network with reliable channels, i.e. with the inequality (7.1) always satisfied. The delay T of an uncoded message in the network is determined by the maximum delay among the K packets of the given message

$$T = \max\{t_1, \ldots, t_K\},$$

where t_i is the delay of the ith packet of the message; i.e., the message delay is equal to the delay of the packet, which arrives last. If we re-denote the packet delays of the message in increasing order $t_{1:K} \le t_{2:K} \ldots \le t_{K:K}$, we have

$$T = t_{K:K}.$$

We can now apply the coding at the transport level of the network and encode the message, which consists of K packets, with the help of an MDS (N, K) code (for example Reed-Solomon code). In the case of the Reed-Solomon code each of the K packets is considered as an element of a field $GF(2^s)$ (s is the packet length), and after encoding the original message consisting of K packets is replaced by a message consisting of N packets. When the encoded messages are transmitting over the network, the traffic increases by a factor of $1/R$ ($R = K/N$ is the rate of the code used). This naturally leads to an increase in the mean packet delay in the network. However, at the node-addressee, to reconstruct the message (in view of properties of MDS codes), only K packets need to be received, as against all N packets. We will show that with some restrictions on the operation of the network this method leads to a decrease of the mean message delay. We will denote this method further on as *transport coding*.

In the case of using transport coding, the delay of the encoded message is

$$T_{cod} = t_{K:N}.$$

Using the apparatus of order statistics [8], the mathematical expectation of the delay of the ith packet (for an overall number of packets N) can be written as follows:

$$E[t_{i:N}] = B_i \cdot \int_{-\infty}^{\infty} t \cdot [P(t)]^{i-1} \cdot [1 - P(t)]^{N-i} dP(t) \tag{7.5}$$

or

$$E[t_{i:N}] = B_i \cdot \int_{0}^{1} P^{-1}(u) \cdot u^{i-1} \cdot [1 - u]^{N-i} du, \tag{7.6}$$

where $B_i = N \cdot \binom{N-1}{i-1}$, $P(t)$ is the distribution function of packet delay and $P^{-1}(u)$ is the inverse function of $P(t)$. In the case of an exponential distribution of the packet delay in the network equations (7.5), (7.6) can be rewritten as follows [8]:

$$E[t_{i:N}] = \bar{t} \cdot \sum_{j=N-i+1}^{N} j^{-1}, \qquad (7.7)$$

where $\bar{t}$ is the mean packet delay in the network (depends on λ and μ). The mean delay of the uncoded message in the network is then

$$\overline{T}_1 = E[t_{K:K}] = \bar{t}(\lambda, \mu) \cdot \sum_{j=1}^{K} j^{-1}, \qquad (7.8)$$

where $\bar{t}(\lambda, \mu)$ is defined by (7.4). The sum on the right-hand side of (7.8) can be represented as follows:

$$\sum_{j=1}^{K} j^{-1} = \varepsilon + \ln K + \frac{1}{2K} - \sum_{i=2}^{\infty} \frac{A_i}{K \cdot (K-1) \cdot \cdots \cdot (K+i-1)},$$

$$A_i = \frac{1}{i} \cdot \int_0^1 x \cdot (1-x) \cdot \cdots \cdot (i-1-x)dx,$$

where $\varepsilon = 0.577\ldots$ is Euler's constant. Hence, we obtain the following estimate for the mean delay of the uncoded message $\overline{T}_1$:

$$\overline{T}_1 \geq \bar{t}(\lambda, \mu) \cdot (\varepsilon + \ln K) = \frac{\bar{l}}{\mu C} \cdot \frac{1}{1-\rho} \cdot (\varepsilon + \ln K). \qquad (7.9)$$

We can write the mean delay of the encoded message $\overline{T}_2$ for given $N > K$, in accordance with (7.7), as follows:

$$\overline{T}_2 = E[t_{K:N}] = \bar{t}(\lambda/R, \mu) \cdot \sum_{j=N-K+1}^{N} j^{-1}, \qquad (7.10)$$

where $\bar{t}(\lambda/R, \mu)$ is the mean packet delay for traffic that has been increased as a result of using encoded messages by a factor of $1/R$; $R = K/N$ is the rate of the code used. The sum on the right-hand side of (7.10) can be represented as follows:

$$\sum_{j=N-K+1}^{N} j^{-1} = \sum_{j=1}^{N} j^{-1} - \sum_{j=1}^{N-K} j^{-1} = \ln\left(\frac{N}{N-K}\right) + \frac{1}{2N} - \frac{1}{2(N-K)}$$

$$+ \sum_{i=2}^{\infty} \left(\frac{A_i}{(N-K)\ldots(N-K+i-1)} - \frac{A_i}{N\ldots(N+i-1)}\right) \leq \ln\left(\frac{N}{N-K}\right)$$

$$(7.11)$$

It is possible to choose the code rate R in such a way as to minimise the mean message delay $\bar{T}_2$. For the best-chosen code, we obtain

$$\bar{T}_2 = \min_R \left\{ \bar{t}(\lambda/R, \mu) \cdot \sum_{j=K/R-K+1}^{K/R} j^{-1} \right\}. \tag{7.12}$$

Then, with the help of (7.11) we can estimate (7.12) as

$$\bar{T}_2 \leq \min_R \left\{ \bar{t}(\lambda/R, \mu) \cdot \ln\left(\frac{1}{1-R}\right) \right\}. \tag{7.13}$$

The mean packet delay for traffic which has been increased by a factor of $1/R$ can be written in accordance with (7.4), as

$$\bar{t}(\lambda/R, \mu) = \frac{\bar{l}}{\mu C} \cdot \frac{R}{R-\rho}, \tag{7.14}$$

where $\bar{l} = \dfrac{\lambda/R}{\gamma/R} = \dfrac{\lambda}{\gamma}$ is the mean path length traversed by a packet along the network and $\rho = \dfrac{\lambda}{\mu \cdot C}$ is the load of the network when using uncoded messages. Minimising (7.13) with respect to R, we obtain the optimal code rate $R_0 = \dfrac{2\rho}{1+\rho}$. Substituting R_0 to (7.13) we obtain

$$\bar{T}_2 \leq \frac{\bar{l}}{\mu C} \cdot \frac{4\rho}{(1-\rho)^2}. \tag{7.15}$$

The gain of using transport coding can be obtained when the following condition is satisfied:

$$\bar{T}_1 - \bar{T}_2 > 0. \tag{7.16}$$

Substituting (7.9) and (7.15) into (7.16), we obtain the following condition for the gain of using transport coding:

$$\varepsilon + \ln K > \frac{4\rho}{1-\rho}. \tag{7.17}$$

The plots of transport coding gain (in the sense of decrease of the mean message delay) are represented in Figures 7.1 and 7.2. As one can see from Figure 7.1 the exact calculation shows that the gain of using transport coding can be obtained with a wider range of network load than follows from condition (7.17). However, the estimation reflects the proper trend of the gain behaviour versus network load. The plot in Figure 7.2 shows that an increase of the number of information packets in a message leads to a gain increase that has logarithmic behaviour.

It is necessary to mention the example of the packet length s being too long. Then it is hard to use, for instance, the Reed-Solomon code for transport coding because of the necessity of operating in a very big field $GF(2^s)$. It is quite easy to eliminate this situation by

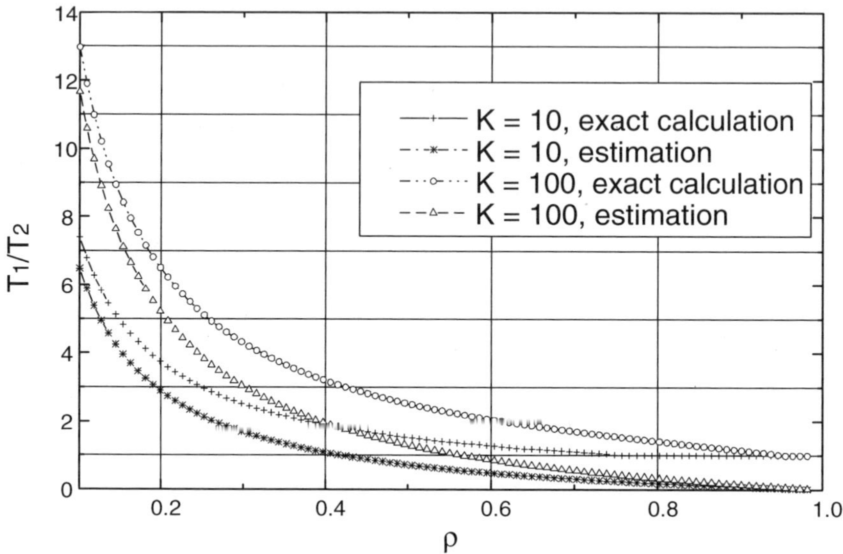

Figure 7.1   A gain of transport coding vs. network load. Reliable network

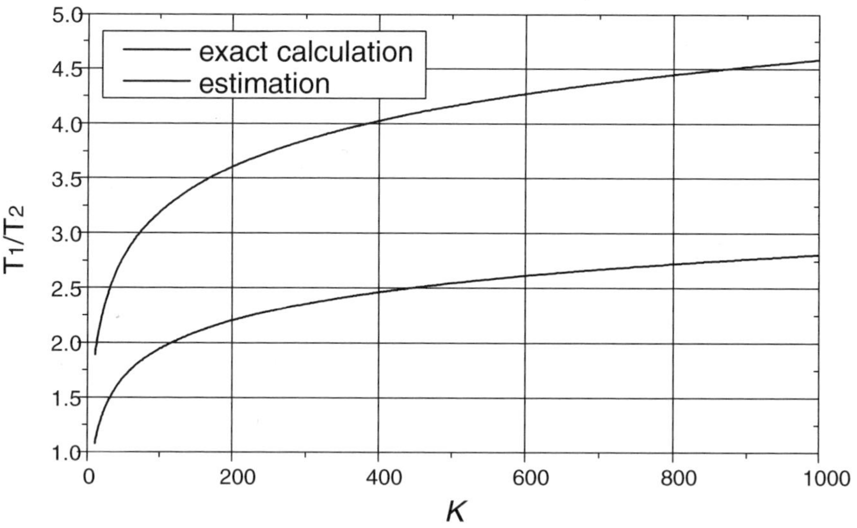

Figure 7.2   A gain of transport coding vs. number of information packets in message. Reliable network

splitting each packet, as is shown in Figure 7.3, into s/m parts, where m is the divisor of s. After that we can consider each m-bits part of packet as a symbol over $GF(2^m)$ and transport coding will produce the corresponding sets of $N - K$ redundant symbols for each of the s/m codewords. So we can change the encoding–decoding procedure over field $GF(2^s)$ to s/m procedures over field $GF(2^m)$. There are no changes in transmission protocols in this case.

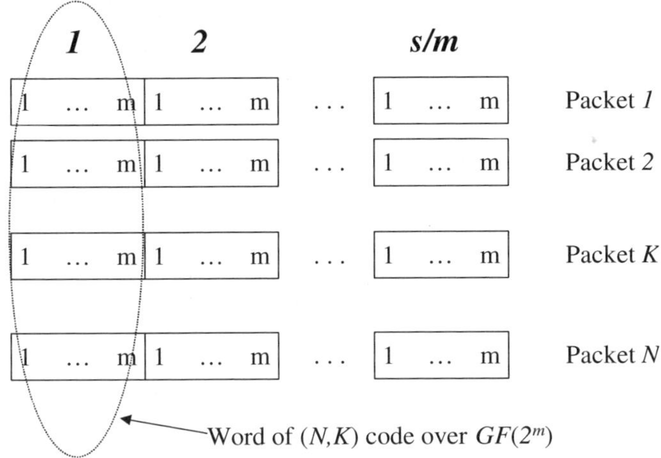

Figure 7.3 Splitting of long packets for encoding in a small field $GF(2^m)$

As before K information packets of the uncoded message will be replaced by N packets after encoding. And the length of each packet remains the same: s bits. In the case when m is not the divisor of s it is necessary to increase the length of the packet to s' (m is the divisor of s').

Let us discuss some assumptions made in this section. The strongest assumption is the independence of the packet delays. This fact restricts us to the consideration of datagram networks. Moreover, even for the datagram network model this assumption is not canonical in spite of the fact that it fits the Kleinrock's assumption of independence and is proven for some network models in [7]. Other assumptions, e.g. the exponential distribution of the packet delay and the dependence of the packet delay on the network load are not fundamental and were made to estimate analytically the possible gain of transport coding.

7.2 TRANSMISSION OF MESSAGE DURING LIMITED TIME

For many data networks the probability $P(T_0)$ of transmission of a message during a time of no more than T_0 has the same importance as the mean message delay. Let us show that in this case transport coding also can be used to increase $P(T_0)$. Let $p = \Pr\{t \le T_0\}$ denote the probability of transmission of a packet during a time of no more than T_0 and let p_R denote the same probability for the encoded message having regard to the increased network load. Then for the uncoded messages we have

$$P(T_0) = p^K \tag{7.18}$$

and in case of using the encoded messages (code length is N) the probability $P^{(R)}(T_0)$ that the encoded message is delivered during a time no more than T_o can be written as follows

$$P^{(R)}(T_0) = \sum_{i=0}^{N-K} \binom{N}{i} \cdot (1 - p_R)^i \cdot p_R^{N-i}. \tag{7.19}$$

Now if we use the same assumptions as in section 7.1 about exponential distribution of packet delay and about dependence of the mean packet delay on the network load (7.4) we obtain the following expression for p:

$$p = 1 - e^{-T_0/\bar{t}(\lambda,\mu)}, \tag{7.20}$$

where $\bar{t}(\lambda, \mu)$ is the mean packet delay defined by (7.4). Let us denote the ratio $T_0/\bar{t}(\lambda, \mu)$ as a, and ratio of the mean packet delay in the ordinary network without transport coding to the mean packet delay in the network with transport coding as $\xi = \bar{t}(\lambda, \mu)/\bar{t}(\lambda/R, \mu) = (1 - \rho R^{-1})/(1 - \rho)$. Then formulas (7.18) and (7.19) can be rewritten as follows:

$$P(T_0) = (1 - e^{-a})^K, \tag{7.21}$$

$$P^{(R)}(T_0) = \sum_{i=0}^{N-K} \binom{N}{i} \cdot (1 - e^{-a\xi})^{N-i} \cdot e^{-ia\xi}. \tag{7.22}$$

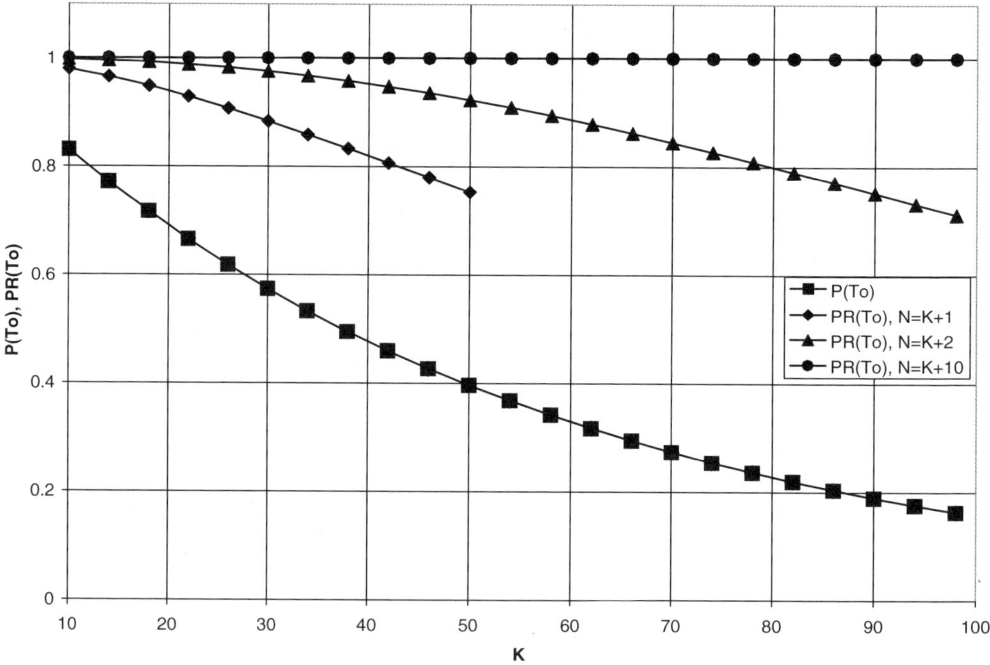

Figure 7.4   A probability of message delivery during time T_0 vs. number of information packets in message, $a = 4$, $\rho = 0.2$

It is easy to verify that $P(T_0) \to 0$ with increasing the number of packets in the message K. On the other hand, in case $(1 - R) > e^{-a\xi}$, $P^{(R)}(T_0) \to c$, $0 < c \leq 1$ with increasing K. The condition $(1 - R) > e^{-a\xi}$ with the restriction $R > \rho$ can be written as the following inequality

$$\rho < R < 1 - e^{-a\frac{1-\rho/R}{1-\rho}}. \tag{7.23}$$

Thus, for any R satisfying (7.23) the addition of N-K redundant packets to the message leads to the fact that the probability $P^{(R)}(T_0)$ tends to the constant, which is greater than zero, whilst the probability $P(T_0)$ tends to zero with increasing K. The plots of $P(T_0)$ and $P^{(R)}(T_0)$ against K are represented in Figures 7.4, 7.5 and 7.6.

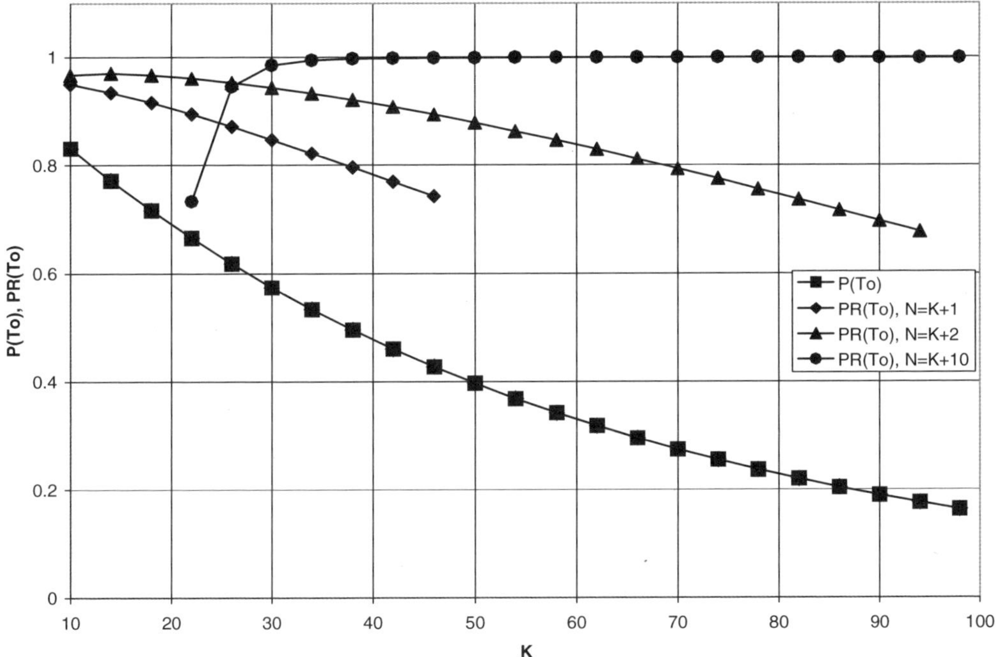

Figure 7.5 A probability of message delivery during time T_0 vs. number of information packets in message, $a = 4$, $\rho = 0.6$

Example 7.1 Let $K = 8$ and $T_0 = 7\,\bar{t}(\lambda, \mu)$. Then $a = 7$ and $P(T_0) = 1 - 7.27 \cdot 10^{-3}$. The probabilities $1 - P^{(R)}(T_0)$ that the message will not be delivered during given time T_0 for different values of the network load ρ and the length of transport code N are shown in Table 7.1. It is obvious that another possible approach is to use the fixed probability that the message will be delivered during a given time, and to decrease this given time with the help of transport coding. Very often we have given time T_0 and given probability $P_{deliv.}$ that the message should be delivered during this time. Then it is necessary to estimate the maximal acceptable network load $\rho_{acc.}(T_0, P_{deliv.}) = \rho_0$ for which these requirements are fulfilled. The results presented in this section show that the region $(T_0, P_{deliv.})$, where the transport coding allows an increase in the network load, is not empty.

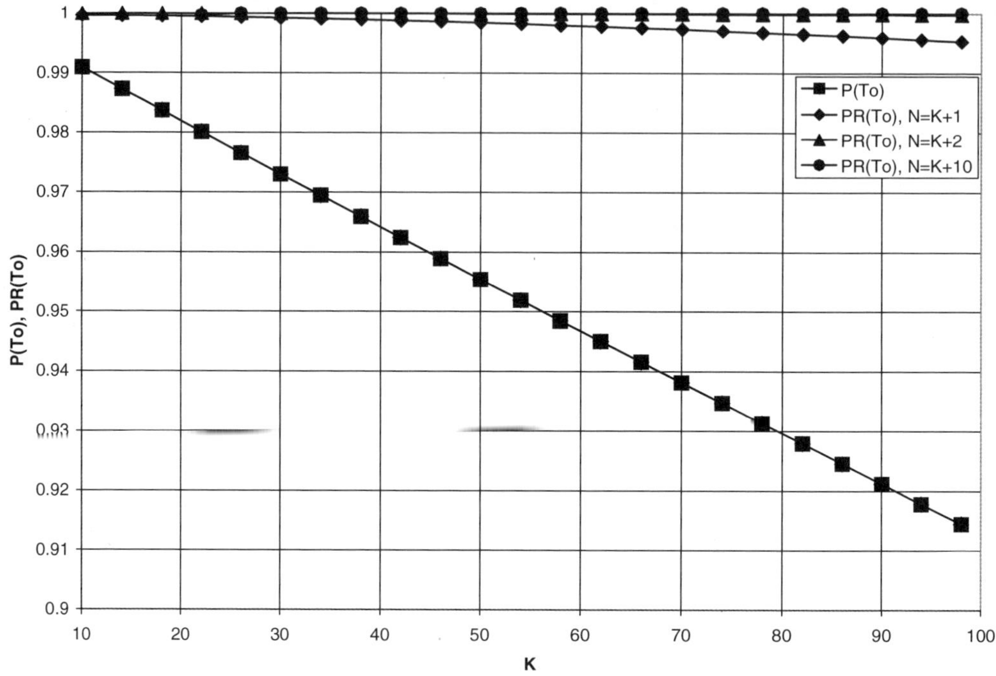

Figure 7.6 A probability of message delivery during time T_0 vs. number of information packets in message, $a = 7$, $\rho = 0.6$

Table 7.1

	$\rho = 0.4$	$\rho = 0.5$	$\rho = 0.6$	$\rho = 0.7$	$\rho = 0.8$
$N = 9$	$9.5 \cdot 10^{-5}$	$1.7 \cdot 10^{-4}$	$4.1 \cdot 10^{-4}$	$1.7 \cdot 10^{-3}$	$2.9 \cdot 10^{-2}$
$N = 10$	$3.0 \cdot 10^{-6}$	$1.7 \cdot 10^{-5}$	$2.2 \cdot 10^{-4}$	$1.4 \cdot 10^{-2}$	

7.3 TRANSMISSION OF PRIORITY MESSAGES WITHOUT USING PRIORITY PACKETS

A useful feature in many networks is the ability to transmit some special messages with less delay. Usually this feature is achieved with the help of priority packets. That means all messages, and correspondingly the packets belonging to these messages, are divided into several different priority classes. Class 1 has the highest priority, class 2 the second highest and so on. There are different queues for different priority classes in each network node and the packet will be served only when all higher class queues are empty. Of course, the presence of some special priority packets in a network leads to an increase in the complexity of network protocols.

We suggest using transport coding for the transmission of priority messages instead of the priority packets [9]. Let us consider the simplest case with only two priority classes: the priority messages form class 1 and the ordinary messages form class 2. Let denote by q, $(q < 1)$ the share of the priority messages in external traffic. Then we can write the external traffic and the network traffic for the priority messages as $\gamma_p = \gamma \cdot q$ and $\lambda_p = \lambda \cdot q$ correspondingly. In the case of transmitting the priority messages by priority packets the transmission of the ordinal information does not influence the priority message traffic. So, we can say that the priority information is transmitted at a state of low network load $\rho_p = \rho \cdot q$. In this case, the mean delay of a priority packet $\bar{t}_p$ and the mean delay of a priority message $\bar{T}_{p1}$, in accordance with (7.4) and (7.8), can be written as follows:

$$\bar{t}_p = \frac{\lambda q}{\gamma q \mu C} \cdot \frac{1}{1 - \rho q} = \frac{\bar{l}}{\mu C} \cdot \frac{1}{1 - \rho q}. \tag{7.24}$$

$$\bar{T}_{p1} = \frac{\bar{l}}{\mu C} \cdot \frac{1}{1 - \rho q} \cdot \sum_{j=1}^{K} j^{-1}. \tag{7.25}$$

For the considered network model the mean packet delay (including both priority and ordinary packets) is the same as for the network without priority packets.

$$\bar{t} = \frac{\bar{l}}{\mu C} \cdot \frac{1}{1 - \rho}. \tag{7.26}$$

Denote by $\bar{t}_{ord}$ the mean delay of the ordinary packets. In accordance with Little's theorem we can write the following equation

$$\bar{t} = \bar{t}_p \cdot q + \bar{t}_{ord} \cdot (1 - q). \tag{7.27}$$

Then from the equation (7.27) we obtain the following expression for the mean delay of the ordinary packet:

$$\bar{t}_{ord} = \frac{\bar{t} - \bar{t}_p q}{1 - q} = \frac{\bar{l}}{\mu C} \cdot \frac{1}{(1 - \rho)(1 - \rho q)}. \tag{7.28}$$

And the mean delay of an ordinary message can be written as follows:

$$\bar{T}_{ord} = \bar{t}_{ord} \cdot \sum_{j=1}^{K} j^{-1}. \tag{7.29}$$

Now let us consider the network without priority packets. As was mentioned above, instead of using priority packets we will use transport coding to transmit the priority information. It means that the priority messages will be replaced by the encoded messages and the ordinary messages will be transmitted without coding. As a result of transport coding, the external traffic and network traffic for the priority messages will increase by a factor of $1/R$, i.e. $\gamma_{p.cod} = \gamma \cdot q/R$ and $\lambda_{p.cod} = \lambda \cdot q/R$; and for the ordinary messages, values will be the same as they were $\gamma_{ord} = \gamma \cdot (1 - q)$ and $\lambda_{ord} = \lambda \cdot (1 - q)$, correspondingly. Then the overall external traffic and network traffic can be written as

$$\gamma_{new} = \gamma_{p.cod} + \gamma_{ord} = \gamma \cdot \left(1 + \frac{q}{R} - q\right),$$

$$\lambda_{new} = \lambda_{p.cod} + \lambda_{ord} = \lambda \cdot \left(1 + \frac{q}{R} - q\right).$$

Then the mean delay of a priority message can be written in accordance with (7.10) as:

$$\bar{T}_{p2}(R) = \frac{\bar{l}}{\mu C} \cdot \frac{1}{1 - \rho(1 + q/R - q)} \cdot \sum_{j=K_1/R\cdot K+1}^{K/R} j^{-1}, \tag{7.30}$$

and the mean delay of an ordinary message can be written as:

$$\bar{T}_{ord2}(R) = \frac{\bar{l}}{\mu C} \cdot \frac{1}{1 - \rho(1 + q/R - q)} \cdot \sum_{j=1}^{K} j^{-1}, \tag{7.31}$$

Let us choose the code rate $R = R_o$ satisfying the following condition:

$$\bar{T}_{ord2}(R) = \bar{T}_{ord}. \tag{7.32}$$

That means the mean delay of the ordinary message has the same value for both networks (with and without using transport coding). With a little manipulation we will obtain R_o from (7.32) as follows:

$$R_o = \frac{1}{2 - \rho}. \tag{7.33}$$

Substituting R_o of (7.33) in (7.30) gives us the value of mean delay of priority massage in the network with transport coding $\bar{T}_{p2}$:

$$\bar{T}_{p2} = \bar{T}_{p2}(R_o) = \frac{\bar{l}}{\mu C} \cdot \frac{1}{(1 - \rho)(1 - q\rho)} \cdot \sum_{j=K(1-\rho)+1}^{K(2-\rho)} j^{-1}. \tag{7.34}$$

Using the same technique to estimate $\bar{T}_{p1}$ and $\bar{T}_{p2}$ as was employed in the section 7.1 to estimate $\bar{T}_1$ and $\bar{T}_2$, we obtain the following:

$$\bar{T}_{p1} \geq \frac{\bar{l}}{\mu C} \cdot \frac{\varepsilon + \ln K}{1 - \rho q}, \tag{7.35}$$

$$\bar{T}_{p2} \leq \frac{\bar{l}}{\mu C} \cdot \frac{1}{(1 - \rho)(1 - \rho q)} \cdot \ln \frac{2\rho}{1 - \rho}. \tag{7.36}$$

The condition for gain in using transport coding to transmit priority messages can be obtained in the same manner as in (7.17) by substituting (7.35) and (7.36) in (7.16):

$$\varepsilon + \ln K > \frac{1}{1-\rho} \cdot \ln \frac{2\rho}{1-\rho}. \qquad (7.37)$$

As can be seen from Figure 7.7 the condition (7.37) is quite accurate in comparison with exact calculations. For case $K = 100$, as we can see from the plot on Figure 7.7, the estimate coincides with the exact calculation. And the gain of using transport coding for priority messages can be achieved in a very wide range of network load values. By this means the use of transport coding allows the elimination of the special procedures for processing the priority packets and simultaneously provides the gain in the mean delay of priority messages without increasing the mean delay of ordinary messages.

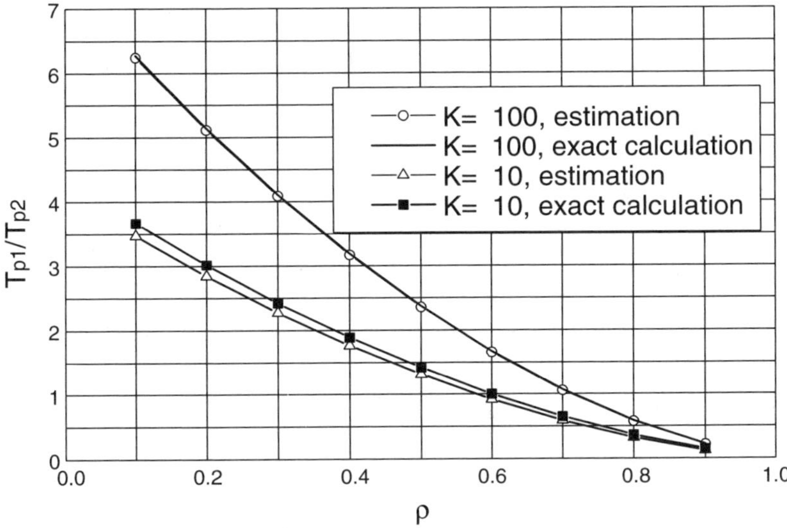

Figure 7.7   A gain of transport coding for priority messages vs. network load

7.4   ESTIMATION OF THE EFFECTIVENESS OF TRANSPORT CODING FOR THE NONEXPONENTIAL MODEL OF PACKET DELAY

As was mentioned in section 7.1 the assumption about the exponential distribution of the packet delay is not fundamental and was made to simplify some analytic estimates. In this section we will try to obtain the estimates of the mean delay of the encoded message in the case of arbitrary distribution of packet delay.

In estimating the moments of rank statistics, different lines of approach can be considered. Most of these allow for the calculation of the estimates of the moments reasonably well. To obtain the gains in using transport coding in an explicit form it is necessary to have the explicit formulas to calculate the estimates of the mean message delay. It is possible to obtain such formulas with the help of estimates of the moments of rank statistics expressed by the quantiles of distribution.

Let $F(x)$ be a cumulative distribution function (cdf) and $\varphi(x)$ be the inverse function of $F(x)$, i.e. $\varphi(F(x)) = x$. Denote by Φ the class of cumulative distribution functions $F(x)$, which have a positive continuous derivative $f(x)$ over some interval I. Then $\varphi(F)$ is uniquely defined for the interval $0 < F < 1$ and has a positive continuous derivative in this range. It is possible to show [8] that for every pair of cdfs F_1 and F_2 in Φ there exists a strictly increasing function $g(x)$ such that, if the random value x has cdf F_1, then $g(x)$ has cdf F_2. And the function $g(x)$ is uniquely defined by the equation

$$g(x) = \varphi_2(F_1(x)), \tag{7.38}$$

where $\varphi_2(x)$ is the inverse function of F_2. There are defined two classes of ordering relation for the functions from Φ [8]. We will say that F_1 *c-precede* F_2

$$F_1 \underset{c}{<} F_2$$

if $g(x)$ is the convex function on the interval I. If $F_1 \underset{c}{<} F_2$ and $F_2 \underset{c}{<} F_1$, then F_1 is equivalent to F_2 $F_1 \sim F_2$.

Let us introduce Θ the subclass of Φ, $\Theta \subset \Phi$ consisting of the symmetric cdfs, i.e.

$$F(x_0 - x) + F(x_0 + x) = 1 \tag{7.39}$$

for some x_0 and all x, where $F(x) \in \Theta$. Now we can define *s-ordering* or *s-comparison*: Let F_1 and F_2 are in Θ, then $F_1 \underset{s}{<} F_2$ if and only if $g(x)$ is convex for $x > x_0$. The letter s stands for symmetry, and F_1 is said to *s-precede* F_2.

The following lemmas give the conditions of c and s-ordering.

Lemma 7.1   [8] If F_1 and F_2 are in Φ, then $F_1 \underset{c}{<} F_2$ if and only if $\varphi_2'(y)/\varphi_1'(y)$ is nondecreasing for $0 < y < 1$.

Lemma 7.2   [8] If F_1 and F_2 are in Θ, then $F_1 \underset{s}{<} F_2$ if and only if $\varphi_2'(y)/\varphi_1'(y)$ is nondecreasing for $\frac{1}{2} < y < 1$.

Let random variates X and X^* have the cdfs F_1 and F_2 correspondingly. Then the following theorems hold true.

Theorem 7.1   [8] If F_1 and F_2 are in Φ, then $F_1 \underset{c}{<} F_2$ implies that

$$F_1(E[X_{K:N}]) \leq F_2(E[X_{K:N}^*]) \tag{7.40}$$

for all K ($K = 1, 2, \ldots, N$) and all N, for which $E[X_{K:N}]$ and $E[X_{K:N}^*]$ exist, where $E[X]$ is the expectation of X.

Theorem 7.2   [8] If F_1 and F_2 are in Θ, then $F_1 \underset{s}{<} F_2$ implies that

$$F_1(E[X_{K:N}]) \leq F_2(E[X_{K:N}^*]) \tag{7.41}$$

for all K in $\frac{1}{2}(N + 1) \leq K \leq N$ and all N, for which $E[X_{K:N}]$ and $E[X_{K:N}^*]$ exist.

Table 7.2

Estimate	Condition, under which the estimate holds true	Method of estimation
1. $E[X_{K:N}] \leq \varphi\left(\frac{K}{N+1}\right)$	$F(x)$ is any convex function	c-comparison with the uniform distribution
2. $E[X_{K:N}] \geq \varphi\left(\frac{K}{N+1}\right)$	$F(x)$ is any concave function	
3. $E[X_{K:N}] \leq \varphi\left(\frac{K-1}{N}\right)$	$I = (-\infty, -1)$, $\dfrac{1}{F(x)}$ is a concave function over I	c-comparison with the distribution $F^*(x) = \dfrac{1}{x}$
4. $E[X_{K:N}] \geq \varphi\left(\frac{K-1}{N}\right)$	$\dfrac{1}{F(x)}$ is a convex function over I	
5. $E[X_{K:N}] \leq \varphi\left(\frac{K}{N}\right)$	$\dfrac{1}{1-F(x)}$ is a convex function over I	c-comparison with the distribution $F^*(x) = \dfrac{x-1}{x}$, $1 < x < \infty$
6. $E[X_{K:N}] \geq \varphi\left(\frac{K}{N}\right)$	$\dfrac{1}{1-F(x)}$ is a concave function over I	
7. $E[X_{K:N}] < \varphi\left(\frac{K}{N+1/2}\right)$	$\dfrac{F'(x)}{1-F(x)}$ is an increasing function over I	c-comparison with the exponential distribution
8. $E[X_{K:N}] \leq \varphi\left(\frac{K}{N+1}\right)$	$F(x)$ is U-shape symmetric distribution	s-comparison with the uniform distribution
9. $E[X_{K:N}] \geq \varphi\left(\frac{K}{N+1}\right)$	$F(x)$ is symmetric unimodal distribution	

These theorems allow us to obtain the upper and lower bounds of expectation of rank statistics of some cdf with the help of c or s-comparison of the given cdf with cdfs for which the expectation of rank statistics is known. The results of such a comparison are listed in Table 7.2. The following lemma shows that c-comparison does not depend on the parameters of shift and scaling. Because of this in Table 7.2 the results of c-comparison with normalised cdfs are listed.

Lemma 7.3   [8] If F_1 and F_2 are in Φ, then $F_1 \sim F_2$ if and only if $F_1(x) = F_2(ax + b)$ for some constants $a > 0$ and b.

These results allow us to obtain in the explicit form the conditions for gain of using transport coding in the case when the explicit formula for the expectation of the rank statistic with the given distribution of the packet delay in the network is not known.

Example 7.2   Consider the gain of using transport coding under the condition of one side normal distribution $F_\rho(t)$ of the packet delay in a network.

$$F_\rho(t) = \begin{cases} 0, & -\infty < t < 0 \\ \sqrt{\dfrac{2}{\pi\sigma^2}} \displaystyle\int_0^t e^{-\frac{x^2}{2\sigma^2}}\, dx, & 0 \leq t < \infty \end{cases} \tag{7.42}$$

where the value σ^2 is defined from the condition of equality of the mean packet delay in this network to the mean packet delay in the Kleinrock model $\bar{t}(\rho)$:

$$\sigma\sqrt{\frac{2}{\pi}} = \bar{t}(\rho).$$

From this equality with regard to (7.4) obtain

$$\sigma_\rho = \sqrt{\frac{\pi}{2}} \cdot \bar{t}(\rho) = \sqrt{\frac{\pi}{2}} \cdot \frac{\lambda}{\gamma\mu C} \cdot \frac{1}{1-\rho} \tag{7.43}$$

The function $F_\rho(t)$ is concave function over $I = [0, \infty)$. Then in accordance with the estimate 5 of Table 7.2 obtain the estimate of the mean delay of the uncoded message

$$\bar{T} = T_{K:K} \geq \varphi_1\left(\frac{K}{K+1}\right), \tag{7.44}$$

where φ_1 is the inverse function of (7.42). If the transport coding is used in the network, the cdf of packed delay becomes as follows

$$F_{\rho/R}(t) = \begin{cases} 0, & -\infty < t < 0 \\ \sqrt{\dfrac{2}{\pi\sigma_R^2}} \displaystyle\int_0^t e^{-\frac{x^2}{2\sigma_R^2}} dx, & 0 \leq t < \infty, \end{cases} \tag{7.45}$$

where R is the code rate of used code and σ_R is defined by the following equality

$$\sigma_R = \sqrt{\frac{\pi}{2}} \cdot \bar{t}(\rho, R) = \sqrt{\frac{\pi}{2}} \cdot \frac{\lambda}{\gamma\mu C} \cdot \frac{1}{1-\rho/R}.$$

It is easy to show that $\dfrac{1}{1 - F_{\rho/R}(t)}$ is the convex function over the I. Therefore, in accordance with estimate 5 of Table 7.2

$$\bar{T}_R = T_{K:N} \leq \varphi_R(R), \tag{7.46}$$

where φ_R is the inverse function of (7.45).

Then to obtain the gain of using transport coding in the given network it is only necessary that, in accordance with (7.44) and (7.46), the following condition holds true

$$\varphi_1\left(\frac{K}{K+1}\right) > \min_R \varphi_R(R), \tag{7.47}$$

The plot of gain of using transport coding in the given network against the network load ρ is represented in Figure 7.8.

Unfortunately, the estimates of the mean delay expressed with the help of distribution quantiles are frequently not accurate enough. However, to estimate the effectiveness of using transport coding it is enough to estimate some differences of the rank statistics rather than statistics themselves.

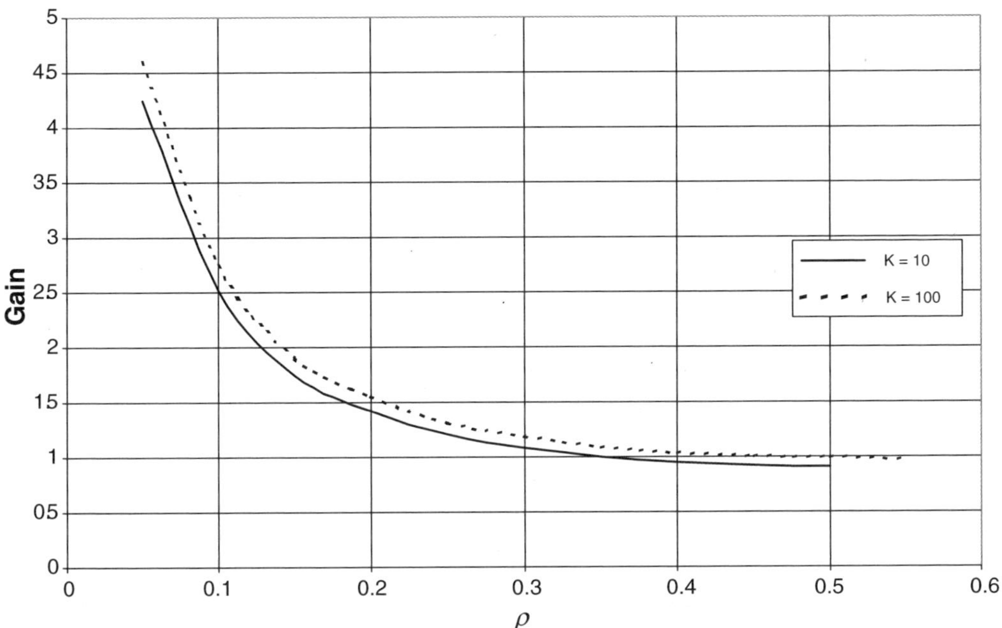

Figure 7.8   A gain of transport coding for one side normal distribution of packet delay

The effectiveness of the transport coding is defined by the sign of the difference $\bar{T} - \bar{T}_{cod}$, where $\bar{T} = E[t_{K:K}]$ is the expectation of the Kth rank statistic of K independent random variables with cdf $F_\rho(t)$, and $\bar{T}_{cod} = E[t_{K:N_0}]$ is the expectation of the Kth rank statistic of N_0 independent random variables with cdf $F_{\rho/R_0}(t)$, $R_0 = K/N_0$ is the value of the code rate that minimizes $E[t_{K:N}]$ over R. To prove the gain of using transport coding it is enough to show that the difference $\Delta = \bar{T} - \bar{T}_{cod}$ corresponding to the distribution $F(t)$ is no less than the difference $\Delta^* = \bar{T}^* - \bar{T}^*_{cod}$ corresponding to some distribution $F^*(t)$, for which this difference Δ^* is positive for some R. It is well known from the theory of rank statistics that for random variable X the expectation of the Kth rank statistic $E[X_{K:N}]$ can be written as follows [8]:

$$E[X_{K:N}] = N \cdot \binom{N-1}{K-1} \cdot \int_0^1 \varphi(u) \cdot u^{K-1} \cdot (1-u)^{N-K} du \qquad (7.48)$$

where $\varphi(u)$ is the inverse function of cdf of X. This formula has already been used in (7.6).

Let $\varphi(u, R)$ be the inverse function of cdf of packet delay in a network with the transport coding rate R ($\varphi(u, 1)$ corresponds to the network without transport coding). Then

$$\Delta = \bar{T}_{K:K} - \bar{T}_{K:N} = \int_0^1 \left[K \cdot \varphi(u, 1) - N \cdot \binom{N-1}{K-1} \cdot \varphi(u, R) \cdot (1-u)^{N-K} \right] \cdot u^{K-1} du$$

$$= \int_0^1 K \cdot \varphi(u, 1) \cdot u^{K-1} \left[1 - \frac{1}{R} \cdot \binom{N-1}{K-1} \cdot \frac{\varphi(u, R)}{\varphi(u, 1)} \cdot (1-u)^{N-K} \right] du \qquad (7.49)$$

In the same way

$$\Delta_0 = \int\limits_0^1 K \cdot \varphi_0(u, 1) \cdot u^{K-1} \left[1 - \frac{1}{R} \cdot \binom{N-1}{K-1} \cdot \frac{\varphi_0(u, R)}{\varphi_0(u, 1)} \cdot (1 - u)^{N-K} \right] du, \qquad (7.50)$$

where $\varphi_0(u, R)$ is the inverse function of the cdf, for which $\Delta_0 > 0$ for given $R = R_0$. Denote $\dfrac{\varphi(u, R)}{\varphi(u, 1)}$ by the $z(u)$ and $\dfrac{\varphi_0(u, R)}{\varphi_0(u, 1)}$ by the $z_0(u)$. Assume that $z_0(u)$ does not depend on u ($z_0(u) = z_0$). Then the function

$$\psi_0(u) = 1 - \frac{1}{R} \cdot \binom{N-1}{K-1} \cdot z_0 \cdot (1 - u)^{N-K} \qquad (7.51)$$

increases over the range $0 \leq u \leq 1$ and has no more than one root. Denote by u_0 the root of the function (7.51), if it exists and assume that $u_0 = 0$ if it does not exist. The function

$$\chi_0(u) = K \cdot \varphi_0(u, 1) \cdot u^{K-1}$$

increases over the $(0, 1]$. Moreover, $\chi_0(u) \geq 0$ since the cdf of packet delay has no sense for the negative value of the argument.

Then Δ_0 can be represented as the difference of the positive values

$$\Delta_0 = I_{01} - I_{02} = \underbrace{\int\limits_0^{u_0} \psi_0(u) \cdot \chi_0(u)du}_{I_{01}} - \underbrace{\int\limits_{u_0}^1 (-\psi_0(u)) \cdot \chi_0(u)du}_{I_{02}}. \qquad (7.52)$$

Split the integral (7.49) in two parts I_1 and I_2 relatively to u_0

$$\Delta = I_1 - I_2 = \int\limits_0^{u_0} \psi(u) \cdot \chi(u)du - \int\limits_{u_0}^1 (-\psi(u)) \cdot \chi(u)du, \qquad (7.53)$$

where $\psi(u) = \dfrac{1}{R} \cdot \binom{N-1}{K-1} \cdot z(u) \cdot (1 - u)^{N-K} - 1$ and $\chi(u) = K \cdot \varphi(u, 1) \cdot u^{K-1}$.
  The inequality

$$\Delta \geq \Delta_0, \qquad (7.54)$$

holds true if the following inequalities hold true

$$I_1 \leq I_{01},$$
$$I_2 \geq I_{02}. \qquad (7.55)$$

Analysing (7.52) and (7.53) it is possible to show many variants of the restrictions on functions $\varphi(u, 1)$ and $\varphi(u, R)$ leading to the case when inequalities (7.55) hold true. In particular, the conditions sufficient for gain in using transport coding are the following

$$\begin{cases} \varphi(u, 1) \geq \varphi_0(u, 1), & \text{for } u \geq u_0 \\ \varphi(u, 1) \leq \varphi_0(u, 1), & \text{for } u < u_0 \\ z(u) \leq z_0. \end{cases} \tag{7.56}$$

Consider the application of this analysis to the network in which the cdf of packet delay corresponds to Weibull distribution. Note, that for this distribution the conditions, which allow applying the results of c and s-comparison listed in Table 7.2, do not hold true.

Example 7.3   Let the cdf of the packet delay in a network be

$$F_\rho(t) = 1 - e^{-\beta(\rho) \cdot t^a}, \tag{7.57}$$

where $\beta(\rho)$ is chosen from the condition of equality of the mean packet delay in this network to the mean packet delay in the Kleinrock model $\bar{t}(\rho)$:

$$\beta(\rho) = \left[\frac{\gamma \mu C}{\lambda} \cdot (1 - \rho) \cdot \Gamma\left(\frac{1}{a} + 1\right) \right]^a = \left[\frac{\gamma \mu C}{\lambda a} \cdot (1 - \rho) \cdot \Gamma\left(\frac{1}{a}\right) \right]^a. \tag{7.58}$$

In the case of using transport coding with code rate R the cdf of the packet delay can be written as follows

$$F_{\rho/R}(t) = 1 - e^{-\beta(\rho/R) \cdot t^a}, \tag{7.59}$$

where

$$\beta(\rho/R) = \left[\frac{\gamma \mu C}{\lambda} \cdot \left(1 - \frac{\rho}{R}\right) \cdot \Gamma\left(\frac{1}{a} + 1\right) \right]^a = \left[\frac{\gamma \mu C}{\lambda a} \cdot \left(1 - \frac{\rho}{R}\right) \cdot \Gamma\left(\frac{1}{a}\right) \right]^a. \tag{7.60}$$

Then the mean packet delay in the network with transport coding is

$$\bar{t}(\rho, R) = [\beta(\rho/R)]^{-\frac{1}{a}} \cdot \Gamma\left(\frac{1}{a} + 1\right) = \frac{\lambda}{\gamma \mu C} \cdot \frac{R}{R - \rho}. \tag{7.61}$$

The inverse functions of cdf are as follows

$$\varphi(u, R) = \left[\frac{\ln\frac{1}{1-u}}{\beta(\rho/R)} \right]^{\frac{1}{a}}, \tag{7.62}$$

$$\varphi(u, 1) = \left[\frac{\ln\frac{1}{1-u}}{\beta(\rho)} \right]^{\frac{1}{a}}. \tag{7.63}$$

The ratio z of (7.62) to (7.63) does not depend on u:

$$z = \frac{\varphi(u, R)}{\varphi(u, 1)} = \left[\frac{\beta(\rho)}{\beta(\rho/R)}\right]^{\frac{1}{a}} = \frac{1 - \rho}{R - \rho} \cdot R. \tag{7.64}$$

Choosing the exponential distribution as a reference

$$F_0(t) = 1 - e^{-\frac{t}{\bar{t}(\rho)}}. \tag{7.65}$$

obtain

$$\varphi_0(u, R) = \bar{t}(\rho/R) \cdot \ln\frac{1}{1 - u},$$
$$z_0 = \frac{\bar{t}(\rho/R)}{\bar{t}(\rho)} = \frac{1}{R - \rho} \cdot \frac{\rho}{R - \rho} \cdot R = z \tag{7.66}$$

Then $\psi(u) = \psi_0(u)$. Now find u_0 the root of the equation $\psi_0(u) = 0$, that is the root of the equation

$$1 - \frac{1}{R} \cdot \binom{N - 1}{K - 1} \cdot z_0 \cdot (1 - u)^{N - K} = 0.$$

Taking in account (7.66) and the equality $N - K = (1 - R) \cdot N$ obtain

$$u_0 = 1 - \left[\frac{R - \rho}{1 - \rho} \cdot \frac{1}{\binom{N - 1}{K - 1}}\right]^{\frac{1}{N - K}} \approx 1 - e^{-\frac{H(R)}{1 - R}}. \tag{7.67}$$

Now find u^* the cross-point of the functions $\varphi(u, 1)$ and $\varphi_0(u, 1)$:

$$\varphi(u^*, 1) = \varphi_0(u^*, 1). \tag{7.68}$$

Substituting (7.63) and (7.66) to (7.68) obtain

$$\frac{a}{\Gamma(1/a)} = \left(\ln\frac{1}{1 - u^*}\right)^{1 - \frac{1}{a}}. \tag{7.69}$$

It follows from (7.69) that

$$u^* = 1 - \exp\left\{-\left(\frac{a}{\Gamma(1/a)}\right)^{\frac{a}{a - 1}}\right\}$$

To make the conditions (7.56) true it is enough to fulfill the following condition

$$u^* \geq u_0,$$

which is equivalent to

$$\frac{H(R)}{1-R} \leq \left(\frac{a}{\Gamma(1/a)}\right)^{\frac{a}{a-1}}. \tag{7.70}$$

For the Weibull distribution with $a = \dfrac{1}{2}$ the condition (7.70) can be written as

$$\frac{H(R)}{1-R} \leq 2. \tag{7.71}$$

The inequality (7.71) holds true if $R \leq \dfrac{1}{2}$. The optimal code rate of transport coding for the exponential distribution of the packet delay, as was shown in the section 7.1, is

$$R = \frac{2\rho}{1+\rho}.$$

It follows from the condition $\dfrac{2\rho}{1+\rho} \leq \dfrac{1}{2}$ that for the case $\rho < \dfrac{1}{3}$ the gain of using transport coding in the network with the Weibull distribution of the packet delay ($a = 1/2$) is no less than the gain in the case of the exponential distribution. Similar calculations for the case $a = 2$ allow us to make the same conclusion for $\rho \leq 0.18$.

The conditions (7.71) are too strict. More accurate calculations done in similar way show that the gain of using transport coding can be obtained over the wider range of ρ values.

REFERENCES

1. Bertsekas, D. and Gallager, R. (1992). *Data Networks*. Prentice Hall, NJ, USA.
2. Kleinrock, L. (1975). *Queuing Systems. Theory, vol.1*, John Wiley & Sons, Chichester, UK.
3. Kleinrock, L. (1964). *Communication nets; stochastic message flow and delay*. Dover, N.Y. USA.
4. Kleinrock, L. and Naylor, W. (1974). *On Measured Behavior of the ARPA Network*, AFIPS Conf. Proc., National Computer Conf., 1974, **43**, pp. 767–80.
5. Kabatiansky, G. A. and Krouk, E. A. (1993). *Coding Decreases Delay of Messages in Networks*. IEEE International Symposium on Information Theory, Proceedings.
6. Maxemchuk, N. F. (1975). *Dispersity routing*, IEEE Conf. Commun., **3**. San Francisco.
7. Vvedenskaya, N. D. (1998). Large Queuing System where Messages are Transmitted via Several Routes, *Problems Of Information Transmission*, **34**, (2) pp. 180–9.
8. David, H. A. (1981). *Order statistics*. John Wiley & Sons, Chichester, UK.
9. Krouk, E. and Semenov, S. (2004). *Transmission of Priority Messages with the Help of Transport Coding*. Proceedings of 10th International Conference on Telecommunications, Tahiti French Polynesia, pp. 1273–8.

8

Providing Security of Data in a Network with the Help of Coding Methods

8.1 PUBLIC-KEY CRYPTOGRAPHY

Traditional cryptography investigated the problem of protection against unauthorised access during transmission of information between two parties. It was based on both parties having the same secret keys for encryption and decryption.

Let F denote an information transformation algorithm (encryption). F maps message u from the set of possible messages U to ciphertext $v = F(u)$ from the set of possible ciphertexts V. Respectively, F^{-1} denotes the transformation reverse to F, i.e., it performs decryption of ciphertext v to the message $u = F^{-1}(v)$ for each $v \in V$. In the context of traditional cryptography, the functions F and F^{-1} are common secret (secret keys) of the legitimate communicating parties, and form a *secret-key* cryptosystem. All attack methods on such cryptosystems are based on attempts to use correlation dependencies in ciphertexts. The encryption task is hiding or weakening these dependencies while satisfying encryption procedure complexity restrictions. These restrictions are the core of the encryption task in traditional cryptography, because without them it becomes easy to construct theoretically unbreakable ciphers [1]. The most common example for such a cipher is a 'one-time pad' cipher, where each encryption transformation occurs only once.

Emergence of open networks with hundreds of thousands of users unacquainted with other users (i.e. users who do not have common secret keys) made it impossible to provide information security by the traditional methods. Two major cryptographic problems presented by the open networks are

1. key distribution, and

2. user authentication.

Detailed consideration of algorithmic, software and organisational challenges posed by the mentioned problems is out of the scope of this section. We focus on the investigation of coding theory approaches to these problems.

Error Correcting Coding and Security for Data Networks   G. Kabatiansky, E. Krouk and S. Semenov
© 2005 John Wiley & Sons, Ltd   ISBN: 0-470-86754-X

The basis of information security in open networks was founded in the paper by Diffie and Hellman [2]. It was the beginning of new *public-key* cryptography with asymmetric keys. This approach is based on the notion of a 'trapdoor function.'

Let F be an encryption function without an inverse one. Thus, a non-legitimate information exchange party that has access to F and ciphertext v is still unable to recover u. Obviously, such a function is useless for the legitimate party as well, because it does not help in obtaining the message u.

However, let's suppose that the function F^{-1} exists, but the knowledge of some additional secret key K is required in order to calculate it. Diffie and Hellman named this function a 'trapdoor function.'

Below the principles of a trapdoor-based cryptosystem are explained in more detail.

Every user of the system has his own function F_K, which serves as his public key. He publishes this function, i.e. makes it available to all users on the network, but does not reveal his secret key K. Anyone who wants to send him a message u should obtain an encrypted message using the published function F_K:

$$v = F_K(u) \tag{8.1}$$

The legitimate user recovers u using his knowledge of secret key K

$$u = F_K^{-1}(v, K),$$

and non-legitimate user has to recover u without knowledge of K

$$u = F_K^{-1}(v, \cdot) \tag{8.2}$$

Note that calculation (8.2) is always possible by performing an exhaustive search over the whole set of keys K. Thus, it is impossible for public key cryptosystems to provide perfect secrecy, and only computational cryptographic resistance can be considered. Nowadays, computational resistance requirements are defined by the work factor (the number of operations required to break the cryptosystem) on the level of 2^{50}, that seems to be able to withstand computing power growth for the next few years.

From the above it follows that constructing an inverse function to a trapdoor function should be equivalent to solving some computationally hard problems.

Further, an informal introduction to the theory of intractable *NP-complete* and *NP-hard* problems is given. First, the notion of an abstract '*non-deterministic (probabilistic) algorithm*' is defined.

An algorithm is usually defined as a sequence of operations, where each subsequent step is unambiguously determined by the result of the previous one. In particular, this means every time the algorithm performs a single operation. Let us elaborate this notion. Consider a finite state machine, which is intended to solve some problem.

Let us represent all initial conditions of the problem (the algorithm's input data) as a bit sequence x. Let n denote the length of this sequence. The solution of the problem is a binary sequence as well (denote it by y). Then a traditional (further referred to as 'deterministic') algorithm is a sequence of bit operations (binary operations) performed by the state machine to obtain y from x:

$$y = f(x),$$

where f is an algorithm.

The complexity of a deterministic algorithm is a number of operations L performed by the finite state machine during calculation of f.

The deterministic algorithm is said to be of polynomial complexity if

$$L = L(n) = O(n^a) \tag{8.3}$$

holds, where a is constant. The deterministic algorithm is said to be of exponential complexity if for any polynomial $p(n)$ the following equation is satisfied:

$$L = L(n) > p(n). \tag{8.4}$$

In the context of the complexity theory it is considered that an algorithm of polynomial complexity is realisable and an algorithm of non-polynomial complexity is not.

The above definitions of an algorithm and its complexity allow us to classify problems by their complexity. The problem belongs to class P if there exists a deterministic algorithm of polynomial complexity for solving this problem. The problem belongs to class $\bar{P}$ if there exists no deterministic algorithm of polynomial complexity for solving this problem.

However, it should be clearly specified that such a classification is a matter of convention, because the problem solving complexity depends on the input data as well. The same problem could have a polynomial complexity with one set of input data and could have no solving algorithm of polynomial complexity with another one. Therefore, when the problem is referred to as non-polynomial, it means that, at least for a single set of initial conditions, no polynomial solutions can be found.

The fruitfulness of problem classification is connected with the fact, that with this classification it is possible to find problem equivalence, and then analyse the whole problem classes instead of separate problems.

From the public key cryptography point of view, the existence of the problems of classes P and $\bar{P}$ would give an opportunity to build trapdoor functions on the basis of problems, which belong to class $\bar{P}$, but with some specific input data have an algorithm of polynomial complexity. Then, the published data of encryption algorithm (public key) should be chosen in such a way that the decryption problem belongs to $\bar{P}$, but there exists a transformation (secret key) of problem input data, enabling a polynomial solution.

However, complexity theory does not offer such an opportunity to cryptography. The majority of problems producing trapdoor functions have the following properties:

1. Deterministic algorithm of polynomial complexity is so far not known.

2. This algorithm has not been proven to be non-existent.

The impossibility of dividing problems into simple and hard according to the P and $\bar{P}$ classification resulted in the introduction of the NP problem class, which is based on the notion of abstract nondeterministic algorithms.

Let the encrypting transformation f be a nondeterministic state machine, i.e. the state machine, which could be in several (generally speaking, optionally not finite) states simultaneously. This state machine can simultaneously perform several actions. We define the step of this state machine as the set of actions which can be executed simultaneously (in other words – a set of states, in which the state machine could be at a single moment of time).

Then as a nondeterministic algorithm we define the sequence of nondeterministic state machine steps, which should be performed to calculate y from x.

Obviously, the described nondeterministic state machine is not realisable and, respectively, the nondeterministic algorithm is abstract.

However, considering a probabilistic state machine, that chooses its next state (of the nondeterministic state machine) according to some probability distribution, we obtain a probabilistic model of the nondeterministic algorithm. This algorithm randomly chooses one sequence of states of nondeterministic algorithm from the set of all possible sequences.

A nondeterministic algorithm is said to be of polynomial complexity if the following conditions are satisfied:

1. The number of steps of the nondeterministic state machine, which executes this algorithm, is polynomial.

2. The number of states of nondeterministic state machine, which follow the given state, is polynomial.

3. The number of steps required to make a decision about the termination of the algorithm is polynomial.

In terms of the probabilistic state machine the polynomial complexity of nondeterministic algorithm means, respectively, that:

1. The number of states that were used by state machine while executing the algorithm, is polynomial.

2. The complexity of choosing the next state is polynomial (the polynomial probability $1/n^a$ of choosing the next state).

3. The complexity of making decision about an algorithm's termination is polynomial.

Hence, polynomial complexity of nondeterministic algorithm means that there exist

$$M(n) = n^{m(n)}$$

deterministic algorithms of polynomial complexity ($m(n)$ is a polynomial of n) among which there is an algorithm that transforms the input sequence x into the required output sequence y. This also means that one can find a deterministic algorithm from the P class solving the problem with probability at least $n^{-m(n)}$.

We should note that the given definitions of a nondeterministic algorithm with the help of probabilistic and nondeterministic state machines are equivalent. In each particular case we shall use the definition that suits best.

The notion of the nondeterministic algorithm permits us to locate the equivalence class for the problems that are especially interesting to applications (in particular, to cryptography).

We define *NP* as the class of problems for which there exists a nondeterministic algorithm of polynomial complexity. It is clear that $P \in NP$. On the other hand, it is not known (and this constitutes the main problem of complexity theory) if $P \neq NP$ or not, i.e., whether all of the problems from *NP* could be solved with a polynomial deterministic algorithm.

The most interesting classes of problems are two special classes of *NP*-problems: *NP-complete* problems and *NP-hard* problems.

We call problem *A* polynomially reducible to problem *B*, if there is an algorithm of polynomial complexity which allows reformulating the problem *A* according to the terminology of problem *B*. Obviously, if there exists an algorithm for *B* from *P* or *NP*, then there exists also an algorithm for problem *A* (from *P* or *NP*, respectively). The problems *A* and *B* are called polynomially equivalent if they are mutually reducible (i.e. *A* is reducible to *B* and vice versa). We call problem *A NP-hard* if all the problems from *NP* are polynomially reducible to *A*. A *NP-hard* problem is said to be *NP-complete*, if it belongs to *NP*, i.e., it can be solved by a nondeterministic algorithm of polynomial complexity.

Hence, all *NP-complete* problems are polynomially equivalent, i.e., if one of them is solved, then all others can also be solved. Therefore, we can use less strict, but possibly more intuitive, interpretation of *NP-hard* and *NP-complete* problems. Problem *A* is *NP-hard* if it is 'not easier' than any problem from *NP* (i.e. it is impossible to solve *A* without solving all *NP* problems) and problem *A* is a *NP-complete* if it is 'not easier' but at the same time 'not harder' than all *NP* problems.

As was mentioned before, there is no proof (or contradiction) of impossibility to solve *NP-hard* or *NP-complete* problems using algorithms of polynomial complexity. But the lasting and, most importantly, intensive search for their solution (beginning with an outstanding book [3], there exists a big list of *NP*-complete problems from various fields of mathematics being researched independently and actively due to their importance for respective fields) gives a certain amount of confidence that these problems have no simple solutions. Thus, they can be used for constructing trapdoor functions. At the same time, using *NP*-complete problems for constructing trapdoor functions is non-trivial. In order to construct such a function, one has not only to specify an encryption method that results in *NP*-complete decryption complexity in the absence of secret key, but also provides a decryption procedure that will avoid solving this problem when the secret key is present. This, of course, grants certain abilities to non-legitimate users as well.

One classical example of such an attempt and corresponding contradiction is the Merkle-Hellman scheme [4] based on the knapsack problem [3].

Below we describe this example in details.

The problem is: given a set of integers $a_1, \ldots, a_n$ and a integer s, find vector $x = (x_1, \ldots, x_n)$, $x_i \in \{0, 1\}$ satisfying the following equation

$$\sum_{i=1}^{n} x_i \alpha_i = S \tag{8.5}$$

The vector $a = (\alpha_1, \ldots, \alpha_n)$ is called a *knapsack vector*.

The general knapsack problem is *NP*-complete. However there are some knapsack vectors, for which the problem could be solved comparatively easy. One of them is a superincreasing vector (used by Merkle and Hellman [4]), i.e. the vector composed of elements, which satisfy the inequality $\alpha_j > \sum_{i<j} \alpha_i$ for any j.

The trapdoor function, which is based on the knapsack problem, is defined as follows:

- Public key: the vector $a = (\alpha_1, \ldots, \alpha_n)$.

- Secret key: the transformation of the vector a to some superincreasing knapsack vector $a' = (\alpha'_1, \ldots, \alpha'_n)$.

- The set of messages to be encrypted is the set of vectors $x = (x_1, \ldots, x_n)$, $x_i \in \{0, 1\}$.

- The encryption method: $x \rightarrow S = \sum_{i=1}^{n} x_i \cdot \alpha_i$.

For mapping of vector a into vector a' the following transformation was used in the paper [5]:

$$\alpha_i = \alpha'_i W \bmod M, \quad (W, M) - 1, \quad i - \overline{1, k},$$

where W, M are the secret key of the system.

Legal user knowing W and M can use received S to calculate

$$S' \equiv SW^{-1} \equiv \sum_{i=1}^{n} x_i \alpha_i W^{-1} \equiv \sum_{i=1}^{n} x_i \alpha'_i,$$

and then solve the system using a superincreasing knapsack vector. But the attacker has to solve the knapsack problem in general form.

Intensive cryptanalysis of this system was directed to avoid necessity of solving the knapsack problem in an explicit form. Below we describe the simple facts on which the successful attack on the Merkle-Hellman cryptosystem is based.

Let $U = W^{-1}$. It is evident from the mapping of vector a, that there is system of equations

$$\alpha_i U = k_i M + \alpha'_i, i = \overline{1, k}$$

Therefore,

$$\frac{U}{M} - \frac{k_i}{\alpha_i} = \frac{\alpha'_i}{\alpha_i M} \qquad \text{for any } i \tag{8.6}$$

As long as α_i forms the superincreasing vector, then

$$\sum_{j=1}^{i} \alpha_j < \frac{1}{2} \sum_{j=1}^{i+1} \alpha_j,$$

with subject to $a_n < M$, we obtain

$$\sum_{j=1}^{i} \alpha_j < M \, 2^{i-n}.$$

And from (8.6) we obtain

$$\left| \frac{k_i}{\alpha_i} - \frac{k_j}{\alpha_j} \right| = \left| \frac{\alpha_i'}{\alpha_i M} - \frac{\alpha_j'}{\alpha_j M} \right|$$

and

$$\left| k_i \alpha_j - k_j \alpha_i \right| = \frac{1}{M} \left| \alpha_i' \alpha_j - \alpha_j' \alpha_i \right| < M \, 2^{i-n} \tag{8.7}$$

As was shown by Shamir [6], the restrictions described above allow breaking the Merkle-Hellman cryptosystem with polynomial complexity.

Of course the Merkle-Hellman knapsack cryptosystem is not the only cryptosystem based on the knapsack problem (in particular, the codebased Niederreiter cryptosystem, considered in the next section, is also classified as a knapsack cryptosystem). The attention paid to the Merkle-Hellman cryptosystem in this section is connected with the clearness of described attack, which avoids solving the original problem used as the base of the cryptosystem.

Thus, in order to construct a trapdoor function a hard (preferably, *NP*-complete or *NP*-hard) problem should be found, that also has subproblems of polynomial complexity. The decoding problem belongs to this very class. The minimum distance decoding problem is *NP*-complete [7]. The analysis performed in [8] shows that for almost all codes even the decoding of errors of a given weight requires, probably, an exponential number of operations. On the other hand, there exist codes (first and foremost, Goppa codes) with polynomial decoding complexity. In the next sections, trapdoor function construction based on combinatorial decoding problem will be investigated.

8.2 CODEBASED CRYPTOSYSTEMS: McELIECE AND NIEDERREITER

As has been discussed in the previous section, in order to construct a public-key cryptosystem it is necessary to use a hard problem (preferably *NP-complete*), which has a rather wide subclass of problems of polynomial complexity. The problem of decoding in an arbitrary linear code satisfies these requirements:

1. The problem of decoding is *NP-complete* [7], and

2. There is an exponentially large class of alternant codes, for which polynomial decoding algorithm exists (if the number of errors is less than the half of a designed code distance).

The McEliece [9] and Niederreiter [5] cryptosystems were the first systems of this type. The Niederreiter cryptosystem is the generalised knapsack system, where the knapsack vector is defined over a finite field. Below we give a detailed description of this cryptosystem.

Its secret key consists of:

1. $\mathbf{H}$ is an $(n-k) \times n$ check matrix of a (n,k)-code C defined over F_q, capable of correcting t-fold errors and for which a 'simple' (realisable) decoding algorithm is known

2. $\mathbf{P}$ is an $(n \times n)$ permutation matrix

3. $\mathbf{M}$ is an $((n - k) \times (n - k))$ nonsingular matrix

And the public key consists of:

1. Matrix $\mathbf{H}' = \mathbf{MHP}$

2. Value t

The set of messages to be encrypted is the set of n-tuples $\mathbf{u} = (u_1, \ldots, u_n)$ defined over F_q with weight equal to t. The encryption is specified as follows

$$\mathbf{v} = \mathbf{H}'\mathbf{u}^T \tag{8 8}$$

To break this cryptosystem one has to solve the decoding problem, i.e., to find an error vector $\mathbf{u}$ according to a syndrome $\mathbf{v}$. A legal user could easily solve this task using the secret key ($\mathbf{P}$ and $\mathbf{M}$):

$$\mathbf{v} = \mathbf{H}'\mathbf{u}^T = \mathbf{MHPu}^T,$$

hence

$$\mathbf{vM}^{-1} = \mathbf{H}(\mathbf{uP}^T)^T.$$

That is a legal user can decode the error vector $\mathbf{u}' = \mathbf{uP}^T$ using the code C, which has a realisable decoding algorithm, and then recover $\mathbf{u} = \mathbf{u}'(\mathbf{P}^T)^{-1}$. The security of the Niederreiter cryptosystem is based on the following assumption: to break this system one has to solve the decoding problem for the code using parity-check matrix $\mathbf{H}'$, which has no special algebraic structure. And this problem is considered to be hard.

As was mentioned above, the Niederreiter cryptosystem can be interpreted as a generalised knapsack problem. This circumstance allows usage of cryptanalysis methods, which were developed for non-code cryptosystems of knapsack type.

However, historically the first codebased public-key cryptosystem was the nonknapsack McEliece cryptosystem, which is also based on the decoding problem. Below we describe McEliece cryptosystem in detail.

The McEliece cryptosystem's secret key consists of:

1. $\mathbf{G}$ is a $(k \times n)$ generator matrix for a (n, k)-code C defined over F_q, capable of correcting t-fold errors and for which a realizable decoding algorithm is known

2. $\mathbf{P}$ is an $(n \times n)$ permutation matrix

3. $\mathbf{M}$ is a $(k \times k)$ nonsingular matrix

The public key consists of:

1. Matrix $\mathbf{G}' = \mathbf{MGP}$

2. Value t

The set of messages that can be encrypted is the set of k-tuples $\mathbf{u} = (u_1, \ldots, u_k)$ defined over F_q. To obtain a ciphertext $\mathbf{w}$ from a corresponding plaintext $\mathbf{u}$ one should compute $\mathbf{w} = \mathbf{u}\mathbf{G}' + \mathbf{e}$, where $\mathbf{e}$ is a n-vector of weight t defined over F_q.

The decryption procedure has three steps:

1. Compute vector $\mathbf{w}' = \mathbf{w}\mathbf{P}^{-1} = \mathbf{u}\mathbf{M}\mathbf{G} + \mathbf{e}\mathbf{P}^{-1} = \mathbf{u}'\mathbf{G} + \mathbf{e}'$, where $\mathbf{u}' = \mathbf{u}\mathbf{M}$ and vector $\mathbf{e}' = \mathbf{e}\mathbf{P}^{-1}$ has weight t.

2. Decode vector $\mathbf{w}'$ using the realisable decoding algorithm for code C and obtain vector $\mathbf{u}'$ as result.

3. Compute $\mathbf{u} = \mathbf{u}'\mathbf{M}^{-1}$.

If we note that in both cases one has to solve the decoding problem, it becomes easy to understand [10, 24], that the McEliece and Niederreiter cryptosystems are polynomially equivalent. In the McEliece cryptosystem one should decode using a syndrome and a check matrix, in the Niederreiter cryptosystem, using an obtained vector and a generator matrix[1]. That is why cryptanalysis of these systems can be performed jointly. For simplicity we will consider the security of the McEliece cryptosystem only.

There are two main directions of any public-key encryption scheme cryptanalysis (attacks of the first and the second type respectively):

1. The direct attack, i.e. search for effective algorithms solving a problem, on which the cryptosystem under analysis is based.

2. The attack on key, i.e. search for transformation of the hard problem on which the cryptosystem is based to some simple one.

Since any cryptosystem is based on a hard, for example NP-complete, problem, the potential of the first type attack is restricted. However it is incorrect to say, that this way is absolutely hopeless, because the real capabilities of algorithms, which solve these problems on finite lengths, essentially exceed the theoretical bounds estimated for infinite length. The considerable progress which was made in the direct attack on the RSA system [11], demonstrates the expedience of cryptanalysis investigations in this direction anyhow.

A direct attack on the McEliece cryptosystem consists of a search for effective methods to decode an arbitrary linear code. However we should note the difference in requirements between cryptography and coding theory. The coding theory requires the probability of correct decoding to be close to 1. A cryptosystem is considered to be compromised if the probability of correct decoding (i.e., decryprion) is not small enough, for instance, is greater than 1/2 or even 1/1000.

---

[1]Nevertheless we should not regard these methods as different ways to define the same cryptosystem, because, for example, the sets of input messages are different.

For estimation of the decoding complexity of McEliece cryptosystem the following result is usually used [12]

$$\chi(n, t) \leq f(n) \frac{\binom{n}{t}}{\binom{n-k}{t}}, \tag{8.9}$$

where $f(n)$ — is a polynomial of n.

The main efforts [12] undertaken inside the direct attack approach were connected with attempts to decrease non-exponential member $f(n)$ in the equation (8.9). On this way some progress was achieved [21], which allowed a decrease in the work factor of the McEliece system. Particularly, the work factor value of about $2^{62.1}$ elementary operations was achieved for the system based on the Goppa $(1024, 524)$-code with 50 correctable errors. This value is less than $2^{80.7}$ achieved in the previous estimation. Nevertheless, such a work factor noticeably exceeds all values acceptable for cryptanalysis.

However there are other opportunities. The random choice of information sets for decoding of almost all linear codes were suggested and analysed by the author in [13]. In further works [14,8] some exponential improvements of estimation (8.9) were obtained.

Taking into account the results mentioned above, it is possible to show, that the complexity of the McEliece cryptosystem can be decreased to 2^{55}. It seems to be real to obtain work factor of order 2^{50}, which lies on the border of encryption standard requirements.

The attack of the first type doesn't use any information about the structure of the code, on which the cryptosystem is based. The attack of the second type is entirely based on this information. Indeed, there are not so many classes of codes suitable for being used in public-key cryptosystems. Required codes should have realisable decoding algorithms, good parameters, and they should be big enough to prevent exhaustive search for appropriate code.

It was supposed (at the moment of publication) to use q-ary alternant codes (generalised Reed-Solomon codes) in the Niederreiter cryptosystem and binary Goppa codes in the McEliece cryptosystem. Moreover, in view of the fact that binary Goppa codes can be described as subcodes over F_2 of alternative codes, it is clear, that we are talking about codes of the same class. The alternant codes have many algebraic and combinatorial properties, which are difficult to conceal even using matrices $\mathbf{M}$ and $\mathbf{P}$. Consequently, these properties could be utilised for identification of original (used by legal user) code.

We will describe two attacks of the second type on the McEliece cryptosystem.

The first attack was proposed in the brilliant work of V.Sidelnikov and S.Shestakov [15]. It leads to breaking the cryptosystem if generalised Reed-Solomon codes are used. This attack is based on the fact that cryptanalyst doesn't need to find exactly the same polynomial as was used by the legal user in order to apply an effective algebraic decoding algorithm (for example, the Berlecamp algorithm). The point is that the same code can be defined by different Goppa polynomials.

The parity-check matrix of generalised Reed-Solomon (n, k)-code (and, in particular, of q-ary Goppa code) over F_q could be defined in the following way:

$$\mathbf{H}(\alpha_1, \ldots, \alpha_n; z_1, \ldots, z_n) = \begin{bmatrix} z_1\alpha_1^0 & \cdots & z_n\alpha_n^0 \\ z_1\alpha_1^1 & \cdots & z_n\alpha_n^1 \\ \vdots & \vdots & \vdots \\ z_1\alpha_1^{r-1} & \cdots & z_n\alpha_n^{r-1} \end{bmatrix} \tag{8.10}$$

where $r = n - k, \alpha_1, \ldots, \alpha_n, z_1, \ldots, z_n \in F_q$.

Let the matrix $\mathbf{H}'$ be known

$$\mathbf{H}' = \mathbf{S}\mathbf{H}(\alpha_1, \ldots, \alpha_n; z_1, \ldots, z_n), \tag{8.11}$$

where $\mathbf{S}$ is nonsingular $(r \times r)$ matrix.

Examine the problem of solving equation (8.11) with respect to $\mathbf{H}$ and $\mathbf{S}$. Matrix $\mathbf{H}'$ is known. Matrix $\mathbf{H}$ is expected to be found in the class of matrices (8.10). This equation has many solutions. In [15] the algorithm, which enables us to find at least one (any) solution of the equation (8.11) with the complexity $O(r^4)$, is proposed.

This result puts in question (or even denies) reliability of the McEliece cryptosystem when q-ary modified Reed-Solomon codes are used. Another important weakness is that these cryptosystems based on concatenated codes are also unsafe [22]. However, the algorithm mentioned requires huge amounts of operations for certain codes. Particularly, for the $(1024, 524)$-code the complexity of attack seems to be quite high. Although the proposed attack puts into question the reliability of the McEliece cryptosystem in the case of using q-ary Goppa codes (and in the general case as well), nevertheless it could not be directly applied for binary Goppa codes. Anyway, the amount of appropriate polynomials for binary Goppa codes is essentially smaller, so a new modification of the searching algorithm is needed.

Another approach to attack the McEliece cryptosystem is based on the fact, that there are many Goppa codes having big semiaffine symmetry group [16].

It is said, that permutation π preserves the code (or the code is symmetric with respect to permutation π), if the result of application of π to any codeword $\mathbf{a}$ of some code is codeword $\mathbf{a}\pi$ of that code as well. Permutations that preserve the code form a group. This group is called a symmetry group of the code.

Lemma 8.1 Let permutation π preserve the code C with generator matrix $\mathbf{G}$. Then permutation $\pi' = \mathbf{P}^{-1}\pi\mathbf{P}$ preserves the code C' with generator matrix $\mathbf{G}' = \mathbf{MGP}$.

Proof If π preserves the code C, then one could find such nonsingular matrix $\mathbf{V}$, which satisfies

$$\mathbf{G}\pi = \mathbf{V}\mathbf{G}.$$

Then we can construct the appropriate matrix $\mathbf{V}'$ for permutation π and the matrix $\mathbf{G}'$ in the following way

$$\mathbf{G}'\pi = \mathbf{G}'\mathbf{P}^{-1}\pi\mathbf{P} = \mathbf{MGPP}^{-1}\pi\mathbf{P} = \mathbf{MG}\pi\mathbf{P} = \mathbf{MVGP} = \mathbf{V}'\mathbf{MGP} = \mathbf{V}'\mathbf{G}',$$

where $\mathbf{V}' = \mathbf{MVM}^{-1}$ is nonsingular $(k \times k)$ matrix.   *Q.E.D.*

Lemma 8.1 implies, that with information about permutation π', which preserves the code C' (public key), and about symmetry group of code C the cryptanalyst could find permutation matrix $\mathbf{P}$ by solving the equation

$$\mathbf{P}\pi' = \pi\mathbf{P} \tag{8.12}$$

with respect to $\mathbf{P}$ and all permutations π from symmetry group of code C.

Using the knowledge of permutation π', which preserves the code C', one could solve the equation (8.12) with polynomial complexity (it polynomially depends on n). The cardinality of symmetry group of the code C is also polynomial. Therefore, the complexity of the whole attack depends mainly on the complexity of finding permutation π' that preserves the code C'. Generally, finding a permutation that preserves the code, is a hard problem. Some interesting algorithms of solving this problem are known [23]. But this problem could be simplified using some considerations, which are presented below.

Lemma 8.2 For any permutation matrix $\mathbf{P}$, the permutation π and the permutation $\pi'\mathbf{P} = \mathbf{P}^{-1}\pi\mathbf{P}$ have the same structure of cycles.

Proof Let π has an order s, i.e., $\pi^s = \mathbf{I}$, then

$$(\pi')^s = (\mathbf{P}^{-1}\pi\mathbf{P}) \cdot (\mathbf{P}^{-1}\pi\mathbf{P}) \cdots (\mathbf{P}^{-1}\pi\mathbf{P}) = \mathbf{P}^{-1}\pi^s\mathbf{P} = \mathbf{I},$$

so π and π' have the same order. It follows from this statement that if permutations π and π' have a different structure of cycles, then one of the permutations includes cycles of length divisible by the length of cycles of the other permutation. But it is impossible, because in this case, one permutation has positions, where moves are more than one number. *Q.E.D.*

Thus, with knowledge of permutation π, we can search for permutation π' with the same structure of cycles.

It is said, that code $C\pi$ is the eigen subcode of permutation π, that preserves code C, if $C\pi$ includes codewords of code C, which are eigen vectors with respect to permutation π.

Lemma 8.3 Eigen subcode of permutation π, that preserves linear code C, is also linear.

Proof Let $\mathbf{a}_1, \mathbf{a}_2 \in C$ be an eigen vectors with respect to permutation π. Then $(a_1\mathbf{a}_1 + a_2\mathbf{a}_2)\pi = a_1(\mathbf{a}_1\pi) + a_2(\mathbf{a}_2\pi) = a_1\mathbf{a}_1 + a_2\mathbf{a}_2$, where $a_1, a_2 \in F_q$. *Q.E.D.*

Lemma 8.4 Let π be a permutation of order s that preserves (n, k)-code C. Then the number of vectors in the eigen subcode of code C is no less, than k/s.

Proof Let $\mathbf{a}$ be a codeword of code C, then $\mathbf{a}\pi, \ldots, \mathbf{a}\pi^{s-1}$ are also codewords of this code, and the codeword $\mathbf{a}' = \mathbf{a} + \mathbf{a}\pi + \ldots + \mathbf{a}\pi^{s-1}$ is the eigen vector of permutation π. The algorithm for construction of eigen subcode of permutation π could be specified as follows:

1. Choose some non-eigen codeword $\mathbf{a}_1$ and construct a space $\mu(\mathbf{a}_1)$ using vectors $\mathbf{a}_1, \ldots, \mathbf{a}_1\pi^{s-1}$ as a basis. Include vector $\hat{\mathbf{a}}_1 = \mathbf{a}_1 + \mathbf{a}_1\pi + \ldots + \mathbf{a}_1\pi^{s-1}$ into the basis of eigen subcode.

2. Choose another non-eigen codeword $\mathbf{a}_2$ that does not belong to space $\mu(\mathbf{a}_1)$, and construct space $\mu(\mathbf{a}_1, \mathbf{a}_2)$ using vectors $\mathbf{a}_1, \ldots, \mathbf{a}_1\pi^{s-1}$ and $\mathbf{a}_2, \ldots, \mathbf{a}_2\pi^{s-1}$ as a basis. Include vector $\hat{\mathbf{a}}_2$ into the basis of eigen subcode.

3. On the step i choose noneigen codeword $\mathbf{a}_i$, that does not belong to space $\mu(\mathbf{a}_1, \ldots, \mathbf{a}_{i-1})$, and construct space $\mu(\mathbf{a}_1, \ldots, \mathbf{a}_i)$ using vectors $\mathbf{a}_1, \ldots, \mathbf{a}_1 \pi^{s-1}, \ldots, \mathbf{a}_i, \ldots, \mathbf{a}_i \pi^{s-1}$ as a basis. Include vector $\hat{\mathbf{a}}_i$ into the basis of eigen subcode. Vector $\hat{\mathbf{a}}_i$ is linearly independent on vectors $\hat{\mathbf{a}}_1, \ldots, \hat{\mathbf{a}}_{i-1}$ by design.

4. If $C \backslash \mu(\mathbf{a}_1, \ldots, \mathbf{a}_i)$ contains only eigen vectors, then construct the basis of these remaining eigen vectors, and include this basis into the basis of eigen code.

Obviously, the space $\mu(\mathbf{a}_1, \ldots, \mathbf{a}_i)$ contains no more than $(q^s)^i$ vectors. Then $|C \backslash \mu(\mathbf{a}_1, \ldots, \mathbf{a}_i)| \geq q^k - (q^s)^i = q^{si}(q^{k-si} - 1)$. Thus, we can conclude, that the constructed eigen subcode has cardinality of no less than $i + \log_q(q^{si}(q^{k-si} - 1))$. The last expression achieves minimum when $k = si$. Thus, $i \geq k/s$.   Q.E.D.

Let $\pi = (j_1 j_2) \ldots (j_{n1} j_n)$ be a permutation of order 2, that has cycles of length 2. The permutation π preserves the code C. The number of vectors in the eigen code of this permutation is no less than $q^{k/2}$. The method of eigen code construction described in Lemma 8.4, gives us generator matrix $\mathbf{G}_0$ of eigen code in the form

$$\mathbf{G}_0 = [\mathbf{G}_{01} \vdots \mathbf{G}_{01}]\mathbf{P},$$

where $\mathbf{G}_{01}$ is an $(i \times n/2)$-matrix and $\mathbf{P}$ is an $(n \times n)$ permutation matrix.

Thus, we can attack the McEliece cryptosystem by finding a set of vectors which can be split in two equal subvectors.

The described attacks on the McEliece cryptosystem based on Goppa codes may not compromise it totally, but nevertheless it is necessary to consider a modification of this cryptosystem based on other classes of codes.

Interesting modification of the McEliece cryptosystem based on codes correcting errors in the rank metric was proposed by E. M. Gabidulin in [17].

Below we describe the modification of the McEliece cryptosystem based on codes with $(\mathbf{x}, \mathbf{x} + \mathbf{y})$-construction [18].

Let X be an (n_1, k_1, d_1)-code over F_q and let Y be an (n_2, k_2, d_2)-code over the same field. Code C, which consists of code words $\mathbf{v} = (\mathbf{x}, \mathbf{x} + \mathbf{y})$, $\mathbf{x} \in X, \mathbf{y} \in Y$, is called $(\mathbf{x}, \mathbf{x} + \mathbf{y})$-construction. Moreover if $n_1 \neq n_2$, then by sum $\mathbf{x} + \mathbf{y}$ we understand the vector obtained as a sum of long vector and short vector expanded by a necessary number of zeros. The resulting code C has the following parameters: $n = n_1 + max(n_1, n_2)$, $k = k_1 \cdot k_2$, $d = min(d_2, 2d_1)$.

Generator matrix G of the described code C is:

$$G = \begin{bmatrix} \mathbf{G}_X & \mathbf{G}_X^* \\ \mathbf{0} & \mathbf{G}_Y^* \end{bmatrix},$$

where $\mathbf{G}_X^*$ and $\mathbf{G}_Y^*$ are the generator matrices of codes X and Y respectively. The superscript $*$ means that matrices G_x and G_y can be expanded by zeros as needed.

Decoding of the code C could be done in the following way. Let $\mathbf{b} = (\mathbf{b}_1, \mathbf{b}_2) = (\mathbf{v}_1 + \mathbf{e}_1, \mathbf{v}_2 + \mathbf{e}_2)$ be a received vector, where $\mathbf{b}_1$ is a subvector of length n_1, $\mathbf{b}_2$ is a subvector of length $max(n_1, n_2)$, $\mathbf{v}_1, \mathbf{v}_2$ and $\mathbf{e}_1, \mathbf{e}_2$ are corresponding subvectors of code word and error vector respectively.

1. Compute $\mathbf{b}' = \mathbf{b}_2 - \mathbf{b}_1$ from the vector $\mathbf{b}$. Before subtraction, an appropriate number of zeroes should be added/removed at/from the end of $\mathbf{b}_2$ in order to obtain vector $\mathbf{b}' = \mathbf{y} + \mathbf{e}'$ as a result of subtraction, where the weight of $\mathbf{e}'$ does not exceed the sum of $\mathbf{e}_1$ and $\mathbf{e}_2$ weights.

2. Decode $\mathbf{b}'$ in the code Y obtaining vector $\mathbf{y}$ as a result (of course, if $wt(\mathbf{e}_1) + wt(\mathbf{e}_2) \leq \lfloor (d-1)/2 \rfloor$).

3. Compute vector $\mathbf{b}'_1 = \mathbf{b}_2 - \mathbf{y} = \mathbf{x} + \mathbf{e}_2$.

4. Decode vectors $\mathbf{b}_1$ and $\mathbf{b}'_1$ in the code X.

5. Choose one of the two variants of vector x obtained on the previous step as a result of decoding such vectors, which corresponds to the error vector of the smallest weight.

The described decoding procedure of code C is reduced to decoding of codes X and Y of smaller lengths. Thus, it has reasonable complexity if the complexities of decoding in codes X and Y are also reasonable.

Assume now that $n_2 > n_1$. Then code word of code C will have the form of $(\mathbf{x}, (\mathbf{x}, \mathbf{0}) + \mathbf{y})$, where $\mathbf{0}$ is a zero vector of length $n_2 - n_1$. Now we can describe a modification of the McEliece cryptosystem based on $(\mathbf{x}, \mathbf{x} + \mathbf{y})$-construction.

Let $\mathbf{G}$ be a $(k \times n)$-matrix obtained from the matrices $\mathbf{G}_X$ and $\mathbf{G}_Y$:

$$
\mathbf{G} = \begin{bmatrix} \mathbf{G}_X & [\mathbf{G}_X \vdots \hat{\mathbf{0}}]\mathbf{P}_1 \\[2ex] \mathbf{0} & \mathbf{G}_Y \mathbf{P}_2 \end{bmatrix},
$$

where $\hat{\mathbf{0}}$ is $(k_1 \times (n_2 - n_1))$ zero matrix, $[\mathbf{G}_X \vdots \hat{\mathbf{0}}]$ is $k_1 \times n_2$-matrix (obtained by adding of matrix $\hat{\mathbf{0}}$ to $\mathbf{G}_X$), and $\mathbf{P}_1, \mathbf{P}_2$ are $(n_2 \times n_2)$ permutation matrices.

Obviously, the code with generator matrix $\mathbf{G}$ has the same parameters as the original code with $(\mathbf{x}, \mathbf{x} + \mathbf{y})$-construction, so the decoding algorithm described for original could also be applied.

Further, a public-key cryptosystem based on the code with generator matrix $\mathbf{G}$ is described.

The secret key consists of:

1. $\mathbf{G}_X, \mathbf{G}_Y$ generator matrices of codes X and Y

2. $\mathbf{P}_1, \mathbf{P}_2$ permutation matrices

3. $\mathbf{P}$ permutation $(n \times n)$-matrix

4. $\mathbf{M}$ nonsingular matrix

The public key consists of:

1. Matrix $\mathbf{G}' = \mathbf{MGP}$

2. Value t

The set of messages to be encrypted is a set of k-vectors over F_q. The encryption algorithm is the same, as for the McEliece cryptosystem

$$\mathbf{w} = \mathbf{uG}' + \mathbf{e} = \mathbf{v} + \mathbf{e}$$

The decryption algorithm for the legal user is specified as follows:

1. Compute

$$\mathbf{w}' = \mathbf{wP}^{-1} = \mathbf{uMG} + \mathbf{e}' = \mathbf{v}' + \mathbf{e}' = (\mathbf{v}'_1, \mathbf{v}'_2) + (\mathbf{e}'_1, \mathbf{e}'_2) = (\mathbf{x}, (\mathbf{x}, \mathbf{0})\mathbf{P}_1 + \mathbf{yP}_2) + (\mathbf{e}'_1, \mathbf{e}'_2).$$

2. Divide the vector $\mathbf{w}'$ into two parts $\mathbf{w}'_1$ and $\mathbf{w}'_2$ with lengths n_1 and n_2 respectively:

$$\mathbf{w}' = (\mathbf{w}'_1, \mathbf{w}'_2) = (\mathbf{v}'_1 + \mathbf{e}'_1, \mathbf{v}'_2 + \mathbf{e}'_2) = (\mathbf{x} + \mathbf{e}'_1, (\mathbf{x}, \mathbf{0})\mathbf{P}_1 + \mathbf{yP}_2 + \mathbf{e}'_2).$$

3. Expand the vector $\mathbf{w}'_1$ by concatenation with zero vector of length $(n_2 - n_1)$. Multiply the resulting vector by $\mathbf{P}_1$. As a result, we obtain a vector $\mathbf{w}''_1 = (\mathbf{w}'_1, \mathbf{0})\mathbf{P}_1 = (\mathbf{x}, \mathbf{0})\mathbf{P}_1 + \mathbf{e}'_1\mathbf{P}_1$.

4. Compute residual $\mathbf{w}'_2 - \mathbf{w}''_1$, and apply permutation $\mathbf{P}_2^{-1}$ to it. As a result we obtain a vector

$$\mathbf{w}''_2 = ((\mathbf{x}, \mathbf{0})\mathbf{P}_1 + \mathbf{yP}_2 + \mathbf{e}'_2 - (\mathbf{x}, \mathbf{0})\mathbf{P}_1 - \mathbf{e}'_1\mathbf{P}_1)\mathbf{P}_2^{-1} = (\mathbf{yP}_2 + \mathbf{e}'_2 - \mathbf{e}'_1\mathbf{P}_1)\mathbf{P}_2^{-1} = \mathbf{y} + \mathbf{e}''_2$$

where $\mathbf{e}''_2 = (\mathbf{e}'_2 - \mathbf{e}'_1\mathbf{P}_1)\mathbf{P}_2^{-1}$ is a vector of weight lower or equal to t.

5. Decode vector $\mathbf{w}''_2$ in the code Y obtaining vector $\mathbf{y}$ as a result.

6. Compute vector $(\mathbf{w}'_2 - \mathbf{yP}_2)\mathbf{P}_1^{-1} = ((\mathbf{x}, \mathbf{0})\mathbf{P}_1 + \mathbf{e}'_2)\mathbf{P}_1^{-1} = (\mathbf{x}, \mathbf{0}) + \mathbf{e}'_2\mathbf{P}_1^{-1}$

7. Cut last $n_2 - n_1$ positions obtaining vector $\mathbf{w}''_1 = \mathbf{x} + \tilde{\mathbf{e}}_2$, where $\tilde{\mathbf{e}}_2$ is a vector of weight no more than of weight of $\mathbf{e}'_2$.

8. Decode vectors $\mathbf{w}'_1 = \mathbf{x} + \mathbf{e}'_1$ and $\mathbf{w}''_1 = \mathbf{x} + \tilde{\mathbf{e}}_2$ in the code X

9. Choose from two variants of vector $\mathbf{x}$ obtained on previous step the one that corresponds to the error vector of the smallest weight

10. Compute vector $\mathbf{uM}$ using the vectors $\mathbf{x}$ and $\mathbf{y}$, and then vector $\mathbf{u}$

The described cryptosystem does not allow attacks that are based on special properties of the codes used, e.g., on search for symmetry group or for appropriate Goppa polynomial.

Application of attack based on the search for subcode with generator matrix $\mathbf{G}'$ seems to be difficult because in the construction described, the columns of matrix $\mathbf{G}_X$ are interleaved with zero columns of matrix $\hat{0}$ (moreover, the construction can be easily modified by substitution of matrix $\hat{0}$ for an arbitrary submatrix of appropriate size).

From our point of view, an attack based on the search for generator matrix with big zero submatrix has more possibilities. However, realisation of this idea is not clear for us yet.

The described modification of the McEliece cryptosystem seems to be more resistant to attacks of the second type than the original one due essentially to less structuredness of codes with $(\mathbf{x}, \mathbf{x} + \mathbf{y})$-construction, than Goppa codes.

However the parameters of $(\mathbf{x}, \mathbf{x} + \mathbf{y})$-codes are worse than the parameters of the best Goppa codes. Thus, the protection against attack of the first type requires longer codes and, consequently, larger public and secret key sizes, than of the original McEliece cryptosystem.

Therefore, the main drawbacks of the McEliece cryptosystem

1. Large sizes of public and secret keys, and

2. Excessive redundancy in ciphertext.

are preserved (and become even stronger due to the worse parameters of used codes).

But all advantages of the original cryptosystem are also preserved: comparatively fast (e.g., when compared to RSA) encryption and decryption algorithms. Additionally, the proposed cryptosystem is more resistant to attacks of the second type.

In the next section we describe a cryptosystem which substantially does not have the mentioned disadvantages but preserves all the advantages of the original McEliece cryptosystem.

8.3 CRYPTOSYSTEMS BASED ON FULL DECODING

Efficiency of attack on the McEliece cryptosystem is substantially determined by the fact, that for success it is only necessary to correct the errors of weight less than or equal to t. The attack becomes much more complex, if the decryption requires an ability to correct all the used code cosets leaders. However generally speaking, the decoding of all leaders (full decoding) is not a trap-door function, but a one-way function, because such decoding is a NP-complete task.

Below we define a key, which allows the construction of a trap-door function based on the full decoding.

Let C be a q-ary (n, k)-code with a minimum distance $d = 2t + 1$ and a generator matrix $\mathbf{G}$. Assume also that the realisable decoding algorithm for code C exists. Denote by $\mathbf{M}$ some nonsingular $(n \times n)$-matrix over F_q, and by $\mathbf{G}'$ matrix

$$\mathbf{G}' = \mathbf{GM} \tag{8.13}$$

The matrix $\mathbf{G}'$ specifies a linear (n, k)-code C'. The minimum distance of that code $d(C')$ depends on $\mathbf{G}$ and $\mathbf{M}$, and can be significantly less than $d(C)$.

Let n-tuple $\mathbf{e}'$ can be represented as

$$\mathbf{e}' = \mathbf{eM}, \tag{8.14}$$

where $\mathbf{e}$ is the n-tuple of weight $wt(\mathbf{e}) \leq t$. Then any vector $\mathbf{b}'$, which is equal to sum of vectors $\mathbf{a}' \in C'$ and $\mathbf{e}'$, can be decoded in the following way. Multiply $\mathbf{b}'$ by $\mathbf{M}$ and decode obtained vector $\mathbf{b}$

$$\mathbf{b} = \mathbf{b}'\mathbf{M}^{-1} = (\mathbf{a}' + \mathbf{e}')\mathbf{M}^{-1} = \mathbf{a}'\mathbf{M}^{-1} + \mathbf{e}'\mathbf{M}^{-1} = \mathbf{a} + \mathbf{e}$$

in code C. According to the choice of vector $\mathbf{e}$ we obtain codeword $\mathbf{a}$, and then can easily compute $\mathbf{a}' = \mathbf{a}\mathbf{M}$.

Denote by $E_{t,n}$, the set of n-vectors of weight no more than t, and by E the set of n-tuples $\mathbf{e}'$.

$$E = \{\mathbf{e}': \quad \mathbf{e}' = \mathbf{e}\mathbf{M}, \mathbf{e} \in E_{t,n}\}$$

The matrices $\mathbf{G}$, $\mathbf{M}$ and the set E specify a trap-door function F.

$$\mathbf{v} = F(\mathbf{u}) = \mathbf{u}\mathbf{G}' = \mathbf{u}\mathbf{G}\mathbf{M},$$
$$\mathbf{w} = f(\mathbf{v}) = \mathbf{v} + \mathbf{e},$$

where $\mathbf{u}$ is the message to be encrypted, and $\mathbf{e} \in E$.

With knowledge of $\mathbf{M}$ and code C the decoding algorithm backward transformation $\mathbf{w} \Rightarrow \mathbf{u}$ is performed using the method described above. Without knowledge of $\mathbf{M}$ the backward transformation is the full decoding in code C'.

The inconvenience of the described trap-door function is in the necessity to store and use the set E consisting of q^r vectors ($\mathbf{M}$ should be kept secret). Thus, to construct a cryptosystem based on the proposed trap-door function a constructive and compact method for specification of E is required. Properly speaking, for the McEliece cryptosystem such specification is given by $E = E_{t,n}$. Note, we do not need to specify the whole set E. By specifying only part of this set we complicate the search for matrix $\mathbf{M}$ for cryptanalyst.

In [13] a method for specification of E was proposed. Let us split $(n \times n)$-matrix $\mathbf{M}$ into submatrices $\mathbf{M}_1$ and $\mathbf{M}_2$ of dimensions $(p \times n)$ and $((n - p) \times n)$, respectively.

$$\mathbf{M} = \begin{bmatrix} \mathbf{M}_1 \\ \mathbf{M}_2 \end{bmatrix}$$

Let E' be the set of n-vectors

$$E' = \{\mathbf{e}': \quad \mathbf{e}' = \mathbf{e}\mathbf{M}_1, \mathbf{e} \in E_{t,p}\}$$

Set E' is specified by matrix $\mathbf{M}_1$ and by values of t and p. Now we can proceed with a full description of new public-key cryptosystem.

The secret key consists of:

1. $\mathbf{G}$ is a generator matrix of q-ary (n, k)-code C with minimum distance $d \geq 2t + 1$

2. $\mathbf{M}$ is an $n \times n$ matrix

The public key consists of:

1. $\mathbf{M}_1$ is a (p, n)-matrix

2. $\mathbf{G}' = \mathbf{GM}$ is a $(k \times n)$-matrix

The set of messages to be encrypted is a set of k-vectors over F_q.
The encryption algorithm is specified in following way:

1. Compute vector $\mathbf{v} = \mathbf{uG}'$, where $\mathbf{u}$ is the original message

2. Randomly choose p-vector $\mathbf{e}$ from set $E_{t,p}$

3. Compute vector $\mathbf{e}' = \mathbf{eM}_1$

4. Compute vector $\mathbf{w} = \mathbf{v} + \mathbf{e}'$

The decryption is done as follows:

1. Compute $\mathbf{w}' = \mathbf{wM}^{-1} = (\mathbf{v} + \mathbf{e}')\mathbf{M}^{-1} = \mathbf{v}' + \mathbf{e}$, where $\mathbf{v}' = \mathbf{vM}^{-1} = \mathbf{uGM}^{-1} = \mathbf{uG}$

2. Decode vector w' in code C obtaining vector $\mathbf{v}'$.

3. Compute vector $\mathbf{u}$ from $\mathbf{v}' = \mathbf{uG}$.

Similar to the McEliece cryptosystem the proposed cryptosystem could be attacked in two ways:

1. by decoding of vector w in code C' (the first type attack)

2. by splitting matrix $\mathbf{G}'$ into matrices $\mathbf{G}$ and $\mathbf{M}$ (the second type attack)

It seems, that realisation of the second type attack for the proposed cryptosystem is essentially more complex than for the McEliece cryptosystem. Actually, to break the original McEliece cryptosystem it is enough to find representation of code C as generalised Reed-Solomon code. This is not true for the proposed cryptosystem, because one also needs to know the mapping between a set of coset leaders of code C' and a set of coset leaders of code C (a leader in code C' is not mandatory the most lightweight vector in coset). In other words it is necessary to find matrix $\mathbf{M}$, and therefore, it is necessary to find exactly the same matrix G, as the one kept in secret by the legal user.

To find $\mathbf{M}$ the cryptanalyst could try to use proportion (8.11). However since $\mathbf{M}$ does not have any special structure, it seems, that the complexity of obtaining $\mathbf{M}$ using the mentioned proportion is determined by the number of unknown elements in it, i.e., by the number of elements in matrix $\mathbf{M}_2$.

We cannot use the estimation presented above for complexity of the first type of attack in case of the proposed cryptosystem, because these estimations are oriented towards error correction in Hamming metric; meanwhile vectors from the set $\mathbf{eM}_1$, that we need to decode, are not coset leaders in terms of Hamming metric and so cannot be corrected by methods described previously.

If we multiply the encrypted vector on matrix $\hat{\mathbf{M}}$ such that $\mathbf{M}_1\hat{\mathbf{M}} = [\mathbf{I} \vdots \hat{\mathbf{0}}]$, where $\mathbf{I}$ is an $(p \times p)$ identity matrix, $\hat{\mathbf{0}}$ is $(p \times (n-p))$ zero matrix, then vector $\mathbf{w}'$

$$\mathbf{w}' = \mathbf{w}\hat{\mathbf{M}} = \mathbf{uGM}\hat{\mathbf{M}} + (\mathbf{e}, \mathbf{0})$$

should be decoded in code $\mathbf{GM}\hat{\mathbf{M}}$, which minimum distance could be $< 2t+1$.

Thus, the best estimation for complexity of direct attack on the proposed cryptosystem is

$$L(n) = O\left(\min\left(2^k, \binom{p}{t} \right) \cdot n^2 \right)$$

The term 2^k inside the minimum corresponds to the exhaustive search among all code words with consequent expansion of error vector inside basis of matrix $\mathbf{M}_1$ rows, and $\binom{p}{t}$ corresponds to exhaustive search among all possible vectors e and comparison of error vector syndromes with the syndrome of the ciphertext.

The most interesting attack on the described cryptosystem was proposed by U. Sorger [19]. It is based on the knowledge about part of matrix $\mathbf{M}$ and about some part of ciphertext without errors. This attack tries to reduce the proposed cryptosystem to the McEliece cryptosystem.

Expand public matrix $\mathbf{M}_1$ by random matrix $\mathbf{M}_2'$ in such way, that

$$\hat{\mathbf{M}} = \begin{bmatrix} \mathbf{M}_1 \\ \mathbf{M}_2' \end{bmatrix}$$

will be nonsingular.

Then vector $\hat{\mathbf{w}}' = \mathbf{w}(\hat{\mathbf{M}})^{-1}$ will be equal to

$$\hat{\mathbf{w}}' = (\mathbf{uG}' + \mathbf{e}')\hat{\mathbf{M}}^{-1} = \mathbf{uG}'\hat{\mathbf{M}}^{-1} + (\mathbf{e}, \mathbf{0})\mathbf{M}\hat{\mathbf{M}}^{-1} = \mathbf{uG}'\hat{\mathbf{M}}^{-1} + (\mathbf{e}, \mathbf{0})$$

where $\mathbf{0}$ is a zero vector of length $n-p$ and $wt(\mathbf{e}) = t$.

Now the decoding problem of decoding transforms to decoding of vector $(\mathbf{e}, \mathbf{0})$ in code $\hat{C}$ with generator matrix

$$\hat{\mathbf{G}} = \mathbf{G}'\hat{\mathbf{M}}^{-1} = \mathbf{GM}\hat{\mathbf{M}}^{-1}$$

The distinction from the McEliece cryptosystem is in the structure of the public key: if for the original cryptosystem the matrix $\mathbf{G}'$ is the generator matrix of code with minimum distance $d > 2t+1$, then in the proposed cryptosystem the code $\hat{\mathbf{G}}$ could have, generally speaking, an arbitrary minimum distance, and with high probability this distance is less than $2t+1$. In this case vector $(\mathbf{e}, \mathbf{0})$ most likely cannot be corrected in the obtained code. However, the existence of known zero subvector inside the error vector allows this difficulty to be avoided to some degree.

Consider the matrix $\hat{\mathbf{G}}$ in details.

$$\hat{\mathbf{G}} = \mathbf{G}\begin{bmatrix} \mathbf{M}_1 \\ \mathbf{M}_2 \end{bmatrix}\begin{bmatrix} \mathbf{M}_1 \\ \mathbf{M}_2' \end{bmatrix}^{-1}$$

Reduce $\mathbf{G}$ to the form

$$\mathbf{G} = \begin{bmatrix} \mathbf{G}_2 & \mathbf{I} \\ \mathbf{G}_1 & \hat{\mathbf{0}} \end{bmatrix}, \tag{8.15}$$

where $\mathbf{I}$ is $(n-p) \times (n-p)$ identity matrix, and $\hat{\mathbf{0}}$ is $p \times (n-p)$ zero matrix. This can be done, if the last $n-p$ columns of matrix $\mathbf{G}$ have rank $n-p$.

$$\hat{\mathbf{G}} = \begin{bmatrix} \mathbf{G}_2 & \mathbf{I} \\ \mathbf{G}_1 & \hat{\mathbf{0}} \end{bmatrix} \mathbf{M}\hat{\mathbf{M}}^{-1} = \begin{bmatrix} [\mathbf{G}_2 \vdots \mathbf{I}]\mathbf{M}\hat{\mathbf{M}}^{-1} \\ [\mathbf{G}_1 \vdots \hat{\mathbf{0}}]\mathbf{M}\hat{\mathbf{M}}^{-1} \end{bmatrix},$$

but

$$[\mathbf{G}_1 \vdots \hat{\mathbf{0}}]\mathbf{M}\hat{\mathbf{M}}^{-1} = [\mathbf{G}_1 \vdots \hat{\mathbf{0}}] \begin{bmatrix} \mathbf{M}_1 \\ \mathbf{M}_2 \end{bmatrix} \begin{bmatrix} \mathbf{M}_1 \\ \mathbf{M}_2' \end{bmatrix}^{-1} = [\mathbf{G}_1 \vdots \hat{\mathbf{0}}] \begin{bmatrix} \mathbf{M}_1 \\ \mathbf{M}_2' \end{bmatrix} \begin{bmatrix} \mathbf{M}_1 \\ \mathbf{M}_2' \end{bmatrix}^{-1} = [\mathbf{G}_1 \vdots \hat{\mathbf{0}}]$$

This means that the code $\hat{C}$ (code with generator matrix $\hat{\mathbf{G}}$), which has a short minimum distance contains the subcode with generator matrix $[\mathbf{G}_1 \vdots \hat{\mathbf{0}}]$, i.e., the subcode, which is also contained in code C. Thus, this subcode had a minimum distance of no less than $2t+1$.

Now consider the following procedure for breaking the considered cryptosystem.

Choose matrix $\hat{\mathbf{M}}$ and construct matrix $\hat{\mathbf{G}} = \mathbf{G}'\hat{\mathbf{M}}^{-1}$. Reduce, if it is possible, matrix $\hat{\mathbf{G}}$ to the form

$$\hat{\mathbf{G}}' = \begin{bmatrix} \hat{\mathbf{G}}_1 & \mathbf{0} \\ \hat{\mathbf{G}}_2 & \mathbf{I} \end{bmatrix}, \tag{8.16}$$

i.e., find matrix $\mathbf{Q}$ such that $\hat{\mathbf{G}} = \mathbf{Q}\hat{\mathbf{G}}'$. Then, obviously, a subcode C' with generator matrix $[\mathbf{G}_1 \vdots \hat{\mathbf{0}}]$ is a subcode C with generator matrix $[\mathbf{G}_1 \vdots \hat{\mathbf{0}}]$ specified in another basis.

Consider the vector

$$\hat{\mathbf{w}}' = \mathbf{w}(\hat{\mathbf{M}})^{-1} = \mathbf{u}\hat{\mathbf{G}} + (\mathbf{e}, \mathbf{0}) = \mathbf{u}\mathbf{Q}\hat{\mathbf{G}}' + (\mathbf{e}, \mathbf{0}).$$

Denote vector $\mathbf{u}\mathbf{Q}$ by $\mathbf{u}'$

$$\mathbf{u}' = \mathbf{u}\mathbf{Q} = (\mathbf{u}_1', \mathbf{u}_2'),$$

where $\mathbf{u}_1'$ is a $(n-p)$-subvector and $\mathbf{u}_2'$ is a p-subvector of vector $\mathbf{u}'$.

From here

$$\hat{\mathbf{w}}' = (\mathbf{u}_1'\hat{\mathbf{G}}_1, \mathbf{0}) + \mathbf{u}_2'[\hat{\mathbf{G}}_2 \vdots \mathbf{I}] + (\mathbf{e}, \mathbf{0}) = (\mathbf{u}_1'\hat{\mathbf{G}}_1 + \mathbf{u}_2'\hat{\mathbf{G}}_2 + \mathbf{e}_1, \mathbf{u}_2').$$

Thus, the last p elements of vector $\hat{\mathbf{w}}'$ constitute $\mathbf{u}_2'$.

By subtracting vector $\mathbf{u}_2'[\hat{\mathbf{G}}_2 \vdots \mathbf{I}]$ from $\hat{\mathbf{w}}'$ we obtain vector $\mathbf{u}_1'\hat{\mathbf{G}}_1 + \mathbf{e}_1$. Then by decoding this vector in code with generator matrix $\hat{\mathbf{G}}_1$, which has a minimum distance of no less than t, we obtain vector $\mathbf{u}_1'$. From vector $\mathbf{u}' = (\mathbf{u}_1', \mathbf{u}_2')$ we can find vector $\mathbf{u}$ with the help of matrix $\mathbf{Q}^{-1}$.

Under the condition that matrices $\mathbf{G}$ and $\hat{\mathbf{G}}$ can be reduced to the form (8.15), (8.16) respectively, we reduce the proposed cryptosystem breaking problem to the McEliece cryptosystem breaking problem.

Note that as a result of transformation we work in code C with a 'simple' decoding algorithm, but the search of this 'simple' decoding using only matrix $\hat{\mathbf{G}}_1$ is exactly the same problem as needs to be solved in the original McEliece cryptosystem.

For protection against this attack we should make two modifications in the proposed cryptosystem:

1. New definition of matrix $\mathbf{G}'$ is $\mathbf{G}' = \mathbf{G}\pi\mathbf{M}$, where π is a permutation matrix. Now vector $\mathbf{w}'$ is equal to $\mathbf{w}' = \mathbf{w}\mathbf{M}^{-1} = \mathbf{a}\pi + (\mathbf{e}, \mathbf{0})$. And the legal user should make one additional transformation $\mathbf{w}'\pi^{-1} = \mathbf{a} + (\mathbf{e}, \mathbf{0})\pi^{-1}$. This transformation hides in the vector $\mathbf{a}$ a part, which corresponds to the zero subvector of error vector. But this, generally speaking, is not enough, because if in Sorger's attack we use a code with the generator matrix $\mathbf{G}\pi$ instead of code with the generator matrix $\mathbf{G}$, then it becomes clear that the code with generator matrix $\mathbf{G}\pi$ contains the subcode, which is also contained in the code with generator matrix $\hat{\mathbf{G}}$. In this case the Sorger's attack leads to success, if a matrix

$$\hat{\mathbf{G}} = \mathbf{G}\pi\mathbf{M}(\hat{\mathbf{M}})^{-1}$$

can be reduced to the form (8.16). Then for full neutralisation of this attack we need the second countermeasure, which is described in the next item.

1. We should put a matrix with defect ν in the positions corresponding to a zero subvector of vector $\mathbf{e}'$ in matrix $\mathbf{G}\pi\mathbf{M}$. Then on the attack phase, where $\mathbf{u}'_2$ is determined, we can determine only the fact that $\mathbf{u}'_2$ belongs to subset of 2^ν sequences. Thus, the problem of decoding in code with generator matrix $\hat{\mathbf{G}}_1$ should be solved 2^ν times.

However, the cryptanalyst has the ability to find the position of the zero subvector in the vector e and to apply the attack using this knowledge. The described countermeasures, seemingly, cannot prevent the disclosure of at least $p - \nu$ information symbols.

Thus, the proposed cryptosystem breaking problem can be reduced to a problem of breaking the McEliece cryptosystem by searching matrices $\mathbf{M}'_2$, but in this case only the direct attack can be applied.

8.4 FURTHER DEVELOPMENT OF CODEBASED CRYPTOSYSTEMS

Sorger's attack forces further investigation of ways to increase the cryptographic resistance of a cryptosystem based on full decoding.

The main weakness of a cryptosystem based on full decoding, which was described in the previous chapter (in future we will refer to it as cryptosystem I), lies in the ability to use information about the position of zero subvector inside the error vector $(\mathbf{e}, \mathbf{0})$. Cryptosystem II is another example of a cryptosystem based on full decoding, which does not have the described disadvantage.

Let $\mathbf{M}_1$ be an $(n \times n)$-matrix of rank $t < n$. This means, that rows of matrix $\mathbf{M}_1$ belong to linear space of dimension t, which is defined over F_q. Then construct a $(t \times n)$-matrix $\mathbf{M}_{11}$ by filling it on the arbitrary basis of this linear space.

Denote by $\mathbf{M}$ an $(n \times n)$-matrix of rank n, which has the following form

$$\mathbf{M} = \pi \begin{bmatrix} \mathbf{M}_{11} \\ \mathbf{M}_2 \end{bmatrix} \tag{8.17}$$

where π is a permutation $(n \times n)$-matrix, and $\mathbf{M}_2$ is an arbitrary $(n - t) \times n$-matrix of rank $n - t$. The rows of matrix $\mathbf{M}_2$ and the rows of matrix $\mathbf{M}_{11}$ are linearly independent.

Cryptosystem II is defined as follow.

The secret key consists of:

1. G is a generator matrix of (n, k)-code C, which has the realisable decoding algorithm

2. Matrix $\mathbf{M}$

The public key consists of:

1. Matrix $\mathbf{G}' = \mathbf{GM}$

2. Matrix $\mathbf{M}_1$

The set of encryptable messages is a set of k-vectors over F_q.

The encryption algorithm is specified as follows:

1. Compute $\mathbf{v} = \mathbf{uG}'$

2. Generate an arbitrary non-zero vector e

3. Compute $\mathbf{w} = \mathbf{v} + \mathbf{eM}_1$

The decryption is done in following way:

4. Compute vector

$$\mathbf{w}' = \mathbf{wM}^{-1} = \mathbf{vM}^{-1} + \mathbf{eM}_1\mathbf{M}^{-1} = \mathbf{uG} + \mathbf{eM}_1\mathbf{M}^{-1}$$

As long as matrices $\mathbf{M}_1$ and $\mathbf{M}_{11}$ have equal rank, the following matrix $\mathbf{Q}$ could be found:

$$\mathbf{M}_1 = \mathbf{Q} \begin{bmatrix} \mathbf{M}_{11} \\ \hat{\mathbf{0}} \end{bmatrix},$$

where $\hat{\mathbf{0}}$ is $(n - t) \times n$-zero matrix.

$$\mathbf{eM}_1\mathbf{M}^{-1} = \mathbf{eQ} \begin{bmatrix} \mathbf{M}_{11} \\ \hat{\mathbf{0}} \end{bmatrix} \mathbf{M}^{-1} = \mathbf{e}'\mathbf{M}_{11}\mathbf{M}^{-1} = (\mathbf{e}', \mathbf{0}) \begin{bmatrix} \mathbf{M}_{11} \\ \mathbf{M}_2 \end{bmatrix} \begin{bmatrix} \mathbf{M}_1 \\ \mathbf{M}_2 \end{bmatrix}^{-1} \pi^{-1} = (\mathbf{e}', \mathbf{0})\pi^{-1},$$

where $\mathbf{e}'$ denotes the first t elements of n-vector $\mathbf{eQ}$. Obviously, the weight of $\mathbf{e}'$ and consequently the weight of $(\mathbf{e}', \mathbf{0})\pi^{-1}$ is not higher than t.

5. Decode vector $\mathbf{w}'$ in code C

$$\mathbf{w}' = \mathbf{uG} + \mathbf{e}'',$$

where $\mathbf{e}'' = (\mathbf{e}', \mathbf{0})\pi^{-1}, \quad wt(\mathbf{e}'') < t$. Then obtain message $\mathbf{u}$.

Missing information about position of zero subvector inside error vector $\mathbf{e}''$ makes Sorger's attack difficult.

The cryptanalyst could find a certain basis $\mathbf{M}'_{11}$ of linear space generated by rows of matrix $\mathbf{M}_1$. Then he could construct a matrix

$$\hat{\mathbf{M}} = \begin{bmatrix} \mathbf{M}'_{11} \\ \mathbf{M}'_2 \end{bmatrix},$$

determine matrix $\mathbf{Q}'$, for which the following proportion holds

$$\mathbf{M}_1 = \mathbf{Q}'\begin{bmatrix} \mathbf{M}'_{11} \\ \hat{\mathbf{0}} \end{bmatrix},$$

and finally obtain the vector

$$\hat{\mathbf{w}} = \mathbf{w}(\hat{\mathbf{M}})^{-1} = \mathbf{uG}\pi\begin{bmatrix} \mathbf{M}_{11} \\ \mathbf{M}_2 \end{bmatrix}\begin{bmatrix} \mathbf{M}'_{11} \\ \mathbf{M}'_2 \end{bmatrix}^{-1} + (\hat{\mathbf{e}}, \mathbf{0})$$

However, Sorger's attack does not ensure any success, because the error positions in the vector $(\mathbf{e}', \mathbf{0})\pi^{-1}$ and in the vector $(\hat{\mathbf{e}}, \mathbf{0})$ coincide with very low probability. So it is impossible to prove the existence of subcode with high minimum distance on the positions of matrix

$$\mathbf{G}\pi\begin{bmatrix} \mathbf{M}_{11} \\ \mathbf{M}_2 \end{bmatrix}\begin{bmatrix} \mathbf{M}'_{11} \\ \mathbf{M}'_2 \end{bmatrix}^{-1},$$

that are not contained in zero subvector of vector $(\hat{\mathbf{e}}, \mathbf{0})$.

In any case, it is possible to find a set of 2^t vectors, which contains the original message, by searching through non-zero positions of vector $(\hat{\mathbf{e}}, \mathbf{0})$. Therefore, below we propose some enhancement of cryptosystem II, which we will call cryptosystem III.

Define matrix $\mathbf{M}_1$ in the following way.

Let $\mathbf{M}_{11}$ be an $(n \times n)$-submatrix composed out of code words of (n, k, t)-anticode. And let $\mathbf{M}_1$ be a product

$$\mathbf{M}_1 = \mathbf{M}_{11}\mathbf{M},$$

where $\mathbf{M}$ is a non-singular $(n \times n)$-matrix.

Cryptosystem III is defined in the following way.

Secret key consists of:

6. $\mathbf{G}$ is a generator matrix of (n, k)-code with 'simple' decoding

7. Matrix $\mathbf{M}$

8. Matrix $\mathbf{M}_{11}$

9. Value t

10. Permutation π

Public key consists of:

11. Matrix $\mathbf{G}' = \mathbf{GM}$

12. Matrix $\mathbf{M}_1 = \mathbf{M}_{11}\pi\mathbf{M}$

The set of messages to be encrypted is the same as in cryptosystem II.
The decryption is done in the following way:

13. Compute vector

$$\mathbf{w}' = \mathbf{wM}^{-1} = \mathbf{uG} + \mathbf{eM}_{11}\pi.$$

Vector $\mathbf{eM}_{11}$ belongs to anticode, which has constraint t on the weight of code words, hence, the vector $eM_{11}\pi$ can be successfully decoded in the code with the generator matrix $\mathbf{G}$.

14. Decode it in the code with the generator matrix $\mathbf{G}$ and obtain message $\mathbf{u}$.

Consider the complexity of possible attacks on cryptosystem III.

The direct attack on the described cryptosystem, which can be applied by decoding of ciphertext in code C' with the generator matrix $\mathbf{G}'$, does not seem to be possible, because the set of coset leaders of code C' is unknown to the attacker.

We cannot say that these leaders are vectors of weight $\leq t$. We cannot even say that these leaders are vectors of weight $\leq t$ multiplied by certain non-singular matrix (such observation allows the determination of a set of errors correctable by code C' of cryptosystem I).

However, it is worth nothing, that the matrix $\mathbf{M}_1$ has rank k_1, so theoretically this set still can be found. The cardinality of this set is $2^{k_1} - 1$. Thus, a direct attack requires at least $2^{k_1} - 1$ multiplication of vector by matrix.

Attack of the second type can be applied in the following way. Construct a matrix $\mathbf{M}_1^+$ such that

$$\mathbf{M}_1\mathbf{M}_1^+ = [\mathbf{I} \vdots \hat{\mathbf{0}}].$$

Then

$$\hat{\mathbf{w}} = \mathbf{wM}_1^+ = \mathbf{uG}(\mathbf{MM}_1^+) + (\mathbf{e}, \mathbf{0})$$

Though we can now easily find the zero subvector inside the error vector, Sorger's attack cannot be applied, because the weight of vector e is close to k_1. To prevent the disclosure of information about part of the information symbols, it is enough to construct a submatrix of matrix $\mathbf{GM}$, which corresponds to the zero subvector in the true error vector in such a way that its rank is bounded by $k - k_1$.

The complexity of this attack is proportional to 2^{k_1}.

The described modifications of the cryptosystem based on full decoding achieves complexity about $2^{50} - 2^{55}$, when the generator matrix of (256, 128)-code is used as a secret key.

This means, the large size of the public key, which is the main disadvantage of the McEliece cryptosystem, can be decreased from $2^{10} \cdot 2^9 = 2^{19}$ bits to $2^8 \cdot 2^7 \cdot (1.5 \div 2) \approx 2^{16}$ bits, i.e., more than 8 times.

8.5   CODEBASED CRYPTOSYSTEMS AND RSA: COMPARISON AND PERSPECTIVES

In the previous section we have discussed the methods, capable of improving codebased cryptosystems. In this section we try to estimate the applicability of cryptosystems. First of all, we briefly describe the most popular Rivest-Shamir-Adleman (RSA) cryptosystem [20].

The public key consists of numbers m and t. The secret key consists of numbers p and q the cofactors in factorisation of m.

$$m = pq$$

The set of messages to be encrypted is a set of binary numbers of size n.

$$x = (x_1, \ldots, x_n), \quad x_i \in \{0, 1\}.$$

The encryption algorithm is specified in the following way:

$$y = x^t \bmod m. \tag{8.18}$$

The decryption algorithm follows from the Eulier theorem:

$$a^{\varphi(m)} \equiv 1 \bmod m, \tag{8.19}$$

where a and m are coprime, and $\varphi(m)$ is the Eulier function, i.e., the number of positive integers, which are less than m and coprime with it.

Let $r = (q - 1)(p - 1)$, r and t be coprime. Then with the help of the extended Euclidean algorithm the unique number s could be found such that

$$st = 1 \bmod r \tag{8.20}$$

As long as

$$\varphi(m) = \varphi(pq) = (p - 1)(q - 1) = r,$$

it follows from (8.19) that

$$st = qr + 1$$

for some positive integer q. Hence,

$$y^s = x^{st} = x^{qr+1} = (x^r)^q x$$

with subject to Euler's theorem

$$y^s \equiv (x^r)^q x \equiv 1^q \cdot x \equiv x \bmod m$$

Thus, if $x < m$ and x and m are coprime then the decryption is specified as

$$x \equiv y^t \bmod m. \tag{8.21}$$

Let us compare the main parameters of RSA and codebased cryptosystems:

1. **Encryption and decryption complexity.** The computations (8.18), (8.21) with 1000–20000-bit numbers (today the recommended size of key m is 2048) is a computationally hard task. On the other hand, in codebased cryptosystem the encryption (multiplication vector by matrix) and even the decryption (decoding in Goppa code) are tasks of essentially lower computational complexity. Under the same cryptographic resistance the codebased cryptosystem outperforms RSA more than 8–10 times in terms of the number of elementary operations.

2. **Cryptographic resistance.** The RSA is based on the integer factorisation problem, i.e., on factorisation of numbers into cofactors. This problem is not NP-complete as it is, as indicated by the progress achieved in solving this problem in recent years. Alternatively, the problem of full decoding of linear code is NP-complete. This gives certain reasons to believe that codebased cryptosystems are more cryptographically resistant in the long term. However, one should keep in mind that attacks of the second type were not found during the lifetime of RSA. For codebased cryptosystems we considered several such attacks. In any case it seems that codebased cryptosystems are more robust to the computational power progress than the RSA.

3. **Ease of use.** The mathematical apparatus of coding theory is less popular and is not so widely used as the apparatus of elementary number theory used in the RSA. However, the codebased cryptosystem is essentially more 'ease of use,' than the RSA. The construction of RSA instance requires the generation of big prime integers. Such generation requires some complex techniques of number theory and could not be done by the layman. In contrast, in the construction of codebased cryptosystems it is enough to know how to generate a non-singular matrix **M**.

4. **Size of the secret key.** The size of the secret key s in the RSA cryptosystem is not greater than the size of m. The size of the secret key in the codebased cryptosystem is much bigger. However, by specifying the method of generation of matrix **M**, we can decrease the size of the secret key to an acceptable value.

5. **Size of the public key.** The main disadvantage of codebased cryptosystem, as was mentioned above is the size of the public key. In the original McEliece cryptosystem the size of the public key was 2^{19}. Such a size together other subjective reasons determines the preference for RSA. The recommended size of the RSA key in that time was 128 bit. The success in solving the integer factorisation problem leads to an increase of RSA key size up to 2048 bits. Under equivalent resistance the appropriate key size for cryptosystems based on full decoding is about $(16 \div 20) \cdot 2048$. Thus, from public key size point of view, the RSA still outperforms the codebased cryptosystems. But from the

point of view of current practice such loss is not catastrophic. And what is more, such size of the public key becomes more or less plausible today.

Thus, the tone used in this section shows the subjective authors' confidence in the perspectives of codebased cryptosystems.

The absolute cryptographic resistance of any public-key cryptosystems has not been proved. The evolution of cryptosystem is the subject of game 'attack—defense,' where the most decisive argument is the inventiveness of competitors. In such a situation one of the most decisive arguments about the question of applicability is the experience of attacking. The duration and intensity of attacks on RSA give additional advantages to this cryptosystem from the practical point of view. So, without the total breaking of RSA such an advantage guarantees that RSA dominates the market of security tools. However, there is no doubt that problems, which should be solved by public-key cryptography, have forced the development of algorithms that are alternatives to RSA (that is, algorithms of codebased cryptography). Otherwise, a successful attack on RSA could lead to a total information catastrophe.

8.6   CODEBASED SIGNATURE

At the beginning of this section we will briefly recall the main principles of the McEliece and Niederreiter cryptosystems. In both cryptosystems a secret key is a q-ary (n, k)-code V randomly chosen from a family of linear error-correcting codes for whose rather simple, polynomial complexity algorithm of decoding t or less errors is known. A public key is a generator matrix $\mathbf{G}' = \mathbf{M}_1 \mathbf{G} \mathbf{P}$ for McEliece cryptosystem [9] and a parity-check matrix $\mathbf{H}' = \mathbf{M}_2 \mathbf{H} \mathbf{P}$ for Niederreiter cryptosystem [5], where $\mathbf{M}_1$ and $\mathbf{M}_2$ are arbitrary nonsingular $(k \times k)$ and $(r \times r)$-matrices, respectively, $\mathbf{P}$ is an arbitrary $(n \times n)$ permutation matrix, $\mathbf{G}$ and $\mathbf{H}$ are generator and parity-check matrices, respectively, of the code V. Matrices $\mathbf{P}, \mathbf{G}$ and $\mathbf{M}_1$ or $\mathbf{H}$ and $\mathbf{M}_2$, respectively, form a secret key. Both cryptosystems (we call them McENi for shortness) are equivalent [8] and based on the unproved conjecture that decoding of arbitrary linear code up to half of its minimal Hamming distance is the *NP*-complete problem (it becomes *NP*-complete for the 'full' minimum distance decoding, i.e., for maximum likelihood decoding). Nevertheless for some classes of good codes the systems can be broken by revealing $\mathbf{G}$ (or $\mathbf{H}$ from the public key (see [15]) Let us recall more about the construction of the Niederreiter cryptosystem [5]. Denote as F_q a finite field of q elements and denote as F_q^n the set of all q-ary vectors of length n. Let the set M of possible messages be the set $F_q^{n,t}$ of all q-ary vectors of length n, whose Hamming weight equal to t (there is well known algorithm of enumeration of this set). The encryption substitutes a ciphertext (a syndrome) $\mathbf{s} = \mathbf{H}'\mathbf{m}^T$ to a message $\mathbf{m} \in F_q^{n,t}$, where $\mathbf{H}' = \mathbf{M}_2 \mathbf{H} \mathbf{P}$. A legal user decrypts $\mathbf{s}$ in the following way: $\mathbf{m} = \varphi_{V,t}(\mathbf{M}_2^{-1}\mathbf{s})\mathbf{P}$, where $\varphi_{V,t}$ is a decoding algorithm of the code V capable of correcting t or less errors. As we mentioned above, the system is based on the assumption that finding an error vector of weight t for the given value s of its syndrome has (for large t) a very large complexity, and on the possible hardness of revealing $\mathbf{H}$ from $\mathbf{H}'$ (this is not always true as it was shown for the generalised Reed-Solomon codes in [15]). There is a natural way of constructing a digital signature scheme in the way similar to the Niederreiter cryptosystem. Let the set of messages M be now the set of correctable syndromes, i.e., $M = S_t(\mathbf{H}') = \{\mathbf{H}'\mathbf{e}^T : \quad \mathbf{e} \in F_q^{n,t}\}$. Then the sender signs a message s by e, where a signature e is the solution of the following equation:

$$\mathbf{H}'\mathbf{e}^T = \mathbf{s} : \quad \mathbf{e} \in F_q^{n,t} \tag{8.22}$$

He evaluates the signature as $\mathbf{e} = \varphi_{V,t}(\mathbf{M}_2^{-1}\mathbf{s})\mathbf{P}$ because of knowledge of the algorithm $\varphi_{V,t}$, but everybody who wants to forge the signature should solve the equation (8.22). A receiver can check validity of a received word $[\mathbf{s}; \mathbf{e}]$ by checking equation (8.22). The description of this scheme is not yet complete because usually a set of messages M is represented either by the segment of natural number $N_M = \{1, \ldots, M\}$ or by q-ary (mainly, binary) vectors of length k. Therefore, users should have an effective algorithm of enumeration of the set of messages, i.e., the set $S_t(\mathbf{H}')$ of correctable syndromes. Consider as the first candidate the following enumeration $\beta(i) = \mathbf{H}'\Psi(i)$, where Ψ is the known enumeration $\Psi : N_M \to F_q^{n,t}$. It is clear that such a scheme will be immediately broken because an opponent can create a false message just by setting $\mathbf{e} = \Psi(\mathbf{m})$ and $\mathbf{s} = \beta(\mathbf{m})$. An 'opposite' candidate is a random choice of such a mapping among all $M!$ mappings $N_M \to S_t(\mathbf{H}')$. To forge such a scheme seems to be as hard as to break a McENi cryptosystem, but the scheme is not effective because it demands a huge public key (an enumeration map becomes a common part of public keys and to keep it one needs roughly $M \log M$ bits). In the considered construction its use is probably unobserved as for every linear code the set of its correctable syndromes contains a linear subspace of relatively large dimension L. We restrict a set of messages only to such syndromes and generate them effectively due to their linear structure. Unfortunately, due to the same linear structure the proposed scheme can be broken after approximately (roughly) L usages. The main difference from the previous attempts of constructing a digital signature based on error-correcting codes as well as from McENi systems is that we use arbitrary linear codes [we do not use some classes of codes (like Goppa codes) with a known decoding algorithm. Instead of doing this we construct a set of syndromes, which could be simply decoded in an arbitrary linear code if someone knows the trapdoors.]. At first we describe the basic scheme, and later we consider some improvements. Let V be a q-ary $(n, n - r)$-code with minimum Hamming distance $d(V) > 2t$ and let C be an equidistant (n', k')-code with minimum Hamming distance $d(C) = t$, where $n \geq n' = \dfrac{q^{k'} - 1}{q - 1}$ and $d(C) = t = q^{k'-1}$. Let $(r \times n)$-matrix $\mathbf{H} = [\mathbf{h}_1, \ldots, \mathbf{h}_n]$ be a parity-check matrix of the code V and let $k' \times n'$- matrix $\mathbf{G}$ be a generator matrix of the code C. Define an $r \times k'$-matrix $\mathbf{F} = \mathbf{H}(J)\mathbf{G}^T$, where J is a subset of the set $\{1, \ldots, n\}$ of the cardinality n', and $\mathbf{H}(J)$ is a submatrix of the matrix $\mathbf{H}$ consisting of the columns $\mathbf{h}_j, j \in J$. Recall that $S_t(\mathbf{H}) = \{\mathbf{H}\mathbf{e}^T : \quad \mathbf{e} \in F_q^{n,t}\}$ is the set of syndromes corresponding to errors of weight t. It is easy to prove the following

Proposition 8.1 $\quad \mathbf{F}\mathbf{x}^T \in S_t(\mathbf{H})$ for any $\mathbf{x} \in F_q^{k'}\backslash\mathbf{0}$.

One can define a signature scheme in the following way. There are two public matrices: $\mathbf{H}$ and $\mathbf{F}$. The set J and the matrix $\mathbf{G}$ are secret (private) keys. The sender signs a message $\mathbf{m} \in F_q^{k'}\backslash\mathbf{0}$ by signature e in such a way that

$$\mathbf{H}\mathbf{e}^T = \mathbf{F}\mathbf{m}^T : \quad wt(\mathbf{e}) = t \tag{8.23}$$

where vector $\mathbf{e}$ equals $\mathbf{m}\mathbf{G}$ on the positions of the set J and equals 0 outside of J.

Such a signature is not resistant to homomorphism attack, since after observation of two signed messages $[\mathbf{m}_1; \mathbf{e}_1]$ and $[\mathbf{m}_2; \mathbf{e}_2]$ an opponent can create a false, but valid, word $[\mathbf{m}_1 + \mathbf{m}_2; \mathbf{e}_1 + \mathbf{e}_2]$. To avoid the homomorphism attack let us do the same as is usually done in such a case (for RSA signature, for instance). Namely, consider any 'good' (i.e. 'enough nonlinear,' simple to evaluate, hard to invert, 'no collisions' etc.) hash function $f : M = F_q^K \to F_q^{k'} \backslash \mathbf{0}$, and modify the definition of the signature equation to

$$\mathbf{He}^T = \mathbf{F}f(\mathbf{m})^T : \quad wt(\mathbf{e}) = t \tag{8.24}$$

For any given message $\mathbf{m}$ the signer evaluates the signature $\mathbf{e} = f(\mathbf{m})\mathbf{G}$ The opponent's attempt to solve the equation (8.24) should fail because of the hardness of decoding an *arbitrary* linear code. To find trapdoors, *i.e.*, the set J and the matrix $\mathbf{G}$ also seems to be as difficult as to decode the code V because every column of the public matrix $\mathbf{F}$ is a linear combination of exactly t columns of the matrix $\mathbf{H}$ (taken from the positions of J) and to find this linear combination is the same as to decode V for some particular value of the syndrome.

Example 8.1 Consider binary codes. We would like to make the working factor for the opponent no less than 2^{50}. It implies that the number of values of possible signatures should be at least 2^{50}. Hence, $k' \geq 50$ and $n \geq n' = 2^{k'} - 1 \approx 10^{15}$, but these values are too large for any practical applications.

In order to improve the parameters of the scheme we replace the equidistant code by any (n', k')-code C, whose nonzero codewords c have weight $t_1 \leq wt(c) \leq t_2$. This leads to the following modification of the signature equation

$$\mathbf{He}^T = \mathbf{F}f(\mathbf{m})^T : \quad t_1 \leq wt(\mathbf{e}) \leq t_2 \tag{8.25}$$

Let us note that the condition $d(V) > 2t_2$, which guarantees the uniqueness of a signature, i.e., a solution of the equation (8.25), is not so important for the proposed scheme as for the McENi scheme (one message can have few signatures, but one ciphertext may correspond to only one message). We only need it to be difficult to solve the equation (8.25), i.e., to decode the code V if the number of errors lies in the interval $[t_1, t_2]$. This leads us to the following modification of the initial scheme.

Consider as an (n', k')-code C the code dual to binary BCH code of length $n' = 2^l - 1$ with the designed distance $2s + 1$. It is known that $k' = sl$ and $\left| wt(\mathbf{c}) - \frac{n'+1}{2} \right| \leq (s - 1)\sqrt{n' + 1}$ for any nonzero codeword $\mathbf{c}$ (see [18]).

Consider a *random* binary $r \times n$-matrix as a matrix $\mathbf{H}$. The simple counting arguments (used in the proof of fact that almost all linear codes lie on Varshamov-Gilbert bound) show that

Proposition 8.2 The probability that a random binary $(n, n - r)$-code V has minimum distance $d_V \geq d$ is at least

$$1 - 2^{-r}\left(\sum\binom{n}{i}\right) \geq 1 - 2^{-r + nh(\frac{d-1}{n})},$$

where $h(x) = -x \log_2 x - (1 - x) \log_2(1 - x)$.

Now the description of the system is as follows. The signer chooses *randomly:* a binary $r \times n$-matrix $\mathbf{H}$, a nonsingular $k' \times k'$ matrix A, and n'-subset $J \subset \{1, \ldots, n\}$. He takes a parity-check matrix of the binary BCH code with the designed distance $2s + 1$ as a (generator) matrix $\mathbf{G}$ and forms an $r \times k'$-matrix $\mathbf{F} = \mathbf{H}(J)(\mathbf{AG})^T$ (i.e., $\mathbf{AG}$ is an arbitrary generator matrix of the code C). The public key of this system consists of the matrices $\mathbf{H}$ and $\mathbf{F}$ that take $r(n + k')$ bits, and the private key is relatively small, namely, $\log_2 \binom{n}{n'} nh\left(\frac{n}{n'}\right) \leq n' \log_2 n$ bits for describing the set J and $(k')^2$ bits for the matrix $\mathbf{A}$. The signer evaluates the signature $\mathbf{e}$ as $f(\mathbf{m})\mathbf{AG}$ on positions of the set J and 0 outside of J. All above given arguments concerning the scheme based on the equidistant codes are valid for the considered modification. Let us illustrate it by

Example 8.2 Choose $l = 8$ and $s = 6$. Then the number of possible signatures is large enough (i.e., $2^{48} - 1$), $t_1 = 48, t_2 = 208$. Choose a *random* binary 927×1200-matrix $\mathbf{H}$. With probability $p \geq 1 - 10^{-9}$ the corresponding code V has a minimum distance $d_V > 256$. The public key of this system consists of $253,000$ bits. It takes much more than 2^{50} operations (it takes roughly $\binom{n}{t}/\binom{r}{t}2^{nH(t/n)-rH(t/r)} = 2^{51}$ elementary operations for the best known algorithm) to solve the equation (8.25) for the number of errors in the range [48,208].

Example 8.3 Choose $l = 10$ and $s = 6$. Then the number of possible signatures is $2^{60} - 1$, $t_1 = 352, t_2 = 672$. Choose a *random* binary $2,808 \times 3,000$-matrix $\mathbf{H}$. With probability $p \geq 1 - 10^{-9}$ the corresponding code V has minimum distance $d_V > 1024$. The public key of this system consists of $538,000$ bits. It takes roughly 2^{54} elementary operations to solve the equation (8.25) for the number of errors in the range [352,672].

This construction can be further randomised-improved by the choice a code C as a random code. The following proposition, similar to Proposition 8.2, guarantees that we can do it.

Proposition 8.3 The probability p_C that for of a random binary (n, k)-code C the weight $wt(\mathbf{c})$ of any nonzero codeword $\mathbf{c}$ lies in the range $\left[\frac{n}{2}(1 - \delta), \frac{n}{2}(1 + \delta)\right]$ is at least

$$1 - 2^{-r+1}\left(\sum_{i=0}^{\delta n} \binom{n}{i}\right) \geq 1 - 2^{-r+nh(\delta)+1}$$

Example 8.4 Let code C be generated by a random 60×280-matrix. Due to Proposition 8.3 this is a code with $t_1 = 50, t_2 = 230$ with probability $p_C \geq 1 - 10^{-9}$. Choose a random 990×1250-matrix $\mathbf{H}$. With probability $1 - 10^{-9}$ the corresponding code can correct 140 errors. The number of possible signatures is 2^{51}.

Now we improve these schemes by using just a little bit more complicated enumeration, which allows us to use as C codes with a poor minimal distance.

The suggested improvement below of the basic scheme based on codes C whose minimal distance is not large, but the number of code words with low weight is small. Therefore we try to avoid these vectors in the enumeration procedure. For instance, let C be a direct sum of two codes C_1 and C_2. Then all code words have a large enough weight except vectors of the form $(\mathbf{c}_1, \mathbf{0})$ or $(\mathbf{0}, \mathbf{c}_2)$. It is clear how such vectors can be removed from an enumeration procedure, but an obvious drawback of such a straightforward application is that a standard generator matrix of the code C contains low weight vectors that help the opponent forge a signature. The following scheme seems to be free of this shortcoming.

Consider the finite field $F_{q^{k'}}$ as k'-dimensional vector space over F_q, fix some basis and denote by $\mathbf{M}(\beta)$ the matrix corresponding to a linear mapping $\mathbf{z} \to \beta\mathbf{z}$ where β is an element of $F_{q^{k'}}$, i.e., $\beta\mathbf{z} = \mathbf{M}(\beta)\mathbf{z}^T$. The signer chooses the number P and constructs the generator matrices $\mathbf{G}_1, \ldots, \mathbf{G}_P$ of (n', k')-codes $C_1, \ldots, C_P$ with the property that Hamming weight of any nonzero codeword belongs to the interval $[t_1, t_2]$. He chooses *randomly*: an $r \times n$-matrix $\mathbf{H}$, P nonsingular $k' \times k'$-matrices $\mathbf{A}_j$, nonintersecting n'-subsets J_j and distinct elements $\beta_j \in F_{q^{k'}}\backslash\mathbf{0}$ He forms (for enumeration) a public $r \times Qk'$-matrix $\mathbf{F} = [\mathbf{F}_1, \ldots, \mathbf{F}_Q]$ where

$$\mathbf{F}_i = \sum_{j=1}^{P} \mathbf{H}(J_j)(\mathbf{M}(\beta_j^{i-1})\mathbf{A}_j\mathbf{G}_j)^T$$

are an $r \times k'$-matrices. To sign a message $\mathbf{m}$ he at first applies a hash function $f : M = F_q^K \to F_q^{(Q-1)k'}\backslash\mathbf{0}$ and represents $f(\mathbf{m}) = (\mathbf{u}_2, \ldots, \mathbf{u}_Q)$, where $\mathbf{u}_i \in F_q^{k'} \equiv F_{q^{k'}}$. Then he evaluates $\mathbf{u}(\beta_j)$ and chooses $\mathbf{u}_1 \notin \{-\mathbf{u}(\beta_j) : j = 1, \ldots, P\}$, where $\mathbf{u}(\mathbf{z}) = \mathbf{u}_2\mathbf{z} + \ldots + \mathbf{u}_Q\mathbf{z}^{Q-1}$ (it is always possible if $P < q^{k'}$). He sets $\mathbf{e}$ equals 0 on positions $\{1, \ldots, n\}/\bigcup J_j$ and equals $\mathbf{U}(\beta_j)\mathbf{A}_j\mathbf{G}_j$ on the positions of the set J_j, where $\mathbf{U}(\mathbf{z}) = \mathbf{u}_1 + \mathbf{u}_2\mathbf{z} + \ldots + \mathbf{u}_Q\mathbf{z}^{Q-1} = \mathbf{u}_1 + \mathbf{u}(\mathbf{z})$. As the result he sends the following word $[\mathbf{m}; \mathbf{u}_1, \mathbf{e}]$ The signature equation has a form $\mathbf{He}^T = \mathbf{F}[\mathbf{u}_1 f(\mathbf{m})]^T : \quad Pt_1 \leq wt(\mathbf{e}) \leq Pt_2$

Lemma 8.5 $\mathbf{F}[\mathbf{u}_1 f(\mathbf{m})]^T \in \bigcup\limits_{t=Pt_1}^{Pt_2} S_t(\mathbf{H})$

Proof Denote $\mathbf{u} = (\mathbf{u}_1, \ldots, \mathbf{u}_Q) = (\mathbf{u}_1, f(\mathbf{m}))$ Then

$$\mathbf{Fu}^T = \sum_{i=1}^{Q} \mathbf{F}_i\mathbf{u}_i^T = \sum_{i=1}^{Q}\sum_{j=1}^{P} \mathbf{H}(J_j)(\mathbf{A}_j\mathbf{G}_j)^T\mathbf{M}(\beta_j^{i-1})\mathbf{u}_i^T$$

$$= \sum_{j=1}^{P} \mathbf{H}(J_j)(\mathbf{A}_j\mathbf{G}_j)^T \sum_{i=1}^{Q} \mathbf{M}(\beta_j^{i-1})\mathbf{u}_i^T = \sum_{j=1}^{P} \mathbf{H}(J_j)(\mathbf{U}(\beta_j)\mathbf{A}_j\mathbf{G}_j)^T$$

Since $\mathbf{u}_1 \notin \{-\mathbf{u}(\beta_j) : j = 1, \ldots, P\}$ one has that all $\mathbf{U}(\beta_j) \neq \mathbf{0}$ Hence, $\mathbf{U}(\beta_j)\mathbf{A}_j\mathbf{G}_j$ is a nonzero codeword of the code C_j and $t_1 \leq wt(\mathbf{U}(\beta_j)\mathbf{A}_j\mathbf{G}_j) \leq t_2$ Therefore $\mathbf{Fu}^T$ is a sum of T columns of the matrix $\mathbf{H}(J)$, where $J = \bigcup\limits_{1}^{P} J_j$ and $Pt_1 \leq T = \sum\limits_{j=1}^{P} wt(\mathbf{U}(\beta_j)\mathbf{A}_j\mathbf{G}_j) \leq Pt_2$ *Q.E.D.*

Example 8.5 Let $Q = 14, P = 12$ and $C_1 = \ldots = C_P$ is a binary equidistant $(15, 4)$-code with $t_1 = t_2 = 8$. Choose randomly $(1100, 335)$-code V, which has $d(V) \geq 193$ with probability at least $1 - 10^{-9}$ (or one can choose a hidden by McENi system Goppa code with $n = 1024, k = 280$). The number of possible signatures is $2^{52} - 1$, and complexity of decoding 96 errors by this random code is at least 2^{53} 'trials' (and the best known attack for Goppa code has the complexity $\approx 2^{66}$). The size of the public key is $256, 000$ bits (approximately the same as for the McENi system) and the size of the secret key $= 12 \times 16 + 3 \times 4 + 12 \times 15 \times 10 = 2004$ bits, where the first summand is responsible for a description of 12 nonsingular matrices $\mathbf{A}_j$, the second for a description of 12 nonzero elements of F_{16} and the last one for a description of 12 nonintersecting 15-elements subsets of $\{1, \ldots, 1100\}$.

The last scheme can be improved-generalised by considering instead of matrices $\mathbf{M}(\beta_j^{i-1})$ and $\mathbf{A}_j$ an arbitrary nonsingular $k' \times k'$-matrices $\mathbf{A}_{ij}(1 \leq i \leq Q, 1 \leq j \leq P)$. Define an $k'Q \times n'P$-matrix $\mathbf{G}$

$$\mathbf{G} = \begin{bmatrix} \mathbf{A}_{11}\mathbf{G}_1 & \cdots & \mathbf{A}_{1P}\mathbf{G}_P \\ \cdots & \cdots & \cdots \\ \mathbf{A}_{Q1}\mathbf{G}_1 & \cdots & \mathbf{A}_{QP}\mathbf{G}_P \end{bmatrix}$$

and consider it as a generator matrix of a code C. Its code words have a form

$$\mathbf{x}\mathbf{G} = \left(\sum_{i=1}^{Q} \mathbf{x}_i \mathbf{A}_{i1}\mathbf{G}_1, \ldots, \sum_{i=1}^{Q} \mathbf{x}_i \mathbf{A}_{iP}\mathbf{G}_P \right) = (\mathbf{y}_1\mathbf{G}_P, \ldots, \mathbf{y}_P\mathbf{G}_P)$$

where $\mathbf{x} = (\mathbf{x}_1, \ldots, \mathbf{x}_Q)$, $\mathbf{x}_i$ are k' dimensional vectors and $\mathbf{y}_j = \sum_{i=1}^{Q} \mathbf{x}_i \mathbf{A}_{ij}$. Therefore, if all $\mathbf{y}_j \neq 0$ (above it was granted by the condition $\mathbf{U}(\beta_j) \neq \mathbf{0}$) then $wt(\mathbf{x}\mathbf{G}) = PT$ where $t_1 \leq T \leq t_2$. If we choose by $\mathbf{A}_{ij}$ the matrix corresponding to a linear mappping $z \mapsto \alpha_{ij}z$, where $\alpha_{ij} \in F_{q^{k'}}$, then the conditions $\mathbf{y}_j \neq 0$ can be reformulated as

$$\sum_{i=1}^{Q} \mathbf{x}_i \alpha_{ij} \neq \mathbf{0}, \quad j = 1, \ldots, P$$

This is equivalent to the condition that the vector $\mathbf{x} = (\mathbf{x}_1, \ldots, \mathbf{x}_Q), \mathbf{x}_i \in F_{q^{k'}}$ does not belong to the union of the corresponding hyperplanes. One of the most natural choices is to set $\alpha_{ij} = \alpha_j^{(i-1)}$, where $\alpha_1, \ldots, \alpha_P$ are distinct elements of $F_{q^{k'}} \backslash \mathbf{0}$.

We suggest that the number of possible signatures be large enough (2^{50} in our numerical examples) to avoid a brute-force attack. On the other hand, even if this number is not so large, it is not clear how to expurgate all possible signatures because they are imbedded in a very large space.

REFERENCES

1. Schnier, B. (1996). *Applied Cryptography: protocols, algorithms and source code in C*, (2nd ed.). John Wiley & Sons Chichester, UK.
2. Diffie, W. and Hellman, M.E. (1976). New directions in cryptography, *IEEE Transactions on Information Theory*, **IT-22**(6):644–54.
3. Garey, M. R. and Johnson, D.S. (1979). *Computers and Intractability*. Freeman, W.H. and Company.
4. Merkle, R. C. and Hellman, M. (1978). Hiding information and signatures in trapdoor knapsacks, *IEEE Transactions on Information Theory*, **IT-24**(5):525–30.
5. Niederreiter, H. (1986). Knapsack-type cryptosystems and algebraic coding theory, *Problems of Control and Information Theory*, **15**(2):157–66.
6. Shamir, A. (1984). A polynomial-time algorithm for breaking the basic Merkle - Hellman cryptosystem, *IEEE Trans. Inform. Theory*, **IT-30**, pp. 609-704.
7. Berlekamp, E. R., McEliece, R.J. and van Tilborg, H.C.A. (1978). On the inherent intractability of certain coding problems, *IEEE Transactions on Information Theory*, **24**(5):384–6.
8. Barg, A., Krouk, E. and van Tilborg, H.C.A. (1999). On the Complexity of Minimum Distance Decoding of Long Linear Codes, *IEEE Trans. Inform. Theory*, **IT-45**, pp. 1392–405.

9. McEliece, R. J. (1978). *A public-key cryptosystem based on algebraic coding theory*. DSN Progress Report, pp. 114–6.

10. Li, Y., Deng, R., and Wang, X. (1994). The equivalence of McEliece's and Niederreiter's public-key cryptosystems, *IEEE Transactions on Information Theory*, **40**(1):271–3.

11. Lenstra, A. K. Lenstra Jr., H. W. Manasse, M.S. and Pollard, J.M. (1990). The number field sieve, In *ACM Symposium on Theory of Computing*, pp. 564–72.

12. Lee, P. J. and Brickel, E.F. (1989). *Observation on the Security of McEliece's Public-Key Cryptosystem*. Advances in Cryptology – Proceedings of EUROCRYPT'88, pp. 275–80.

13. Krouk, E. New Public-Key Cryptosystem, in *Proc. 6th Joint Soviet–Swedish Int. Workshop Information Theory* (St.Petersburg, Russia), pp. 285–6.

14. Krouk, E. A. and Fedorenko, S. V. (1995). Decoding by generalized information sets, *Problems of Information Transmission*, **31**(2):143–9.

15. Sidelnikov, V. M. and Shestakov, S. O. (1992). On Insecurity of Cryptosystems Based on Generalized Reed-Solomon Codes, *Discrete Mathematics and Applications*, **2**(4), pp. 57–63.

16. Barg, A. (1998). Complexity issues in coding theory, in (eds V. Pless and W. C. Huffman) *Handbook of Coding Theory*, vol. 1, Elsevier Science, Amsterdam, The Netherlands. pp. 649–754.

17. E. Gabidulin, A. Paramonov, and O. Tretjakov. Ideals over a non-commutative ring and their application to cryptology. *In Advances in Cryptology* - EUROCRYPT'91, LNCS-547, Springer-Verlag, 1991, pp. 482–489.

18. MacWilliams, F. J. and Sloan, J. J. (1977). *The Theory of Error-Correcting Codes*. North-Holland, Amsterdam.

19. Krouk, E. and Sorger, U. (1998). A public key cryptosystem based on total decoding of linear codes, Sixth International Workshop on Algebraic and Combinatorial Coding theory (*ACCT-VI*), Pskov, Russia.

20. Rivest, R., Shamir, A., and Adleman, L. (1978). A method for obtaining digital signatures and public-key cryptosystems, *Commun. Ass. Comput. Mach.* **21**(2).

21. A. Canteaut and F. Chabaud. A new algorithm for finding minimum-weight words in a linear code: Application to McEliece's cryptosystem and to narrow-sense BCH codes of length 511. *IEEE Transactions on Information Theory*, 44(1): pp. 367–378, January 1998.

22. N. Sendrier. On the concatenated structure of a linear code. In *AAECC*, 9(3), pp. 221–242, 1998.

23. N. Sendrier. Finding the permutation between equivalent codes: the support splitting algorithm. *IEEE Transactions on Information Theory*, 46(4), pp. 1193–1203, July 2000.

24. V. M. Sidelnikov. A public-key cryptosystem based on Reed-Muller codes. *Discrete Mathematics and Applications*, 4(3) pp. 191–207, 1994.

9

Reconciliation of Coding at Different Layers of a Network

9.1 TRANSPORT CODING IN A NETWORK WITH UNRELIABLE CHANNELS

In chapter 7 we considered a model of a network with reliable channels; i.e., in the condition that inequality (7.1) is always satisfied. For convenience we repeat it here:

$$P_{err.} < P_{err.a.}, \tag{9.1}$$

where $P_{err.}$ is the probability of obtaining a distorted message at a destination node, and $P_{err.a.}$ is the acceptable error probability of the message.

Let us denote the probability of obtaining a distorted packet at a destination node by p. We will also call the variable p – the network unreliability parameter. Generally speaking this probability depends on different factors, in particular on the length and 'quality' of the route of the packet in a network, but for simplicity we will not take account of these dependencies and will consider the unreliability parameter p as a constant for a given network. Assuming that the message consists of K packets, the probability of obtaining an erroneous message at the node-addressee is

$$P_{err} = 1 - (1 - p)^K \leq K \cdot p. \tag{9.2}$$

The use of transport coding provides an acceptable error probability for the message in the case where the value of p does not satisfy the inequality (9.1). This is possible because of the decoding of the received set of packets by the decoder of the (N, K) code which is capable of correcting errors and erasures. In this case, in order to reduce P_{err} we must receive $K + L$ packets and not K packets in order to restore the message. Then, even if there are erroneous packets among those received, using the correcting capability of the Reed-Solomon code we can correctly restore the message. The value of L is then determined by the value of p.

Suppose that $K + 2\Delta$ packets of an encoded message are received. Then, using the Reed-Solomon code, this message will be decoded correctly if the number of erroneous packets

Error Correcting Coding and Security for Data Networks   G. Kabatiansky, E. Krouk and S. Semenov
© 2005 John Wiley & Sons, Ltd   ISBN: 0-470-86754-X

does not exceed Δ. Hence, the probability of an erroneous decoding of the message P_{err} for a given value of p can be estimated by the following inequality:

$$P_{err} \leq 1 - \sum_{i=0}^{\Delta} \binom{K+2\Delta}{i} \cdot p^i (1-p)^{K+2\Delta-i}. \tag{9.3}$$

Let us denote by Δ_0 the minimum value of Δ for which the inequality (9.1) is satisfied. We will denote the ratio Δ_0/K by δ_0. It is obvious that when coding in a network when the network unreliability parameter p is used, and when messages are assembled from the first $K + 2\Delta_0$ arriving packets out of $N = K/R$, the acceptable error probability $P_{err.a}$ is achieved for a mean message delay of

$$\bar{T}_3 = \min_{R} \left\{ \bar{t}(\lambda/R, \mu) \cdot \sum_{j=K/R-2\Delta_0+1}^{K/R} j^{-1} \right\}. \tag{9.4}$$

The expression (9.4), like (7.10), is obtained as the mean value of the order statistics with number $K + 2\Delta_0$ of the set of packet delays of the message $t_{1:N}, \ldots, t_{N:N}$ for the best chosen code. Then the value of the mean message delay $\bar{T}_3$ can be estimated as follows:

$$\bar{T}_3 \leq \min_{\rho < R < \frac{1}{1+2\delta_0}} \left\{ \frac{\lambda}{\gamma \mu C} \cdot \frac{R}{R-\rho} \cdot \ln \frac{1}{1-R-2\delta_0 R} \right\} \tag{9.5}$$

Using in (9.5) the estimation $\ln x \leq x - 1$ obtain

$$\bar{T}_3 \leq \min_{\rho < R < 1/\sigma_0} \left\{ \frac{\lambda}{\gamma \mu C} \cdot \frac{R}{R-\rho} \cdot \frac{\sigma_0 R}{1-\sigma_0 R} \right\}, \tag{9.6}$$

where $\sigma_0 = 1 + 2\delta_0$. The function in the braces on the right-hand side of (9.6) is a minimum when $R = \dfrac{2\rho}{1+\sigma_0\rho}$ and the condition $\rho < 1/\sigma_0$ is satisfied. Substituting this value of R into (9.6) obtain

$$\bar{T}_3 \leq \frac{\lambda}{\gamma \mu C} \cdot \frac{4\sigma_0 \rho}{(1-\sigma_0\rho)^2}, \tag{9.7}$$

$$\rho < 1/\sigma_0.$$

Comparing (9.7) with (7.15) we can write the following expression

$$\bar{T}_3 \leq \bar{T}_2 \cdot \frac{\sigma_0 \cdot (1-\rho)^2}{(1-\sigma_0\rho)^2} = \bar{T}_2 \cdot \frac{(1+2\delta_0) \cdot (1-\rho)^2}{(1-\rho-2\delta_0\rho)^2}, \tag{9.8}$$

$$\rho < 1/\sigma_0 = 1/(1+2\delta_0).$$

where $\bar{T}_2$ is the mean message delay in the network with the reliable channels and under the condition of using the transport coding. Usually the number of packets in the message is not very large. However, the analysis of expression (9.8) with $K \to \infty$ allows us to decide on the limit capabilities of the transport coding. With large K it is possible to provide a small error probability P_{err} in (9.3) by choosing $\delta = p + \chi$, where χ is the small value decreasing with increasing K. Then it follows from (9.8) that it is possible to provide an arbitrary small value of message error probability P_{err} in the network with unreliability parameter p due to increasing the mean message delay in comparison with the reliable network by the coefficient

$$\frac{(1 + 2\delta_0) \cdot (1 - \rho)^2}{(1 - \rho - 2\delta_0\rho)^2} < 1 + 2\delta_0. \tag{9.9}$$

Notice that for $p > 0$ the error probability of the uncoded messages tends to 1 with increasing K.

9.2 RECONCILIATION OF CHANNEL AND TRANSPORT CODING

The use of unreliable channels in a network leads to the increase of the mean message delay. On the other hand if we are speaking about reliable channels, we assume that some error-correcting coding is used when information is transmitted over each channel of the network, i.e. we assume that powerful codes, capable of making the error probability a fairly small quantity when data is transmitted over the channel, are used. Then we can say that the unreliability parameter p is the function of the mean time of transmission of the packet over the channel $\frac{1}{\mu}$ and by decreasing this time or increasing μ (by means of changing the channel protocols) it is possible not only to increase the unreliability parameter p, but also to decrease the mean packet delay $\bar{t}(\lambda, \mu)$. In this section we consider the problem of choosing the parameter μ in order to decrease the mean message delay and to provide the acceptable message error probability.

Let $\frac{1}{\mu_0}$ be the minimum value of the mean time of transmission the packet over the channel corresponding to the unreliability parameter $p_0 \leq \frac{P_{err.a.}}{K}$ in accordance with (9.1) and (9.2). That means, for μ_0 the network channels can be regarded as reliable ones and for any $\mu = \mu_0 \cdot \nu \ (\nu > 1)$ the network channels are unreliable. Consider p as a function of ν, then $p(1) = p_0, p(\nu) > p_0, \nu > 1$. The mean packet delay in the network with $\mu = \mu_0 \cdot \nu$ and the rate of transport coding R can be written in accordance with (7.14) as

$$\bar{t}(\lambda/R, \mu_0 \cdot \nu) = \frac{\bar{l}}{\mu_0 C} \cdot \frac{R}{R\nu - \rho_0}, \tag{9.10}$$

where $\bar{l} = \frac{\lambda/R}{\lambda} = \frac{\lambda}{\gamma}$ is the mean path length traversed by a packet along the network and $\rho_0 = \frac{\lambda}{\mu_0 \cdot C}$ is the load of the network without transport coding, $\rho_0 < R\nu$.

Substituting the value of the mean packet delay (9.10) in (9.4) and (9.5) obtain the estimate of the mean message delay $\bar{T}_4$ for the network with transport coding and with the unreliability parameter $p(\nu) > p_0$ as follows

$$\bar{T}_4 = \min_{\nu > 1} \bar{T}_3(\nu) \leq \min_{\nu > \frac{p_0}{R}, R < \frac{1}{1+2\delta_0}} \left\{ \frac{\lambda}{\gamma \, \mu_0 \, C} \cdot \frac{R}{R\nu - p_0} \cdot \ln \frac{1}{1 - R - 2\delta_0 R} \right\}, \qquad (9.11)$$

where $\delta_0 = \Delta_0/K$ and Δ_0 is the minimum value of Δ for which the inequality $P_{err}(\Delta_0) \leq P_{err.a.}$ holds true. The application of the same technique as was used for estimation of $\bar{T}_3$ gives us the optimum value for transport code rate $R = \dfrac{2p_0}{\nu + p_0 + 2\delta_0 p_0}$ and the estimate of $\bar{T}_4$ can be written as follows

$$\bar{T}_4 \leq \min_{\nu > p_0(1 + 2\delta_0)} \left\{ \frac{\lambda}{\gamma \, \mu_0 \, C} \cdot \frac{4(1 + 2\delta_0)p_0}{(\nu - p_0 - 2\delta_0 p_0)^2} \right\}, \qquad (9.12)$$

We can define δ_0 as a function of ν by estimating the value P_{err} as follows

$$P_{err} < \sum_{i=\delta \cdot K}^{K/R} \binom{K/R}{i} \cdot p^i (1 - p)^{K/R - i} \leq \exp\left\{ -\frac{K}{R} \cdot h(\delta, p) \right\}, \qquad (9.13)$$

where $h(\delta, p) = \delta \cdot \ln \dfrac{\delta}{p} + (1 - \delta) \cdot \ln \dfrac{1 - \delta}{1 - p}$. Then the condition (9.1) holds true if

$$h(\delta, p) \geq \frac{R}{K} \cdot \ln \frac{1}{P_{err.a.}}. \qquad (9.14)$$

Substituting in (9.14) the optimum value of transport code rate $R = \dfrac{2p_0}{\nu + p_0 + 2\delta_0 p_0}$ obtain the following inequality

$$h(\delta_0, p) \geq \frac{2p_0}{K \cdot (\nu + p_0 + 2\delta_0 p_0)} \cdot \ln \frac{1}{P_{err.a.}}, \qquad (9.15)$$

from which it is possible to estimate the function $\delta_0(\nu)$. However, the use of this function in explicit form obtained from (9.15) is too cumbersome for the further analysis of (9.12).

Obviously the use of transport coding in a network with unreliable channels is more advantageous (in the sense of reducing the mean message delay) than over a reliable network without coding at the transport level, if the following conditions are satisfied:

$$\frac{\varepsilon + \ln K}{1 - p_0} > \min_{\nu} \left\{ \frac{4(1 + 2\delta_0)p_0}{(\nu - p_0 - 2\delta_0 p_0)^2} \right\}, \qquad (9.16)$$

$$\nu > p_0(1 + 2\delta_0)$$

To compare the message delays in a reliable, as against an unreliable, network we have to calculate the parameter ν. We will assume that both networks have the same error probability per bit in channel p_{ch}. In channels of the reliable network, more powerful code with a greater number of redundancy symbols is used. So, the parameter ν can be written as

$$\nu = \frac{n_r}{n_u},\tag{9.17}$$

where n_r and n_u are the lengths of channel codes used in the network with reliable channels and the network with unreliable channels, respectively. The length of channel code is determined by the network unreliability parameter p and by p_{ch} from the equation

$$p = \sum_{i=\frac{d+1}{2}}^{n} \binom{n}{i} \cdot p_{ch}^i (1 - p_{ch})^{n-i},\tag{9.18}$$

where d is the minimum distance of corresponding channel code, and for n the value of n_r or n_u from the equation (9.17) can be used. For calculation, the table of best codes was used. It was assumed that the initial length of packet $s = 10$ bits (number of information bits for channel code). The results of calculation are shown in Figures 9.1–9.4. Here we can see that using transport coding gives an advantage at all values of the initial network load. Thus if the inequalities (9.15) are satisfied it will be possible not only to decrease the mean message delay with the help of transport coding, but also to simplify the procedure of data transmission in the channels. Moreover it will also be possible to provide the same error probability per message as in the network with reliable channels [1].

The data presented in Figures 9.1 and 9.2 are optimised with respect to the parameter ν. The jump in the gain in Figure 9.3 for the number of the information packets in the range of

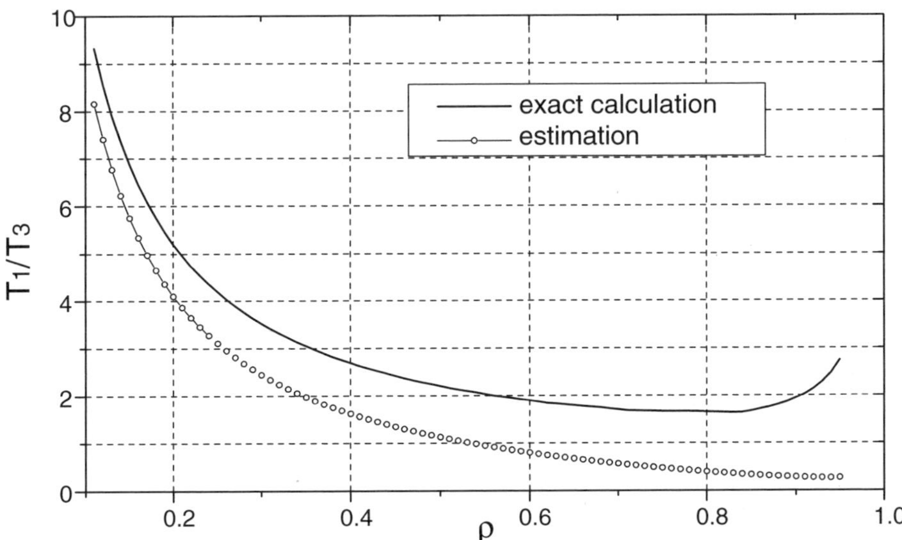

Figure 9.1   A gain of transport coding as function of network load. Unreliable network. $K = 10$, $p_{ch} = 10^{-3}$, $P_{err.a.} = 10^{-9}$

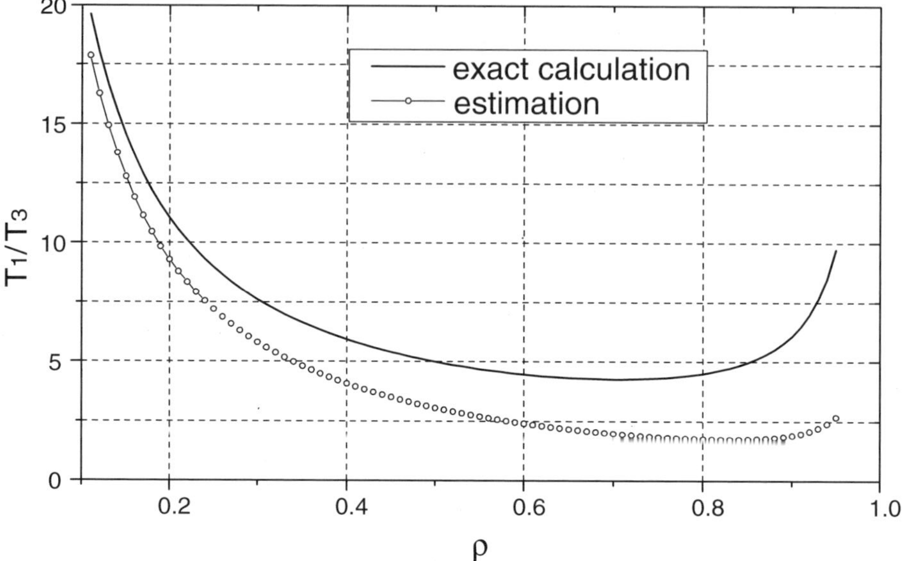

Figure 9.2 A gain of transport coding as a function of network load. Unreliable network. $K = 100$, $p_{ch} = 10^{-3}$, $P_{err.a.} = 10^{-9}$

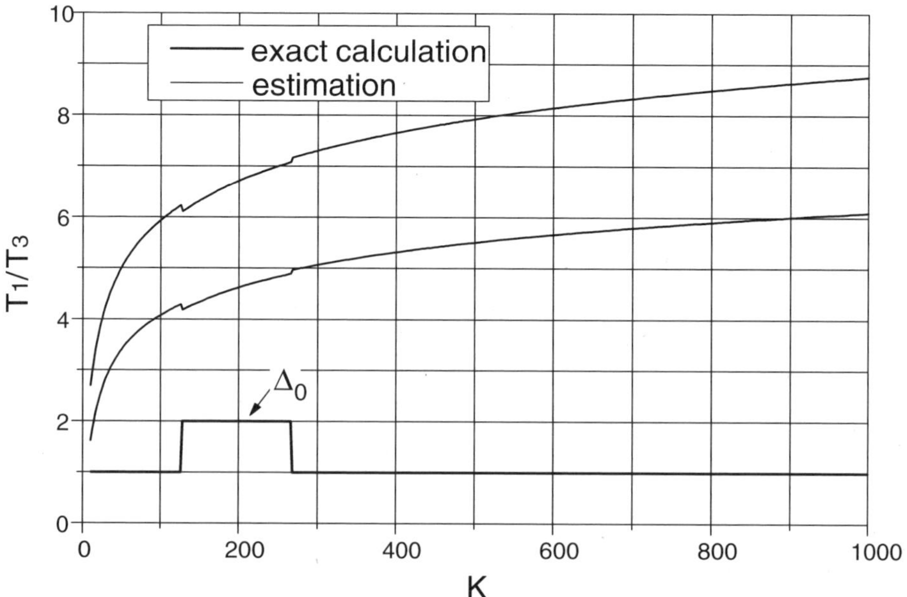

Figure 9.3 A gain of transport coding vs. number of information packets in message. Unreliable network. $\rho = 0.4$, $\nu = 1.29$, $p_{ch} = 10^{-3}$, $P_{err.a.} = 10^{-9}$

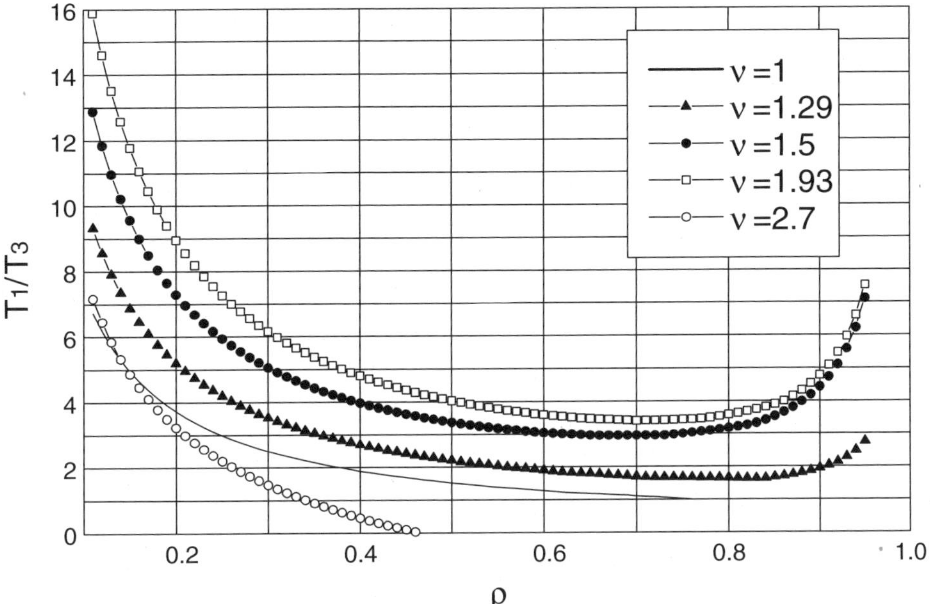

Figure 9.4   A gain of transport coding as a function of network load. Unreliable network. $K = 10$, $p_{ch} = 10^{-2}$, $P_{err.a.} = 10^{-9}$

128–268 can be explained by changing the value of the parameter Δ_0. In Figure 9.4 one can see a gain of transport coding for different values of the parameter ν. It is interesting to note that the value $\nu = 1$ corresponds to the network with reliable channels. So, the data in Figure 9.4 shows that it is possible to choose a value of parameter ν for which the transmission of information over a network with unreliable channels provides less mean message delay than the transmission over a reliable network.

Of course, it is also possible to organise the transmission of priority information in the network with unreliable channels in the same way as in the reliable network. The sole difference will be the fact that priority messages will be encoded by the transport code using a lower code rate than with ordinary messages.

9.3   USE OF TORNADO CODES FOR RECONCILIATION OF CHANNEL AND TRANSPORT CODING

In some cases it is possible to use some suboptimal codes at the transport level of a network rather than MDS codes and in particular Reed-Solomon codes. A good example of such class of suboptimal codes is the class of Tornado codes [2], [3]. The attractive feature of Tornado codes is the low complexity of the encoding and decoding procedures. Unlike the MDS codes, which are capable of recovering the whole message of length N on any K received packets, the Tornado code needs in $(1 + \sigma)K$ packets to reconstruct the message with high probability. And the coefficient $(1 + \sigma)$ is called the *decoding inefficiency*, $\sigma > 0$. Tornado codes are the binary codes. So if the length of packet is s bits we need to use s Tornado encoders and decoders to provide the transport coding. The complexity of the encoding and decoding procedures is the same and is proportional to $N \cdot \ln\left(\frac{1}{\sigma}\right)$ XOR operations. Then the complexity of encoding or reconstruction of a message is proportional to $N \cdot s \cdot \ln\left(\frac{1}{\sigma}\right)$ XOR operations, which is significantly less than the complexity of Reed-Solomon codes encoding and decoding procedures for large values of N and K.

The construction and analysis of Tornado codes are based on bipartite graphs. Let us define a code $V(B)$ with k information bits and $\beta \cdot k$ parity-check bits, by associating these bits with a bipartite graph B. The graph B has k left nodes, corresponding to the information bits and $\beta \cdot k$ right nodes, corresponding to parity-check bits. As shown in Figure 9.5, each parity-check bit c_i, $0 \leq i \leq \beta \cdot k - 1$ on the right side of graph B is obtained as the sum of several information bits from the left nodes of graph B. All operations are done in $GF(2)$. Thus, the encoding complexity is proportional to the number of edges in B. Now, if some of the information bits are missing it is possible to recover them with the help of the parity-check bits. Then the decoding complexity is the same as (or even less than) the encoding complexity.

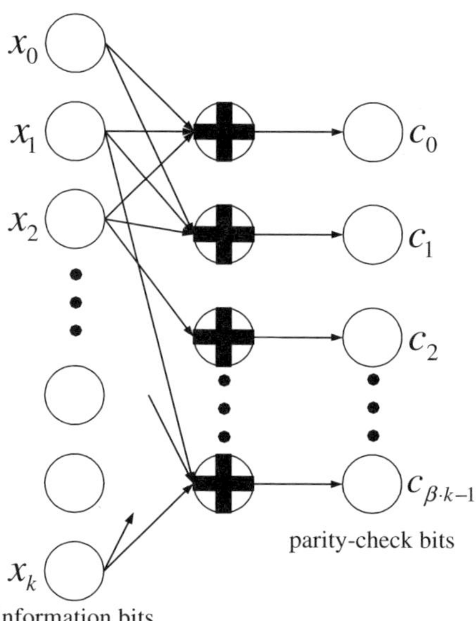

Figure 9.5   A bipartite graph and corresponding calculation of parity-check bits

To produce codes that can recover from losses regardless of their location, it is possible to cascade codes of the form $V(B)$: first code $V(B)$ generates $\beta \cdot k$ parity-check bits for the original k information bits, then a similar code is used to generate $\beta^2 \cdot k$ parity-check bits for the $\beta \cdot k$ parity-check bits produced by the first code, and so on. At the last level, as is shown in Figure 9.6, some conventional loss-resilient code can be used. That means the sequence of bipartite graphs $B_0, B_1, \ldots, B_m$, where B_i has $\beta^i \cdot k$ left nodes and $\beta^{i+1} \cdot k$ right nodes, is used to construct a sequence of codes $V(B_0), V(B_1), \ldots, V(B_m)$. The parameter m is selected in such a way that $\beta^{m+1} \cdot k$ is roughly equal to $\sqrt{k}$ and the last loss-resilient code V' is chosen

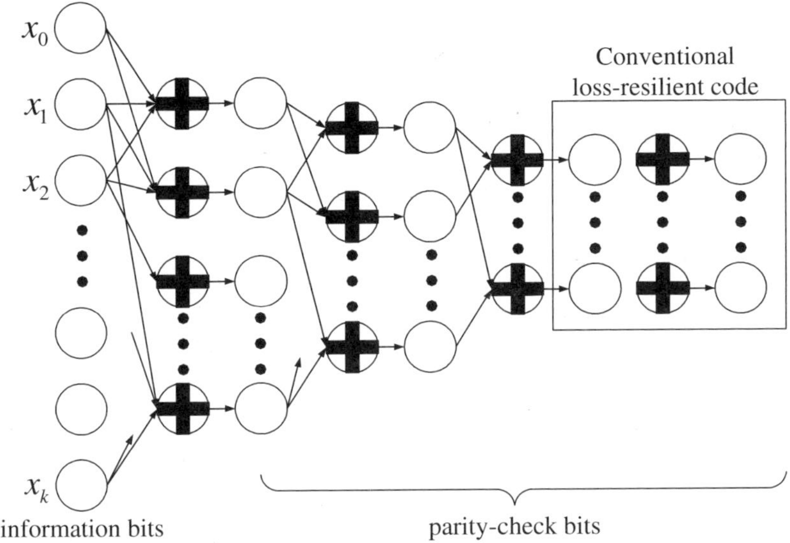

Figure 9.6 An encoding procedure of Tornado code

to be a code of rate $1 - \beta$ with $\beta^{m+1} \cdot k$ information bits for which it knows how to recover from the random loss of β fraction of its bits with high probability. Then the Tornado code can be defined as the code $V(B_0, B_1, \ldots, B_m, V')$ with k information bits and

$$\sum_{i=0}^{m} \beta^{i+1} \cdot k + \frac{\beta^{m+2} \cdot k}{1 - \beta} = \frac{\beta \cdot k}{1 - \beta}, \tag{9.19}$$

parity-check bits formed by using $V(B_0)$ to produce $\beta \cdot k$ parity-check bits for k information bits, using $V(B_i)$ to generate $\beta^{i+1} \cdot k$ parity-check bits for the $\beta^i \cdot k$ parity-check bits produced by $V(B_{i-1})$, and finally using V' to produce an additional $\dfrac{\beta^{m+2} \cdot k}{1 - \beta}$ parity-check bits. As $V(B_0, B_1, \ldots, B_m, V')$ has k information bits and, in accordance with (9.19), $\dfrac{\beta \cdot k}{1 - \beta}$ parity-check bits, it is a code of rate $1 - \beta$.

Then, if code V' can recover from the random loss of a $\beta \cdot (1 - \sigma)$ fraction of its bits with a high probability, and if each code $V(B_i)$ can recover from the random loss of a $\beta \cdot (1 - \sigma)$ fraction of its bits with a high probability, then $V(B_0, B_1, \ldots, B_m, V')$ is a code of rate $1 - \beta$ that can recover from the random loss of a $\beta \cdot (1 - \sigma)$ fraction of its bits with a high probability. In [3] the method of construction of such codes is shown. Thus, it is enough to receive $k \cdot (1 - \sigma)$ undistorted bits to recover all k information bits with a high probability.

Let us consider networks with different values of the unreliability parameter p. As in the previous section we will consider p as a function of ν, i.e. $p(1) = p_0$, $p(\nu) > p_0$, $\nu > 1$, where p_0 corresponds to the network with reliable channels. The problem of using Tornado codes at the transport level of the network with unreliable channels is that Tornado codes are designed to cope with erasures only, not with errors. However, we can avoid this problem by adding some *circular redundancy check* (CRC) to every data packet. The use of some error-correcting code instead of CRC is the same as the use of more powerful channel codes, i.e. it returns us to the case of the network with reliable channels. In the case of CRC we can easily distinguish the corrupted packets. Then to reconstruct the original message it is necessary to receive more than $K \cdot (1 + \sigma)$ packets, e.g. $K \cdot (1 + \sigma) + \Delta$ packets (or that is the same, just to increase the value of σ). The message error probability in this case is the probability that there will be less than $K \cdot (1 + \sigma)$ undistorted packets among $K \cdot (1 + \sigma) + \Delta$ received ones:

$$P_{err} = 1 - \sum_{i=0}^{\Delta} \binom{K \cdot (1 + \sigma) + \Delta}{K \cdot (1 + \sigma)} \cdot p^{\Delta - i} \cdot (1 - p)^{K \cdot (1+\sigma)+i}. \tag{9.20}$$

This value can be estimated as follows

$$P_{err} \leq 1 - \binom{K \cdot (1 + \sigma) + \Delta}{K \cdot (1 + \sigma)} \cdot (1 - p)^{K \cdot (1+\sigma)+\Delta}. \tag{9.21}$$

Let Δ_0 be the minimum value of Δ for which the inequality (9.1) is satisfied. That means we should choose the value of Δ in such a way as to compensate for the change in the value p. The increase of the parameter ν defines the change of the unreliability parameter p. As a rough estimate of (9.8) we can say that the value of ν is proportional to $\dfrac{\ln p_0}{\ln p(\nu)}$. Then the inequality (9.21) can be written as follows

$$P_{err} \leq 1 - \binom{K \cdot (1 + \sigma) + \Delta}{K \cdot (1 + \sigma)} \cdot (1 - e^{\frac{\ln p_0}{\nu}})^{K \cdot (1+\sigma)+\Delta}. \tag{9.22}$$

In accordance with (9.4) and (9.10) the mean message delay $\bar{T}_5(R, \nu)$ in the network with $\mu = \nu \cdot \mu_0$ under the condition of using a Tornado code with rate R at the transport level can be written as follows

$$\bar{T}_5(R, \nu) = \frac{\lambda}{\gamma \mu_0 C} \cdot \frac{R}{R\nu - \rho_0} \cdot \sum_{j=K/R-K(1+\sigma)-\Delta_0+1}^{K/R} j^{-1}, \tag{9.23}$$

where values of ρ_0 and μ_0 correspond to the network with reliable channels. Using the same technique as in chapter 7 and in section 9.2 we obtain the following estimate

$$\bar{T}_5(\nu) \leq \min_{\frac{\rho_0}{\nu} < R < \frac{1}{1+\sigma+\delta_0}} \left\{ \frac{\lambda}{\gamma \mu_0 C} \cdot \frac{R}{R\nu - \rho_0} \cdot \frac{R(1 + \sigma + \delta_0)}{1 - R(1 + \sigma + \delta_0)} \right\}, \tag{9.24}$$

where $\delta_0 = \dfrac{\Delta_0}{K}$. The value of $R = \dfrac{2\rho_0}{\nu + \rho_0(1 + \sigma + \delta_0)}$ gives the minimum of function in the braces on the right-hand side of (9.24) under the condition $\rho_0 < \dfrac{\nu}{(1 + \sigma + \delta_0)}$. Substituting this value of R into (9.24), we obtain

$$\bar{T}_5(\nu) \leq \frac{\lambda}{\gamma \mu_0 C} \cdot \frac{4\rho_0(1 + \sigma + \delta_0)}{(\nu - \rho_0(1 + \sigma + \delta_0))^2}. \tag{9.25}$$

As can be seen from (9.22) the parameter δ_0 depends on ν and we can estimate δ_0 with the help of the following inequality

$$P_{err.a.} \leq 1 - \binom{K \cdot (1 + \sigma + \delta_0)}{K \cdot (1 + \sigma)} \cdot (1 - e^{\frac{\ln p_0}{\nu}})^{K \cdot (1+\sigma+\delta_0)}, \tag{9.26}$$

where $P_{err.a.}$ is the acceptable error probability of the message. Then optimising by ν we can find the optimum balance between the error-correcting capability of channel code and the parameter $\sigma_0 = \sigma + \delta_0$ of the Tornado code used at the transport level of the network. The mean message delay $\bar{T}_6$ in the network with unreliable channels under the condition of using the Tornado code at the transport level can be written as follows

$$\bar{T}_6 = \min_{\nu} \bar{T}_5(\nu) \leq \min_{\nu > \rho_0(1+\sigma_0)} \left\{ \frac{\lambda}{\gamma \mu_0 C} \cdot \frac{4\rho_0(1 + \sigma_0(\nu))}{[\nu - \rho_0(1 + \sigma_0(\nu))]^2} \right\}. \tag{9.27}$$

As in section 9.2 the explicit form of the function $\sigma_0(\nu)$ is too cumbersome to allow effective analysis of the expression (9.27).

As in the previous section, we will compare the results of the calculations for the mean message delay in the network with reliable and unreliable channels. The results of this comparison are represented in Figure 9.7. As one can see, the results obtained when using Tornado codes are a little bit worse than for Reed-Solomon codes. These results can easily be explained by the fact that Tornado codes need a relatively higher number of received symbols to reconstruct the message than do RS codes and, unlike the RS codes, Tornado codes are only capable of coping with the erasures not with the errors. However, the difference in

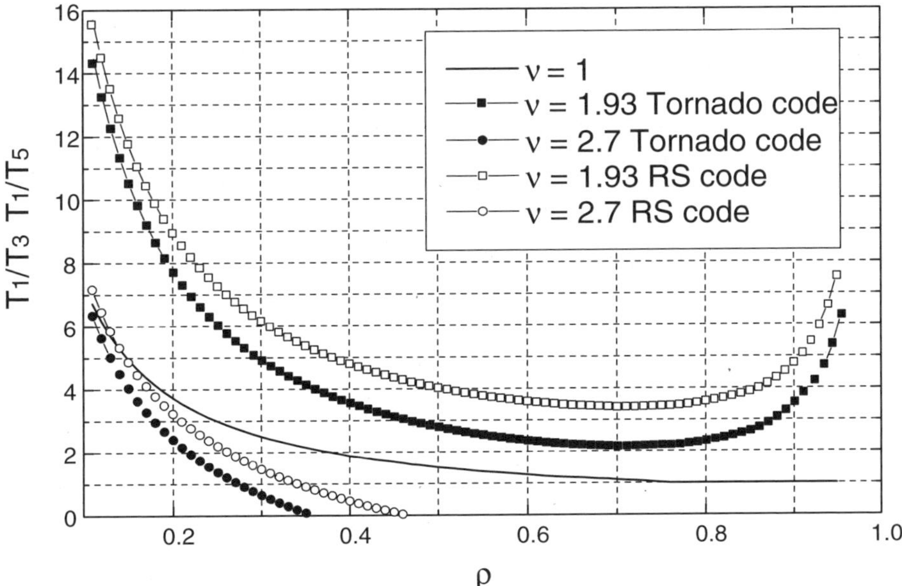

Figure 9.7 A gain of transport coding as a function of network load. Comparison of RS codes and Tornado codes. Unreliable network. $K = 10$, $p_{ch} = 10^{-2}$, $P_{err.a.} = 10^{-9}$

the gain provided by Tornado and RS codes is not dramatic and Tornado codes can be regarded as good candidates for use in transport coding, especially taking into account the low complexity of the encoding and decoding procedure for these codes. It is interesting to note that if we compare the optimised results with respect to the parameter ν the difference between the performance of RS codes and Tornado codes is almost negligible. It can be explained by restrictions on the values of ν, which in some cases are stricter for RS codes. Another good candidates for use in transport coding are LT codes [4]. As in the case of Tornado codes, LT codes also need slightly more than K undistorted symbols in order to reconstruct the message, but the most attractive feature of LT codes is their ability to change the code rate on fly. Thus, it is possible to tune the number of packets to be sent depending on the channel condition or on the network load.

9.4 DEVELOPMENT OF CODING METHODS AT THE PRESENTATION LAYER

In chapter 8 we considered the methods of coding cryptography, which corresponds to coding at the presentation layer (L6) of network. The coding methods to counteract unauthorised access do not exhaust the possibilities of the use of error control codes at the presentation layer. Another example is the application of coding for image compression.

Many image compression techniques use the splitting of an image into domains (some regions of equal size), and then each domain is compressed with loss of quality. Most methods use some spectral transformation, and then the scalar quantisation of the obtained spectral coefficients. The term *"scalar quantisation"* means that some numbers form the input data for this method, and the simplest example of such a method is rounding. The possible output data are the so-called *quantisation levels* or *reconstruction levels*. Usually the quantisation levels are defined with the help of tables. As a result each domain is processed independently.

However, very often the image contains many similar regions, and splitting into domains results in large groups of similar or even equal domains. Obviously, it is possible to obtain some gain from this similarity if, during the compression process, the domain is considered as a whole object rather than a set of independent pixels. For better understanding of the background of possible gain consider the following example.

> ***Example 9.1***   Consider the standard test image of 512×512 pixels, where each pixel is defined by 256 gray scale. Split the image into domains of 8×8 pixels. In this case each domain can be represented as a vector of length 64 or as a point in 64-dimension space. The number of points in this vector space is equal to $256^{64} = 2^{512}$. The image is split into $4096 = 2^{12}$ points, hence, not all the points of vector space are used. If the set of 2^{12} domains is known to both encoder and decoder, each domain can be represented only by 12 bits. Moreover, if compression with loss of quality is used it is possible to represent all similar domains by only one domain, which approximates to these similar domains with some error. Then even fewer than 12 bits can be used to transmit or store each domain. For example, if the set of used samples contains only 256 domains, only 1 byte is needed for the transmission of each domain, and the compression rate in this case is 16 times.

The usage of this feature of domain similarity is not new. There is a lot of work devoted to application of vector quantisation to image compression [5–12].

The general method consists in splitting the image into domains, then these domains are considered as vectors or some points in N-dimensional space and in the process of compression with losses in quality similar domains are regarded as the same sample. All the samples are kept in the codebook. Traditionally the generalised Lloyd's algorithm [1], [11] is used for these purposes. As a result each domain corresponds to the codeword, which is the best approximation of this domain with some restrictions. With this method the whole image corresponds to the codebook (the set of codewords), which is used for quantisation of the image. Vector quantisation itself provides the compression $\dfrac{N \cdot L}{\log(K)}$ times, where N is the vector length (number of pixels in domain), L is the number of bits per symbol of input vector, K is the number of vectors in the code book. The size of the codebook is not taken into account here.

After application of vector quantisation, the reconstructed image differs from the original. The extent to which the image is distorted depends not only on the compression level but also on the set of vectors in the coding book $\{W\} = \{\mathbf{w}_1, \mathbf{w}_2, \ldots \mathbf{w}_K\}$. That means that for one particular image the codebook provides an acceptable level of distortion but for another image the same codebook may introduce unacceptable distortion. For this reason the codebook is built for each particular image, and then it is necessary to transmit the codebook for the reconstruction of the image. As this takes place, the codebook itself is big and forms a significant part of the whole amount of data transmitted or stored after compression. It is possible to increase the compression if storing of the codebook is refused, i.e. if the same codebook is used for the different images.

Let us say that the domain a of size $n \times n$ is covered in code W with radius R if there is $\mathbf{w} \in W$ for which $\sum (w_i - a_i)^2 \leq R$. Then each code can be considered as covering some set of domains with radius R, and the process of vector quantisation can be considered as mapping of the set of domains to the codewords $\mathbf{w} \in W$, as is shown in Figure 9.8.

For the adaptive codebook built with the help of a generalised Lloyd's algorithm, it is possible to obtain a minimal coverage radius because in this case the code is built specially for the given domains of the particular image rather than the coverage of the whole vector space. However, in this case the code has no structure (unlike, for example, the linear code), and, therefore, there is no method of compression for the codebook itself.

If we use some special code for quantisation with the properties of some structure and capable of covering the whole vector space, the covering radius will be significantly more than in the previous case (see Figure 9.9), but now it will be possible to transmit the codebook in some compact form or even not to transmit it at all if we use the same code for all processed images.

Let V be a linear (n, k) code over $GF(q)$, $q = 256$ for the case where a pixel is represented by the 256 level gray scale, and let the domain size be $n_d \times n_d$, $n = n_d \cdot n_d$. If the points of the covered vector space correspond to the domains of some image, then the compression technique can be described as follows: each domain of n pixels is substituted by the closest in the Euclidean metric codeword of code V. Since the codeword is defined by the information set of length k, it is enough to transmit only k elements in order to reconstruct the whole codeword. As a result the compression rate is n/k times. The losses in quality in this case will be caused by the fact that substitution of the original domains by the codewords does not lead to one-to-one correspondence. In this case there is no need to transmit the codebook at all. This method for the case of using Reed-Solomon code is described in [13].

The quality of the reconstructed image depends on the covering radius: the quality of image increase with the decrease of radius. The covering radius in turn depends both on the

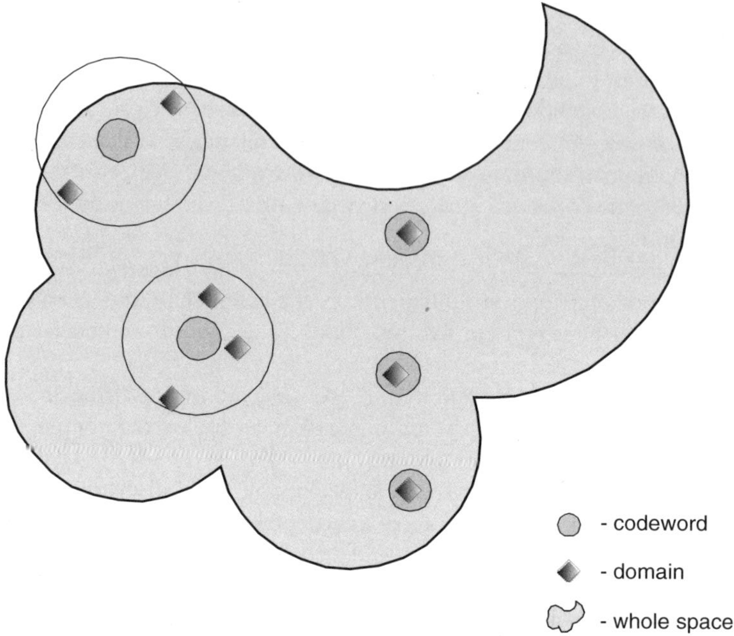

Figure 9.8   An adaptive coverage of given points of space

number of words in code V and on the position of these codewords in the vector space relative to the position of points of the covering set, hence the better is the choice of code, the better is the quality of the reconstructed image. If the linear (n, k) code is used for these purposes it can be transmitted in the compact form of the generator matrix. Then it is

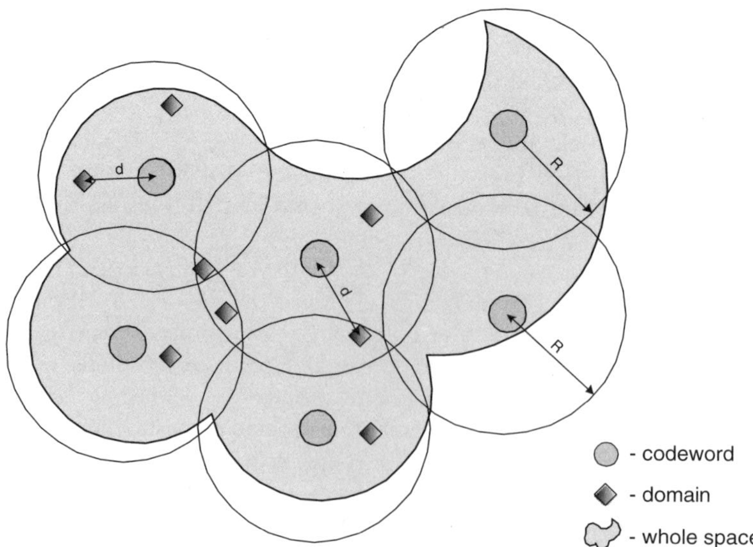

Figure 9.9   A code covering the whole vector space

possible to use different codes for the compression of different images, which leads to an increase in quality of the reconstructed image.

However, even in the case when the linear error-correcting code is not transmitted at all, the overall compression is less in comparison with the compression rate provided by the adaptive code. The adaptive way of building the code allows us to obtain the code with significantly fewer of codewords with the same quality of reconstructed image. Therefore, we need in some way to adapt the code to the particular image.

Let C be the vector space and W be the code covering the whole space C with radius R. That means that for any point of space C there is a word of code W such that the Euclidean distance between this codeword and the given point is no more than R:

$$\exists \mathbf{w} \in W : \quad \forall \mathbf{c} \in C, \quad d(\mathbf{c}, \mathbf{w}) \leq R, \text{ where } d(\mathbf{c}, \mathbf{w}) = \sum (w_i - c_i)^2.$$

Let C' be the vector space, the points of which are the domains of the original image, and let M be the set of invertible transformations defined over the points of C' such that the result of transformation belongs to space C:

$$\forall \mathbf{c}' \in C', m \in M : \mathbf{c} = m\mathbf{c}', \mathbf{c} \in C.$$

Thus, any point from space C' can be mapped with the help of transformation $m \in M$ to the point of space C, where it is covered by the code W with radius R. Hence, the application of the codeword from W as a quantiser for points of form $m\mathbf{c}', \mathbf{c}' \in C'$ leads to the quantisation error of no more than R, and with the help of transformations M it is possible to adapt the code to a particular image, as shown in Figure 9.10.

In the process of quantisation of the points of the vector space, the problem of finding a codeword closest to an arbitrary point of space arises. In general this problem can be solved with the help of an exhaustive search. But in accordance with the assumption that the code covering the space has some specific properties like the error-correcting code, it is possible to consider any point $\mathbf{c} \in C$ as a codeword $\mathbf{a}$ in addition with an error vector $\mathbf{e}$:

$$\mathbf{c} = \mathbf{a} + \mathbf{e}, \quad \mathbf{c} \in C.$$

Let us denote the error weight, i.e., the distance between error vector and all-zero vector, as $wt(\mathbf{e}) = d(\mathbf{e}, \mathbf{0})$. Then the problem of finding a codeword that is at a distance no more than R from the given point, can be reduced to a problem of searching for the error vector of minimal weight:

$$\mathbf{e} : \mathbf{a} = \mathbf{w} + \mathbf{e}, \quad \mathbf{w} \in W, wt(\mathbf{e}) \leq R.$$

This problem can be described as a problem of decoding code W in a radius of no more than R, i.e., the search of the codeword $\mathbf{w} \in W$ such that $\mathbf{w} = \mathbf{c} - \mathbf{e}, \quad wt(\mathbf{e}) \leq R$. Now the general method of image compression with the help of error-correcting code can be described as follows:

- Each domain $\mathbf{x}_i$ of the image is mapped to the point $\mathbf{c}_i$ of the vector space C covered by the code W.

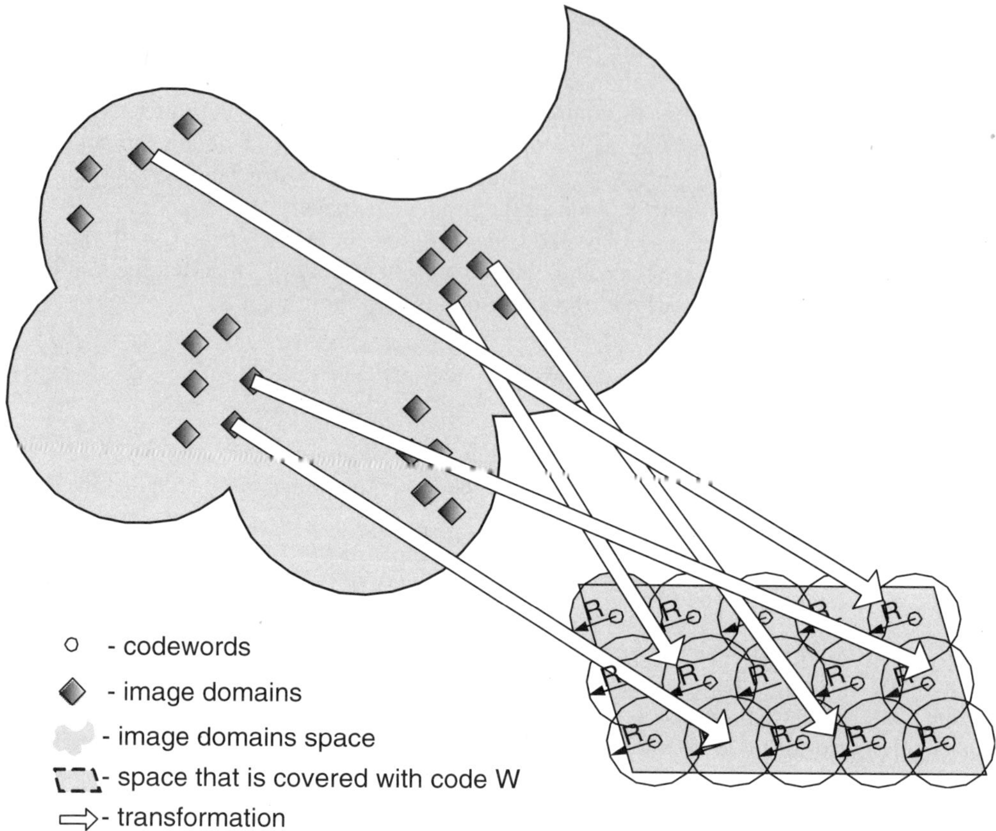

Figure 9.10 Adaptation of given code to the image with the help of special transforms

- The vector $\mathbf{c}_i$ is decoded with the help of code W in space C to the nearest codeword $\mathbf{w}_i \in W$, and $\mathbf{w}_i$ is used as a quantiser for the original domain $\mathbf{x}_i$.

- The information set γ_i of codeword $\mathbf{w}_i$ is transmitted, i.e., instead of the original domain $\mathbf{x}_i$ we transmit the information set γ_i and the compression rate is defined by the rate of lengths of vectors γ_i and $\mathbf{x}_i$.

However, the adaptation of the image for coding can also be done with the help of some transformations. There exists a method of adaptation of an image for a fixed error-correcting code at the cost of some matched losses in the quality. It can be described as follows: each domain is split into bit planes, then each bit plane should be quantised with the help of decoding and given LDPC code. Thus, splitting the domain into bit planes is used as the transformation that maps the domain in another vector space, and the LDPC codes are used as the covering codes.

The main idea of this method consists in the fact that the different bit planes have different significance for the reconstruction of image. For example, if we are going to quantise the most significant bit plane consisting of the most significant bits (MSB), it is obvious that quantisation errors for this plane lead to the maximum distortion in the reconstructed image. However, it is possible to take into account some of these errors in the process of quantisation of the next bit planes.

Example 9.2 Let the luminescent components X_i of pixels in domain be split into 8 bit planes $X_i = \{x_i^1, x_i^2 \ldots\}$. Let $X_i = 255$ and $X_j = 128$. Then the values of both corresponding components of the most significant bit plane x_i^1 and x_j^1 are equal to 1. The errors in quantisation of both bits x_i^1 and x_j^1 lead to the same quantisation error for numbers X_i and X_j. However, the quantisation error of X_i caused by the quantisation error of bit x_i^1 can be only increased after the quantisation of less significant bit planes, whereas the quantisation error of X_j caused by the quantisation error of bit x_j^1 can be compensated during the quantisation of the less significant bit planes up to the unit error if all other less significant bits in X_j will be quantised into the unit bits. In this case the reconstructed value of X_j will be equal to 127 instead of 128, which is indistinguishable to the naked eye.

This example shows that it is important for the quantisation procedure to take into account the significance of bits x_i^1 and x_j^1. Since we are using the decoding of error-correcting code as a quantisation procedure, it is possible to use different reliability parameters for different bits in the decoding procedure. Then the soft-decision decoding can be applied. For this purpose the LDPC codes, which have the simple soft-decision decoding schemes, were chosen. The soft-decision decoding can be applied to image compression in the following way.

Let n be the number of pixels in the domain, $\{X_i\}$, $1 \le i \le n$ be the luminance components of image, which should be quantised, $\{W_j\}$, $1 \le j \le N$ be the LDPC codes, N is the number of bits needed for binary representation of $\{X_i\}$. The search for quantisers will be called the *quantisation step* (recall that quantisers are the codewords corresponding to the bits, from which the luminance components $\{\hat{X}_i\}$ will be restored). Let the number of quantisation steps be equal to N.

- At the first quantisation step the set of bits $\{x_i^1\}$ and set of reliability values $\{l_i^1\}$ are formed, where x_i^1 are the most significant bits (MSB) of numbers X_i. The reliability values are calculated in accordance with formula $l_i^1 = (X_i - 2^{N-1} + x_i^1)^2$. The pairs (x_i^1, l_i^1) form vectors, which are fed to the soft-input decoder of code W_1. The result of decoding is the bit vector $\hat{\mathbf{x}}^1 = (\hat{x}_1^1, \ldots, \hat{x}_n^1)$.

- At each later quantisation step j, $2 \le j \le N$, the differences $r_i^j = X_i - \sum_{m=1}^{j-1} \hat{x}_i^1 \cdot 2^{N-m}$ are calculated, and the sets $\{x_i^j\}$ and $\{l_i^j\}$ are formed, where $x_i^j = \begin{cases} 1, & r_i^j \ge 2^{N-j} \\ 0, & r_i^j < 2^{N-j} \end{cases}$, $l_i^j = (r_i^j - 2^{N-j} + x_i^j)^2$. The pairs (x_i^j, l_i^j), as at the first quantisation step, form vectors, which are fed to the soft-input decoder of code W_j. The results of decoding are bit vectors $\hat{\mathbf{x}}^j$.

After all N quantisation steps the quantised values $\hat{x}_i^j$, $1 \le i \le n$, $1 \le j \le N$, will be obtained. These values are regarded as the elements of image bit planes. Since the codes W_j are the algebraic structures (linear codes) used to reconstruct the whole codeword $\mathbf{w} = (w_1, \ldots, w_n)$ it is necessary to know the information set of codeword $\mathbf{w}$, i.e., vector of length k_j for code W_j. Denote by n_j the code length and by k_j the number of information symbols of code W_j. Then the compression rate due to quantisation is $\frac{\sum_{i=1}^{N} k_i}{\sum_{i=1}^{N} n_i}$ times. Since the codes are fixed for all images, there is no need to transmit them.

In general it is possible to use any LDPC codes for this procedure (recall that the LDPC codes were chosen due to the simplicity of the soft-decision decoding of these codes). However, it is obvious that there is no reason to use the same LDPC code at all quantisation steps because, as was mentioned above, the quantisation errors at different quantisation steps have different significance for the overall quantisation error. Therefore, this should be taken into account in choosing a particular code.

It seems that the simplest way of doing so is by increasing the redundancy (the same the decreasing of code rate) at each quantisation step: $\frac{k_1}{n_1} > \frac{k_2}{n_2} > \cdots > \frac{k_N}{n_N}$. The result of application of the described method of compression to the test image 'LENA' is represented in Figure 9.11. The LDPC codes which were used were (16,8), (16,8), (16,8), (16,8), (64,14), (256,29), (256,29), (256,29).

Figure 9.11 The reconstructed test image. Code quantisation with LDPC codes. (SNR $= 23.61$ dB, PSNR $= 30.25$ dB)

This method of code quantisation of images gives different results for different types of domain, and there are a lot of domain types for which the quality of the restored image outperforms the quality of JPEG. Since the code quantisation of bit planes operates with the domains it is possible to use it combined with some other compression algorithm, which is based on splitting the image into domains, e.g. with JPEG in such a way that for each domain there is the possibility of choosing the best algorithm for this particular domain.

- Fix the acceptable quality of the reconstructed image.
- Compression of image domains with the help of LDPC code quantisation.

- The differences of reconstructed and original domains are compressed with the help of a JPEG algorithm for a different coding to provide the given quality of image.

- If for some domains the LDPC code quantisation, in combined with JPEG differents compression, results in a greater size of compressed data than does the traditional JPEG algorithm at the given quality level, then use the traditional JPEG algorithm to compress this particular domain.

With the described method of compression of 256 gray-scale images under the conditions of providing high quality in the restored image, the gain of using the code quantisation was obtained in 11% of domains on average. The results for particular test images are represented in Table 9.1.

Table 9.1.   The results of test images compression with the help of the LDPC code quantisation and JPEG algorithm.

Test image	Percentage of domains where suggested method outperforms JPEG	Overall compression gain	SNR
g_airplane.bmp	4.37%	500	37.25
g_arctichare.bmp	0.07%	0	41.84
g_baboon.bmp	51.10%	16500	33.02
g_barbara.bmp	14.72%	3000	33.42
g_boat.bmp	21.00%	4000	33.38
g_cat.bmp	10.25%	1800	35.35
g_fruits.bmp	4.83%	800	35.29
g_lena.bmp	5.27%	800	32.19
g_peppers.bmp	4.52%	800	32.89
g_pool.bmp	0.76%	0	30.96

It is also possible to relate this algorithm to the coding methods used on the lower layers of a network. For example, as was shown in section 7.3, it is possible to organise the priority transmission of some messages over the network without the use of special techniques other than transport coding. In this case we can transmit the different quantisation levels on demand in such a way that first it will be possible to deliver a rough approximation of the image with less delay, and then, if needed, to deliver other quantisation levels in order to reconstruct the image with high quality.

9.5   RECONCILIATION OF CODING AT NEIGHBOUR LAYERS OF A NETWORK

In this section we discuss the possibility of joint coding at different network layers and building the common network quality of service system based on coding.

Hereafter we will use some terms which are not in traditional use. By *channel code* we denote an error-correcting code used at physical (or data link) layer of a network; by *transport code* we denote an error-correcting code used at the transport layer of a network; an error-correcting code used for the construction of codebased cryptosystem we call a *cryptocode*; and the term *compression code* we use to denote an error-correcting code used in a code-based method of image or video compression. Also we use the concept of communication channel corresponding to a particular application. The process of encryption we denote as transmission over the *cryptochannel*; the process of data transmission at the transport layer we denote as transmission over the *transport channel*; the process of video data transmission is denoted as transmission over the *videochannel*.

We start the discussion with joint coding at the physical and presentation network layers or, what amounts to the same thing, the joint usage of physical and crypto channels.

The coding cryptosystems, like any coding systems, implement the procedure of introducing redundancy in the information. In this case the redundancy is introduced to cope with unauthorised access. The coding in the lower network layers corresponds to the introduction of redundancy in order to cope with natural errors. The general mathematical method – the coding theory used in this book to control the redundancy in different network layers – allows us to pose the problem of joint coding in different layers.

This problem is quite far from the solution, in fact it is only at the stage of formulating the general problem of controlling the information redundancy in different network layers.

In Section 9.3 we considered joint coding in the data link and transport layers. The interesting problem is the joint coding in the physical or data link layers and in the presentation layer, transport layer and presentation layers.

The code cryptosystems are based on the idea of masking information with the help of an error vector. The transmitted data is encoded by the error-correcting code, and then it is added to the error vector, which can be corrected with the help of this code. On the other hand, during the information transmission in the physical layer the code is used to cope with errors in the physical channel.

In classic transmission schemes the protection of information against artificial and natural errors is provided independently. First the information is encrypted, and then the encrypted data is encoded with an error-correcting code to cope with the natural errors. It is clear that the development of the joint procedure of encryption and encoding is a difficult problem but on the other hand, it is also clear that such a procedure cannot be lost to the classic scheme. In the case of using code methods of protection against unauthorised access, development of the procedure of joint encoding-encryption is especially attractive. In fact, the part of the masking transformation (with some probability) in this case is undertaken by the communication channel, which generates errors during the transmission. Then it is possible to solve the problems of encryption and reliable transmission with the help of the same code that allows us not only to simplify the devices providing the transmission, but also decreases the general amount of redundant data to transmit. The implementation of this idea requires the development of special codes aimed at correcting the specific errors that appeared with superposition of natural errors generated by the communication channel and artificial errors introduced with the encryption.

For simplicity's sake, consider the binary discrete channel (BSC) as the model of the physical channel. Consider confidential data transmission with the help of the McEliece cryptosystem. Let the channel error probability be p. Then the channel capacity C is

$$C = 1 - H(p),$$

where $H(p) = -p \log_2 p - (1-p) \log_2(1-p)$. With independent coding for the physical channel and encryption, the overall transmission rate is equal to

$$R_{ec} = R_c \cdot R_e,$$

where R_e is the rate of channel error-correcting code, R_c is the encryption rate, i.e., the ratio of original data length to the length of encrypted data.

Obviously,

$$R_{ec} \leq R_c \cdot (1 - H(p)). \tag{9.28}$$

For the McEliece cryptosystem (see section 8.2) R_c is the rate of code used in the cryptosystem. In particular, traditionally for this cryptosystem code rate $R_c = 1/2$, R_{ec} does not exceed half of the channel capacity

$$R_{ec} \leq \frac{1}{2} \cdot (1 - H(p)).$$

Now consider the following modification of the McEliece cryptosystem. Let $\mathbf{G}_0$ be a generator matrix of (n, k) code with the simple decoding procedure, capable of correcting no more than $p_0 \cdot n$-fold errors, where $0 < p_0 < 1$, $\mathbf{G} = \mathbf{M}\mathbf{G}_0\mathbf{P}$ be a public key of McEliece cryptosystem, where $\mathbf{M}$ is a nonsingular matrix, $\mathbf{P}$ is a permutation matrix.

The ciphertext $\mathbf{y}$ is calculated on the message $\mathbf{x}$ with the help of the following formula:

$$\mathbf{y} = \mathbf{x}\mathbf{G} + \mathbf{e_c},$$

where $\mathbf{e_c}$ is an arbitrary vector of weight $p_c \cdot n$, $0 \leq p_c \leq 1$.

The ciphertext $\mathbf{y}$ is transmitted over the communication channel without using additional error-correcting coding. Then the received vector $\mathbf{b}$ can be written as follows

$$\mathbf{b} = \mathbf{y} + \mathbf{e}_0 = \mathbf{x}\mathbf{G} + \mathbf{e_c} + \mathbf{e}, \tag{9.29}$$

where $\mathbf{e}$ is the channel error vector.

Without loss of generality we can assume that the error vector $\mathbf{e}_0$ is generated by the BSC with the probability per symbol $p_0 = p_c + p$. The capacity of this 'superchannel' is

$$C_0 = 1 - H(p_0). \tag{9.30}$$

This value is the upper bound on the transmission rate with the described choice of code used both for data transmission over the communication channel and for encryption. The value of this *overall rate* R_0 depends on the choice of p_c.

The value of p_c should be chosen in such a way that the probability of breaking the system does not exceed the acceptable breaking probability P_b. With the considered choice of common code, the system breaking probability is defined by the event when the weight of vector $\mathbf{e}$ $wt(\mathbf{e})$ is less than some value εn, which in turn is defined by the requirements of the complexity of system breaking

$$\Pr\{wt(\mathbf{e}) \le \varepsilon n\} < P_b.$$

For BSC with $\varepsilon < p_0$ this inequality can be written as follows

$$\Pr\{wt(\mathbf{e}) \le \varepsilon n\} \le \sum_{i=0}^{\varepsilon n} \binom{n}{i} \cdot p_0^i \cdot (1 - p_0)^{n-i} \le 2^{-nh(\varepsilon, p_0)}, \tag{9.31}$$

where $h(\varepsilon, p_0) = \varepsilon \cdot \log_2\left(\dfrac{p_0}{\varepsilon}\right) + (1 - \varepsilon) \cdot \log_2\left(\dfrac{1 - p_0}{1 - \varepsilon}\right)$.

Taking in account (9.31) the value of p_0 can be found from the equation

$$h(\varepsilon, p_0) = \frac{1}{n} \cdot \log_2\left(\frac{1}{P_b}\right). \tag{9.32}$$

As was mentioned above, the value of ε defines the decoding complexity. Let us use the following estimate of complexity (in number of operations) T of information set decoding (see Chapter 3):

$$T = n^2 \cdot \frac{\dbinom{n}{\varepsilon n}}{\dbinom{(1 - R_0)n}{\varepsilon n}} \le 2^{n[H(\varepsilon) - (1 - R_0)H(\varepsilon/(1 - R_0))]}. \tag{9.33}$$

Then from the inequality $T \ge T_b$, where T_b is the acceptable breaking complexity obtain ε as the root of equation

$$H(\varepsilon) - H\left(\frac{\varepsilon}{1 - R_0}\right) = \frac{1}{n} \cdot \log_2 T_b. \tag{9.34}$$

Recall that in Chapter 3 we considered the methods of decoding the linear codes to be more effective than information set decoding. With the use of these algorithms a value of ε should be chosen greater than that obtained from (9.34).

Equations (9.32) and (9.34) define the joint choice of code for both data transmission and data encryption. To estimate the achievable results consider this choice in asymptotic as $n \to \infty$.

The BSC can be considered as the error source $\{E, p(\mathbf{e})\}$, where E is a set of error vectors of length n, and

$$p(\mathbf{e}) = p^{wt(\mathbf{e})} \cdot (1 - p)^{n - wt(\mathbf{e})}$$

is the probability distribution on set E.

The *high-probability set $Q_n(\delta)$* is the set of vectors

$$Q_n(\delta) \subset E,$$

for which the sum vectors probabilities no less than given value $1 - \delta$

$$\sum_{\mathbf{e} \in Q_n(\delta)} \Pr\{\mathbf{e}\} \geq 1 - \delta.$$

The high-probability set of BSC consists of vectors $\mathbf{e}$, the weight of which is in the interval

$$(p - \nu)n \leq wt(\mathbf{e}) \leq (p + \nu)n,$$

where ν is some small value depending on δ and decreasing with n increase. This means that in the communication channel the transmitted vector is added to an error vector of weight pn with a probability close to unity.

If ε satisfying (9.34) is less than p, then to satisfy (9.32) it is possible to choose $p_c = 0$. In this case there is no need to add the transmitted ciphertext with vector $\mathbf{e}_c$ as is done in (9.29). Then the achievable code rate is equal to the channel capacity $1 - H(p)$.

In a general case the achievable overall transmission rate with joint coding R_{ec}^c can be written as follows

$$R_{ec}^c = \min\{1 - H(p),\, 1 - H(\varepsilon)\}.$$

With independent error-correcting coding and encryption, the achievable overall transmission rate is

$$R_{ec} = (1 - H(p)) \cdot (1 - H(\varepsilon)) < R_{ec}^c.$$

Thus, application of joint coding in the physical and presentation layers gives a significant gain in overall transmission rate in the case of codebased cryptosystems. This can be explained in a very simple way. In fact the independent error-correcting coding and encryption means using the cascade coding scheme: the inner code is the code used in the McEliece cryptosystem, and the outer code is the error-correcting code protecting the data against the channel noise. In the case of joint coding in physical and presentation layers we use only one code instead of the cascade scheme. Of course, it is easier to optimise the parameters of only one code rather than the parameters of two codes if we can utilise the properties of this one code for both encryption and error-correcting coding.

Obviously, the gain for particular codes is less than in asymptotic. On the other hand, the choice of (1024, 524) Goppa code in the McEliece cryptosystem was not reasonable enough. In particular, there is no need to add ciphertext with error vectors (keys) of weight 50 in order to provide the required level of system resistance to attacks. The (1024, 524) code in accordance with the design distance estimate is capable of correcting 50-fold errors. However, the required breaking complexity has already been achieved by using keys of weight $38 \div 40$. Then redundancy can be used to provide protection against communication channel noise.

Analysis of the joint choice of code for encryption and data transmission over the communication channel becomes more difficult with the use of cryptosystems based on full decoding (see section 8.3).

The error vector **e'** added with the ciphertext is randomly chosen from set E. Set E for these cryptosystems is a set organised in some special way rather than a set of vectors of given weight. For example, it can be a set of vectors obtained from the multiplication of vectors of given weight by some public key matrix $\mathbf{M}_1$:

$$\mathbf{y} = \mathbf{x} \cdot \mathbf{G} + \mathbf{e_c} = \mathbf{x} \cdot \mathbf{G}_0 \cdot \mathbf{M} + \mathbf{e}' \cdot \mathbf{M}_1 \cdot \mathbf{M}, \tag{9.35}$$

where $\mathbf{M}$ is the nonsingular $n \times n$ matrix, $\mathbf{M}_1$ is the special $n \times n$ matrix, the choice of which is described in section 8.3, $\mathbf{e}'$ is the vector of weight $p_0 n$. In the process of decryption $\mathbf{y}$ is multiplied by $\mathbf{M}^{-1}$ and is then decoded in the code with generator matrix $\mathbf{G}_0$.

During the transmission of $\mathbf{y}$ over the communication channel it is added to the error vector $\mathbf{e}$ of weight $wt(\mathbf{e})$. For BSC the weight of error vector has a value close to pn, where p is the error probability. Then to obtain $\mathbf{x}$ it is necessary to decode vector

$$\mathbf{b} = \mathbf{y} + \mathbf{e} = \mathbf{x} \cdot \mathbf{G}_0 \cdot \mathbf{M} + \mathbf{e}' \cdot \mathbf{M}_1 \cdot \mathbf{M} + \mathbf{e}.$$

Multiplying the vector $\mathbf{b}$ by $\mathbf{M}^{-1}$ obtain vector

$$\mathbf{b} \cdot \mathbf{M}^{-1} = \mathbf{x} \cdot \mathbf{G}_0 + \mathbf{e}' \cdot \mathbf{M}_1 + \mathbf{e} \cdot \mathbf{M}^{-1},$$

decoding of which in the code defined by $\mathbf{G}_0$ can be quite a difficult task. Because of this fact we should use some special codes and take into account matrix $\mathbf{M}^{-1}$ for the choice of common code for data transmission over the communication channel and encryption with the help of a cryptosystem based on full decoding.

We will not discuss here the methods of construction of such codes. We aim to discuss only the possibility of joint coding in different network layers.

The modification of the McEliece cryptosystem considered above allows the consideration of processes of data encryption and transmission as a process of data transmission over the joint superchannel using one common code.

This approach allows the reconciliation of this superchannel with coding in the transport layer. It can be done in the same manner as the reconciliation of channel and transport coding shown in section 9.3.

The idea of reconciliation of channel and transport coding consists in using at data-link layer the high-rate code, which does not provide the required error probability. In this case the code in the transport layer is used not only to decrease the mean message delay but also to decrease the error probability to an acceptable level.

An increase in the effectiveness of coding in a network can be achieved with the help of adaptive methods of transmission. Transport coding uses the resources of the whole network rather than particular channels (data links) providing some 'averaging' over the set of routes, which are used by the packets of transmitted messages. Even in the case when adaptive procedures of transport coding (i.e., the adaptation of the number of redundant packets) is not used in an explicit way, there exists conventional adaptation consisting of the following.

With (N, K) code used for transport coding, N packets of encoded message goes over the N not necessarily distinct routes (some packets are going over the same route sequentially). Since it is enough to deliver K packets in order to reconstruct the message, transport coding, in fact, consists in the choice of the best K from N possible routes, i.e., there is adaptation over K from N possible routes.

Notice that the effectiveness of transport coding increases if the error distribution at the output of the superchannel decoder is not uniform. The grouping of errors at the decoder output leads to errors in a rather small number of packets of encoded messages. These errors can be corrected with the transport code.

In the case of reconciliation of transport code with the channel and cryptocode, the transport coding can also be used to increase the resistance of the cryptosystem to attack. Actually, the error at the output of the cryptochannel means that both legal and unauthorised users cannot extract the correct information from the packet. Moreover, if the legal user can deal with the problem by receiving additional numbers of undistorted packets of encoded message, the ability of an unauthorised user to do so is restricted by the necessity of intercepting enough of these additional packets. The crypto-resistance of the system can be improved with the help of inserting in the network some false (erroneous) packets.

Joint coding in different network layers can be used not only to protect information. In Section 9.4 we described the application of coding to compress images or video (video-data). The idea of code compression works because the set of domains is covered in some areas by the error-correcting code (compress code). The words of this code are domains, and instead of original domains the closest codewords to them are chosen for the transmission. The particular codeword to transmit is found by decoding the compression code in some (usually Euclidean) metrics. Instead of the codeword itself the information set (the index) of this codeword is transmitted over the channel. This leads to the compression of information. Let C_{com} be the compression code with code rate R_{com}. Then the data can be compressed by the factor of $1/R_{com}$. With independent channel coding this data after compression is encoded by the channel error-correcting code with code rate R_e. Then the overall transmission rate $R_{com,e}$ is

$$R_{com,e} = \frac{R_e}{R_{com}} \leq \frac{1 - H(p)}{R_{com}}, \tag{9.36}$$

where p is the error probability per bit of BSC.

Let code C_{com} be the optimal compression code for the given quality of reconstructed image. Then (9.36) defines the maximal possible transmission rate.

Now use the same code C_{com} as the channel code, i.e., we will use code C_{com} for the protection against channel noise rather than for data compression. If $R_{com} \leq 1 - H(p)$ then we provide 'error-free' transmission with rate R_{com}. Of course, the transmission rate R_{com} is less than that is guaranteed by (9.36). Notice that we do not need 'error-free' transmission of video data since the transition of the transmitted code domain to the erroneous code domain may lead to a better approximation of the original domain by this erroneous code domain. This means that it is possible to provide a transmission rate greater than $1 - H(p)$.

If we transmit the data with a rate greater than the channel capacity, then there is a high probability that the result of decoding will not be the same codeword that was transmitted. However, if this incorrect codeword defines the domain, which does not differ significantly from the transmitted one, this error will not be significant to the human eye. Of course, in this case we need code and a decoding procedure with the following property: possible (from the decoding procedure point of view) distortions inserted by the decoding should map the received domain so as to be visually close to its domain. In particular, if decoding procedures consist in decoding to the closest word in Hamming metrics, then the code should be constructed in such a way that the closest Hamming metrics codewords should define visually close domains.

The construction of such 'video oriented' codes is a poorly known problem. It is possible to use the compression code C_{com} to solve it.

Let C_{com} be a linear $(n_{com}, R_{com} \cdot n_{com})$ code. With independent channel and compression coding for each domain $R_{com} \cdot n_{com}$ information symbols are to be transmitted. These information symbols are encoded with the independent channel $(n_e, R_e \cdot n_e)$ code. With the joint coding choose the shortened C_{com} code as channel code C_{com}^{Δ} with the following parameters:

- the code length is $n_{com} \cdot \dfrac{R_{com}^2}{R_e}$;
- the number of information symbols is $R_{com} \cdot n_{com}$.

The transmission rate provided by this code is defined as in (9.36) but as obtained from C_{com}, the code C_{com}^{Δ} will be 'video oriented'. This fact makes the described method interesting for further investigation.

As earlier we deal with some superchannel. In this case it is the superchannel compression-transmission. Reconciliation of this superchannel with transport coding can be done in the same way as in case of superchannel encryption-transmission. This means we can distribute the code redundancy between the superchannel code and the transport code. With this, using the transport coding, we can 'correct' the transmitted image, e.g., by transmitting in the transport layer some differences between the original and reconstructed images (those that exceed the acceptable level).

With the transmission of video or speech data in real time, the reconciliation of the superchannel with transport coding should take into account the inadmissibility of big transmission delays.

Some part of the data to be transmitted is priority, i.e., it should be transmitted with minimal delay. In this case, adaptive transport coding can be used to restrict message delay. By changing the rate of transport coding we can change the mean message delay. Then with the help of adaptation of the transport code rate we can provide restriction of the delay dispersion. With uniform distribution on the network channels we can increase the set of routes used by priority messages just by increasing the number of redundant packets (decreasing the rate of transport code) for these messages. Then the probability that the packet of the priority message chooses the low-noise route increases, hence the probability of a big delay for this packet decreases. On the other hand, decreasing the rate of the transport code we are faced with an increase in network load. If this decrease is significant, it is possible to compensate by increasing the rate of transport code for another part of data (which is not as critical as the delays).

All of the preceding allows us to create a new model of network quality of service based on the representation of a network as a system of interacting service superchannels. Control of this system is provided by controlling the redundancy of codes used in the different layers of the network.

Let the network be represented by m types of services $S_1, \ldots, S_m$ (e.g., service of transmission of the encrypted data, service of transmission of video-data, etc.). We described above the construction of some superchannels (encryption-transmission, compression-transmission, etc.). Assume that a particular superchannel is built for each service

$$K_1 = K(S_1), \ldots, K_m = K(S_m).$$

Each superchannel corresponds to its code $V_1, \ldots, V_m$ with the parameters $(n_1, R_1 n_1), \ldots, (n_m, R_m n_m)$. The quality of each service is a function of code rate in the corresponding superchannel. Assume that given codes provide the required quality of service (QoS) for one service request in each superchannel. Let $\lambda_1, \ldots, \lambda_m$ be the mean number of requests for the corresponding service. The transport code V_T with parameters $(N, R_T N)$ is used for improving the quality of each service.

Then the process of services assignation can be treated as the transmission of m users over the generalised superchannel $\hat{K}$ with the capacity restricted by $\hat{C}$. The users of channel $\hat{K}$ are the network services. The intensity of channel use by each service is defined by the corresponding ratio $\dfrac{\lambda_1}{R_1}, \ldots, \dfrac{\lambda_m}{R_m}$.

With the help of such a model it is possible to solve different problems of service distribution optimisation, e.g., distribution of channel capacity between services against service costs with some restrictions on QoS.

The application of transport coding to this model with the help of code V_T allows:

- decreasing the mean delay of service assignation by introducing the redundant packets (requests to service);

- providing QoS control by redistributing the redundancy between the corresponding superchannel and transport coding.

The construction of each superchannel requires a special coding problem to be solved. In this section we have restricted ourselves to the simplest case of BSC, the practical value of which is not very important nowadays. In many cases we considered the simplest methods of coding at higher network layers. Nevertheless, generally speaking we need the construction of new codes with specific properties for superchannel encryption-transmission and compression-transmission. The new codes will probably also be needed to solve the problem of reconciliation of coding for other sets of network layers.

However, it seems that the methods of code construction developed in the coding theory and presented in this book allow us to be optimistic about the development of a system of service control that provides compensation for losses in information protection in one network layer by means of protection in other layers.

REFERENCES

1. Krouk, E. and Semenov, S. (2002). Application of Coding at the Network Transport Level to Decrease the Message Delay. *Proceedings of Third International Symposium on Communication Systems Networks and Digital Signal Processing.*, Staffordshire University, UK, pp. 109–12.
2. Alon, N. and Luby, M. (1996). A linear time erasure codes with nearly optimal recovery. *IEEE Trans. On Inf. Theory*, **42**, (6), pp. 1732–6.

3. Luby, M., Mitzenmacher, M., Shokrollahi, M., A. Spielman, D. A., and Stemann, V. (1997). Practical loss-resilient codes, in *Proc. 29th Ann. Symp. Theory of Computing*, **42**, (6), pp. 150–9.

4. Luby, M. (2002). LT Codes, in *Proc. of the 43rd Annual IEEE Symp. on Foundations of Comp. Science*.

5. Gray, R. M. Fundamentals of Vector Quantisation. http://www-isl.stanford.edu/~gray/compression.html

6. Li, J., Gray, R. M., Olshen, R., (1999). Joint Image Compression and Classification with Vector Quantisation and Two Dimentional Hidden Markov Model in *Data Compression Conference, IEEE Computer Society TCC*, pp. 23–32.

7. Hung, A. C., Tsern, E. K., and Meng T. H., (1998). Error-resilient pyramid vector quantisation for image compression. in *IEEE Trans. on Image. Process.*, **7**, Oct. 1998, pp. 1373–86.

8. Lin, J.-H. and Vitter J. S., (1992). Nearly Optimal Vector Quantisation via Linear Programming in *Data Compression Conference, IEEE Computer Society TCC.* pp. 22–31.

9. Cosman, P. C., Oehler, K. L., Riskin, E. A., and Gray, R. M. (1993). Using vector quantisation for image processing. in *Proc. of the IEEE*, **81**, (9), pp. 1326–41.

10. Bayazıt, U. and Pearlman, W. A., (1999), Variable-Length Constrained Storage Tree-Structured Vector Quantisation. *IEEE Trans. Image Processing*, **8**, (3), pp. 321–31.

11. Garey, M. R., Johnson D. S., and Witsenhausen H. S., (1982). The complexity of the generalized Lloyd-Max problem. In *IEEE Trans. Inform. Theory*, **28**, (2), pp. 255–6.

12. Gersho, A. and Gray, R. M., (1992). *Vector Quantisation and Signal Compression*, Kluwer Academic Publishers, Amsterdam, The Netherlands.

13. Belogolovy, A. V., (1999). Application of error-correcting codes to video compression, *Proc. of Sec. Intern. Youth Workshop BICAMP'99*, St.Petersburg, Russia, p. 119 (in Russian).

Index

Error Correcting Coding and Security for Data Networks   G. Kabatiansky, E. Krouk and S. Semenov
© 2005 John Wiley & Sons, Ltd   ISBN: 0-470-86754-X